the self-build book

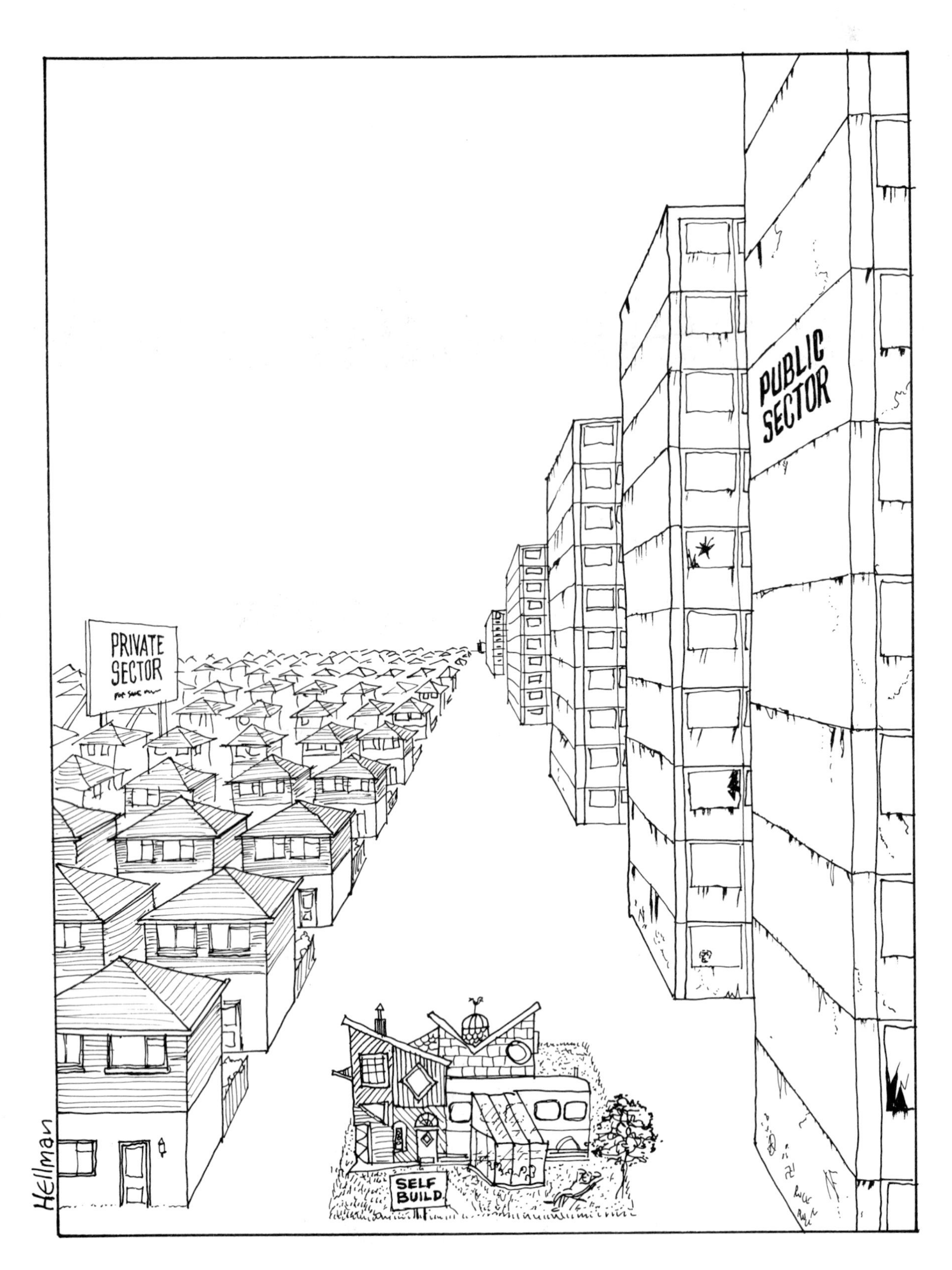
PRIVATE SECTOR
PUBLIC SECTOR
SELF BUILD
Hellman

the self-build book

JON BROOME

BRIAN RICHARDSON

revised and updated edition

GREEN EARTH BOOKS

Revised edition published in 1995
by Green Earth Books, an imprint
of Green Books, Foxhole, Dartington,
Totnes, Devon TQ9 6EB, UK

Second impression 1996

Distributed in the USA
by Chelsea Green Publishing Co., Vermont

Original edition published in 1991 by
Green Books

Cover photograph by Bruna Fionda
kindly supplied by Mike Powell

Designed, photoset in 10 on 12 pt Plantin
and assembled at Five Seasons Press, Hereford
in collaboration with Brian Richardson

Printed by Hartnolls Ltd
Victoria Square, Bodmin, Cornwall

A catalogue record for this book
is available from
The British Library

ISBN 1 900322 00 5

Glenn Storhaug at Five Seasons Press dedicates
his involvement in The Self-Build Book
to the memory of Bob Kindred
who died in France on July 14th 1991
just as this book (whose growing he watched
with interest) was going to press.
In 1969 Bob worked on the Walter Segal house
in Islip, Oxfordshire—one of the
earliest Segal timber-frame designs.

In homage to Walter Segal

Acknowledgements

Our thanks must go first to *John Seymour*, for suggesting to us that we write *The Self-build Book* and for persuading Green Books that they should publish it. We are honoured that he should also have contributed an Afterword.

There are very many other people who have been an inspiration and help to us. Self-build is a convivial business—we have built with and learned from a goodly company—and we could not possibly find space to mention everybody we would like to.

There are our immediate self-building partners, for instance—*Maureen Richardson* in Herefordshire and *Jenny Broome* and *Nick Ackroyd* in Lewisham—but we were also joined by countless friends and relations whose help was unlimited and who made building enjoyable.

For their involvement with this book that comes out of our self-build experiences we particularly thank:

Louis Hellman for his frontispiece

Ken and *Pat Atkins* for the Lewisham story

Chris Gordon for providing all our insights into Zenzele

the Lightmoor pioneers for showing us the Project, particularly *Margaret Wilkinson* and *Robin Heath* for facts, figures and drawings

Margaret and *Cedric Green* and *Derek Leary* for their hospitality at Netherspring and for photographs and drawings

Charlotte Ellis for her portrait of Walter Segal (kindly supplied by The Architectural Press Ltd)

Peter Cook, Nigel Corrie, Mike Trevillion, Phil Sayer & *Martin Charles* for all the remaining photographs not provided by the authors

Jenny Broome for her drawing of the living room at 6 Segal Close

Renz Pijnenborgh for unstintingly sharing his special knowledge of biological building and providing information (not all of it in Dutch!) about MW2 at Maaspoort

Pat Borer, for providing much of the information about environment-friendly materials, energy use and conservation; and with his co-author Cindy Harris for allowing us to quote freely from their book *Out of the Woods*

Vic Sievey, for contributing the chapter on Brighton Diggers

John Willoughby, who gave advice on energy costs

Peter Neale, quantity surveyor, for help with advice on building costs

Alan Jacques, valuer, for land prices

Ross Fraser for sharing his knowledge of organizational forms and finance for self-build groups

Rona Nicholson for helpful suggestions on the text

John Elford and *Lesley Levene* for editorial help and encouragement

Glenn Storhaug for book design and typesetting

The Architects Journal for allowing us to use material originally prepared for their magazine

finally, *Colin Ward* for providing so much good advice and example and for his Foreword.

Contents

Part Three (continued)

Part Four
The Segal method

We explain why we believe this method has much to offer self-builders and then examine the documentation, working methods and also the tools required. Step-by-step building instructions are provided for each element of the construction.

Part Five
Why don't we all self-build?

We examine the history behind the way housing is provided, note some of the shortcomings and suggest a new approach.

Preface to the revised edition

Even more gratifying to the authors than the reviews *The Self-Build Book* received at first publication in 1991 has been the interest generated among a wide readership, and the selling out of the first edition. We are delighted to have been invited by Green Earth Books to re-present our case—that there is great enjoyment to be had in designing and building your own home.

Much of what we said before has been retained in the second edition. Our enthusiasm remains undimmed, as witnessed by the intense building activity we have both been engaged in since, and our vigorous campaigning through the agency of the Walter Segal Self Build Trust and elsewhere to bring self-build more into the housing mainstream. But we could not simply reprint the entire original text; too much has gone on that demands reappraisal.

The huge increase in concern for achieving far higher standards of energy conservation in the construction and use of buildings has led us to update and comprehensively revise the practical advice we give. In particular the whole of Part Four, which is an exploration of the 'Segal approach' to timber-frame building, has been rewritten.

The account we gave in the first edition is now of only historical interest. It described in detail the way a typical house designed by Walter Segal and Jon Broome in the 1970s was put together. Comfortable and economic to run as those houses continue to be, our aspirations for thermal efficiency are now very much higher.

The Building Regulations now in force, although they do not yet require the levels of performance that we now advocate as 'best practice', would not permit the early 'Lewisham' house to be built that way today. Consequently, much thought and development work has gone into devising a building fabric that is still as simple and economical to make as Walter would have wished, but comes up to the very highest standards that can now be achieved.

Other updates have been to Part Three to make sure all our advice in the Action Guide is still applicable in the light of changes to financial and other procedures.

To undertake a self-build project can daunt the bravest spirit and it is reassuring to know that other people have taken it on with success. We have added further examples. Jon has taken on a new ambitious task in succession to his building of the 'classic Segal' design at 6 Segal Close. In the new chapter *Building Again* he describes in diary form the challenges and rewards of aiming high—the result is a house (in Brian's perception at any rate) that both fully satisfies his family's needs and stands as a work of art.

We had little difficulty in persuading enthusiastic self-builder Vic Sievey to tell us of his experiences as a member of the Brighton Diggers. They have produced a most attractive and widely praised group of houses—a fine example of a number of schemes recently completed that owe something to Walter Segal's influence.

In the final part of the book, *Why don't we all self-build?*, our observations on the shortcomings of the conventional approach to dealing with housing, and our constructive suggestions for making self-build a significant component of housing strategy, have unfortunately had to remain largely as we first wrote them.

Brian Richardson & Jon Broome
3 August 1995

Foreword
by Colin Ward

Reader, you will soon realize that the book you have bought or borrowed is in fact a love story.

The authors begin in Part One by describing their own love affair with housing themselves, and the way it, for them, brought the joy back into building. But falling in love is a universal experience and the authors compare notes with half a dozen similar adventures in building, where people with many different motives found ways of housing themselves. They might be badly housed local council tenants, or people from the council's waiting list. They might be single unemployed young people, or they might be pioneers of different styles of earning a living. They may have had visions of making a good life for themselves, by energy-saving, by using the sun to keep warm or by building their own version of a friendly environment. This is where Part Two reassures us.

Every love affair has its grim shadows, and Part Three, as a guide to action, pilots us through the thickets, hazards, and sheer bureaucratic barbed wire that stands between us and our simple, self-justifying and timeless aim. I'll take a bet that you will skip over this third part of the book, simply because, until you are facing the problems for which it provides solutions, you won't feel the need for the advice it gives. But you will return to it.

Fifty years ago the Russian poet Vladimir Mayakovsky killed himself, leaving a note that said 'Love boat smashed against *mores*.' Now if you are a would-be self-builder, these *mores* or social conventions are enormous and suffocating. Access to land, to finance, to planning permission and to approval under the building regulations are enough to wreck any love-boat. They have ruined many a self-build venture.

It is not the function of the authors of this book to conduct a campaign of abuse against these obstacles: their task is simply to help you to find your way through them. This is why the first chapter of Part Three is called *Organizing Yourselves*. Co-operation is the key, and reading the book I am tempted to see Mozart's *Magic Flute* as a self-build allegory. The pair of young hopefuls in that story have to submit themselves to every kind of ordeal before learning that co-operative effort is the key and that

> Those whom this bond cannot unite
> Are all unworthy of the light.

I wouldn't go as far as that, but I do know that collective lobbying of the city hall in Sheffield by members of a housing co-op there achieved results that their architect, who had to depend on day-to-day amicable relations with the bureaucracy, could not achieve. And I do know that the chairholder of an outstandingly successful housing co-operative in Liverpool had to storm the Adelphi Hotel there and to stick his thumb in the soup of the Housing Corporation's chairman, in order to assert that his whole function was to help, and not to hinder people like the Liverpool 8 citizen at the other end of that thumb.

The authors of this book steer an even course between group and family self-build. Circumstances and places profoundly alter cases. I myself have met self-builders who reflected that, given what they now knew, they would work all hours for six months to buy the time to build a house on their own, rather than do it as members of a group. On the other hand, the authors describe group schemes where the self-builders benefited from mutual aid and remained perfectly amicable. This is because they discarded the clauses in the usual Self-Build Housing Association Articles of Association which penalize failure to co-operate fully, which are intimidating rather than helpful.

It is also important to record that the case histories they give us are full of examples of public officials who have 'gone beyond the line of duty' in bending the rules to suit the circumstances of self-builders. In other words they have perceived what their duty really is: to help and not to hinder.

How devastating to learn that the difficulties facing self-builders are not to do with the acquisition of building skills, or the endless hard work. For the authors tell us that in fact, 'These planning stages are the most difficult part of self-build: by comparison, the actual building is straightforward.'

I am reminded of the celebration of the triumph of self-builders throughout history and all over the world, by Bernard Rudofsky in his book *The Prodigious Builders* (Secker and Warburg 1977). Rudofsky observed that

> An African villager looking for the first time in his life at a European house does not suspect the travail and anguish that go into building it—the ritual of buying the land with the help or hindrance of agents, lawyers, and local authorities; securing a bank loan or mortgage; preparing plans, estimates, and documents indispensable for the construction of the house; and paying taxes and insurance policies attached to it forever after. To him the result may look elementary.

When we reach Part Four of this book, we are on happier ground. Walter Segal was an architect, a respected friend of Jon Broome and Brian Richardson, and of mine, who spent years in developing a cheap, simple and dry method of building. He made no claims to originality, his task was to simplify. The great joy of his life was to see this realized in a London borough, and it happened thanks to the support of the authors of this book. In his 'View from a Lifetime' given ten years ago at the Royal Institute of British Architects (to which he did not belong), Walter Segal encapsulated the message of this book:

> Help was to be provided mutually and voluntarily—there were no particular constraints on that, which did mean that the good will of people could find its way through. The less you tried to control them the more you freed the element of good will—this was astonishingly clear. Children were of course expected and allowed to play on the site. And the older ones also helped if they wished to help. That way one avoided all forms of friction. Each family were to build at their own speed and within their own capability. Which meant that we had quite a number of young people in their twenties and late twenties. But we had some that were sixty and over who also managed to build their own houses. They were told that I would not interfere with the internal arrangement. The time it has taken me not to be the normal hectoring architect that lays down everything and persuades everybody to follow his own taste! I let them make their own decisions, therefore we had no difficulties. What I found astonishing with these people was the direct personal friendly contact that I had with them and which they had among themselves. And quite beyond that, the tapping of their own ideas—countless small variations and innovations, and additions were made by them which we have not yet tabulated because the whole thing was not sufficiently organized. But it is astonishing that there is among the people that live in this country such a wealth of talent.

This splendid observation explains the importance of Part Four, a clear and lucid explanation of the method of building that Segal evolved. Anyone can do it. At the same time, the authors of this book carefully stress that the Segal method is one of a variety of techniques of building that people might adopt. Building is a means, not an end. The purpose of Jon Broome and Brian Richardson, as they say from the outset, is to bring the *joy* back into building.

There are people to whom this is irrelevant, just as there are those who will see the advocacy of self-build as a despairing reaction to the collapse of housing policy in Britain. Part Five of this book is an answer to arrogance and prejudice. Firstly the authors nowhere claim that self-build is the answer to everyone's problems. They simply demonstrate that it is one stream of a rational and humane approach to housing. Secondly they contrast this with the manifest failure of housing policy in Britain, whether in the public provision of rented housing or in the speculator's free market in providing houses for profit. Few people get any joy in the process and those in the greatest need not only have the fewest choices, but are usually left out altogether.

Hints and tips on many aspects of building construction and insulation occur all through the book, and this is an account of how to build that we can all learn from, whether we are self-builders or not. The authors are, from their own direct experience, concerned with environmentally-friendly, energy-saving construction and use of buildings.

As the next century approaches these will be issues that concern everyone, even though today they are virtually unknown in the training either of builders or architects. This book, that starts as a love story about the pleasures of designing and building your own home, grows imperceptibly into an environmental manual for the future.

Introduction

Our main aim is to inspire you to build for yourself. This book will show you that designing and building your own house is within the reach of everyone, is enjoyable and can have great economic, practical and social benefits. You could feel the satisfaction of having made something really useful with your own hands and the excitement of dreaming about what your house should be like, how it will be laid out and what you are going to put in it. You will see it slowly taking shape and will imagine the next steps in your mind and have a vision of the finished house. A handmade house is a pleasure to live in and you will know every corner intimately. You will be able to afford a bigger, better house arranged to suit your needs and desires and could be free of the financial burdens that so many people suffer because of their housing costs.

We have worked with groups of self-builders from the local council housing waiting-list in south-east London and have shared their excitement about and enjoyment of their individually designed houses. The self-builders were from very different backgrounds and had very different personalities. They included young people, older people building for their retirement, a single mother, unemployed people and people with low incomes, and almost all were without any experience of building. We have absolutely no doubt that there is an enormous number of people in all walks of life who, like them, would jump at the chance to build a house for themselves if they thought that it was a real possibility. Anyone can do it if they have the determination, are prepared to work hard and, most importantly, if they have the right opportunities.

Rising costs have put good housing beyond the reach of more and more people. The problems of large estates suffering from vandalism, poor maintenance, inadequate heating, lack of shops and community facilities demonstrate to us, among other things, the dangers of ignoring peoples' real needs and wishes. Self-build offers one way to avoid this by allowing people to be in control of their housing.

We do not believe that political expediency and market forces should force anyone to build their own house, and this should not be the only way for them to obtain a good one, but we do believe that people should have the *opportunity* to build their own houses if they wish.

The structure of this book

The book is divided into five parts. In the first we describe our own experiences of building for ourselves to give a flavour of what it can be like and to outline some of the ideas that are involved in designing one's own house.

In Part Two we describe a number of other examples of self-build projects. They have been selected to show the diversity that exists within the idea of self-build and to demonstrate some of the things that people have achieved in various circumstances. From Britain and abroad, they are drawn from town and country, with some people building in brick and others in timber. Many of the projects chosen embody a particular idea: building a place both to live and to work; providing an alternative to council housing for rent; improving peoples' employment opportunities; conserving energy, or building with the ecology of the planet in mind.

We have not chosen examples from the mainstream of self-build activity in Britain—private individuals building conventional houses typical of a suburban estate alone or, less frequently, as members of a group—as this has been covered in other books. Instead we intend to show that there are other ways of doing it that can offer not only financial rewards but also individual control of the design, ease of construction and opportunities for people without much money to build their own houses.

Part Three outlines how ideas of self-building can be turned into practical action. It describes in detail the ingredients necessary for a self-build enterprise:

how to organize, what professional help you may need, how to obtain land and finance, how to approach matters of design and energy conservation, what documents you will need and how to obtain the official permissions necessary. This is a much more comprehensive description than has been available until now and includes the first full treatment of the rapidly evolving methods being devised to enable people on low incomes and without building experience to build for themselves.

When we come to the actual construction on site we do not cover conventional building methods, because these are fully described in many books on the subject. Instead, in Part Four, we describe the Segal method of timber construction in detail. It is not widely known and has many particular advantages for self-builders: it is quick and easy to build, it is economical and allows great freedom of design and group organization. This is also the most complete account available and includes basic step-by-step instructions on how to build a Segal house.

The final part of the book covers what we believe to be some of the shortcomings of the present housing situation in Britain and in it we suggest that self-build should have a much larger part to play, outlining how this could be achieved.

Each part of the book has a different character because we believe not only that self-build activity must be guided by theory but also that theory is not useful unless it can be turned to practical use. One of the strengths of self-build is that it is a very individual affair, but there are basic principles that apply. This book therefore combines the theoretical with the practical and the personal with the general.

Who will find this book useful

We want to inspire people to build for themselves and, in particular, we want to see more opportunities for people without much money and without any previous knowledge of building to self-build. We want this book to be useful to anyone who wishes to build his or her own house, whether individually or as a member of a group, whether using the Segal method or some other more common form of construction.

This book explains the issues, methods and benefits of self-build and should also be useful to those involved in the housing field, whether as councillors, local government officers or housing association staff.

We hope for an increasing awareness of the wider possibilities of self-build among the public at large and among people in the housing world, and hope that the ideas contained in this book will influence the people who control the resources of land and finance.

The description of the Segal method will be of interest to those in the building industry and to architects in particular. We hope that they will want to know more (for we have not attempted to go into all the detail) and will feel moved to work in a participatory manner with self-builders, developing innovatory approaches to design.

There has been a steady increase in self-build over the last decade. It is estimated that in 1991, 15,000 individual builders built on their own land. Of these around 5,000 made all the choices for themselves although they had a formal contract with a builder to do all the work. The remaining 10,000 self-built to some degree. They managed the project but may have employed sub-contractors to do all or part of the work. Of these self-built houses only 150 were built as part of a self-build group. We believe that there is enormous potential in this self-build movement. It has changed a lot since we first became involved in Lewisham in 1975, when self-build was a small fringe activity and not as topical as it is now. Ideas are developing rapidly at the present and we look forward to a couple of small groups building in every town. If this were to happen in just one town in each local authority area in the country, self-build would double and become a significant part of housing in Britain, making a real alternative available to many people.

Part One

Why did we self-build?

It is enjoyable—It is beneficial—It is possible

In this part of the book we trace our own introduction to self-build, tell you of two great men who have influenced us and describe how we went on to enjoy the process of building our own houses, in the hope that it will encourage you to do likewise.

Chapter 1

Bringing the joy back into building

In the context of the tradition of self-building in Britain, we relate what building our own houses has meant to us and suggest how we think you can get the most out of it when you do the same.

In the distant past most people used to build for themselves. When Britain became industrialized, the picture changed and most people rented private and later municipal housing, built in large quantities and still forming a large part of our towns and cities. Throughout, people have continued to build their own houses, often against the odds. For example, the first building societies, founded in the early nineteenth century, were set up as mutual savings organizations to be wound up when all the members had built their own houses. There have been times of particularly high levels of activity, such as the period after the First World War when people from London self-built dwellings in Essex and along the south coast and founded what are now the towns of Basildon and Seaford.

The great majority of self-builders these days are individuals building perfectly conventional houses in the countryside. There are also people building as members of Self-Build Housing Associations in the outer suburbs and in the small towns. They tend to erect large, well-built houses—typically a four-bedroom detached house with garage. They are self-building not because it is the only way that they can get a house but because in this way they can afford the house that they want. There is also a range of individual involvement in self-build. At one end of the scale are the self-builders who carry out all the organizing and building themselves; others employ a self-build consultant to do all the planning and organizing for a group, the self-builders doing a large part of the building but employing subcontractors for some of the work. At the other end of the scale is the person who undertakes the management of the project but employs a series of subcontractors to carry out the building work itself.

The high cost and poor quality of housing have led to a steady increase in self-build recently. In the last few years there have also been a small number of projects in the cities—London, Bristol, Liverpool, for example—for people in housing need.

Should *you* take the momentous decision to self-build, we want you to get the most from it. In Chapter 26 we mention some of the shortcomings of our housing stock in Britain. It would be a shame to fall into any of the common errors and build yourself a house that is not really worthy of you. What we are saying is that you can build yourself a much better home than anyone else could do for you. We are not alone in this. Murray Armor has written in his invaluable *Building Your Own Home* (updated over the years; we quote from the 1987-8 edition) that '[the many thousands of houses built] by D.I.Y. housebuilders are better constructed than average with savings of up to forty per cent on builders' prices'.[1]

The sheer economic advantage of getting into the housing market by contributing your own labour has led to a resurgence of interest in self-build. But as Armor says, 'the real motivation in self build is far deeper than costs per sq ft... The new home is an expression of the individual himself, and will be the highest standard to which he can aspire. At the very least the standard will be several layers above the estate developer's lowest common denominator.'[2] However, he goes on to remark that his conventional self-builder 'is trying to do his thing within the system, not trying to escape from it. His boast is of what his house is worth, and he usually builds a marketable house of conventional appearance and layout.'[3]

It has always been easy to borrow money on the security of a finished product that fits a known slot in the estate agents' books. Consequently, there has been money to be made by people building (cheaply but well) exact replicas of utterly undistinguished houses which they can then trade in to move on up-market. This has been done often enough; there are books telling you how to do it and firms who will undertake the management of self-build groups who have this aim in view. But these management consultants themselves cost money, and with the efficient service they provide come certain important limitations for the intending self-builder. The delight of designing 'the dream house' is largely removed, and the poor substitute is selection from a limited range of pre-designed house types.

Armor tells us that the design of a house in a group scheme 'is agreed by the group as a whole, and once settled there is no opportunity for individual variations'.[4] Both the National Federation of Housing Associations (NFHA) and the management organizations advise strongly against variations between different houses. He observes that in this kind of self-build group, 'standardization and personal choice is much the same as on a developer's housing estate'.[5]

Entry to such a group is selective in that the management consultants require an acceptable mix of skilled tradesmen. Armor observes that the NFHA and the Housing Corporation support this limitation, looking for 'a group that has fifty per cent of its members with building skills, such as bricklayers, carpenters, plasterers, etc., to give a professional feel to the operation. They expect the majority of members to be married and to possess a steady outlook that gives confidence that they will stick a year of unrelenting toil. They look for a balance between men (*sic*) over thirty and under thirty. They look for membership established in their employment and therefore good mortgage risks.' Managed associations, Armor says, 'carry this to extremes, making up lists of possible members from applicants as if they were selecting an England football team'.[6]

Yet another disadvantage, in our view, is the degree of group discipline imposed on this model of a self-build group by the necessity of working to a strict programme. This has led to rules being enforced by fines for non-performance or even to expulsion.

While admitting limitations of freedom of choice in design and way of working inherent in professionally managed, conventional self-build groups, Armor considers that 'these constraints of philosophy and outlook are really side issues'.[7] Where our views differ from his, and the main reason we have written this book, is that these limitations loom much larger *to us* and we seek to overcome them. We wish to widen still further the range of people eligible for self-build and to allow them greater freedom. In our book the emphasis is on the enjoyment of building a house that goes beyond being simply a good investment. It will also be exactly the home you want. As well as being economical to build, it should be an expression of your personal desires, and we include in that the aspiration to make it beautiful. All of us share the same need to live harmoniously in what we sense as a lovely environment. We should all have the opportunity of shaping it, particularly in the fashioning of our own dwellings.

We make clear that there are ways of building that make it easy for people who do not have any building experience to build for themselves and that there are ways of financing schemes that can enable people on the lowest incomes—out of work even—to build their own houses.

We hope that people who have never thought about building their own houses will see that the idea has something for them, that it is something they can do. And having offered some inspiration, we will back it up with good, practical advice on how to get going.

But before you decide to undertake this great task —one that will fundamentally affect your whole life —you need to be sure you are embarking on the right course. We must begin by suggesting answers to your question, 'Why should I do it?' You will certainly be saddling yourself with a great deal of hard work. You will be taking risks. Are the rewards commensurate with the effort and even hardship you will be called upon to endure?

Our own experience

We all need a roof over our heads, and if that were all we got from self-building we would have to admit that there are easier ways of getting one. But we have achieved much more. We took matters into our own hands and found we had done something to the quality of our lives. The roof over our heads is more to us than a roof; it is our roof and it is a good roof. We have made our own decisions about design and quality instead of having them imposed on us.

Greenways: Brian and Maureen's first house self-built at Knockholt, Kent in the 1950s

We have both built our own houses and we have both derived a great deal of satisfaction from doing it. We have experienced that feeling of physical tiredness and mental satisfaction that can come from hard manual work (building makes you very fit). We both get great enjoyment from living in these hand-made houses with their particular characters and idiosyncrasies. We can look forward to reaping the benefits of living in good, warm dry houses that are cheap to live in and do not take a great deal of money or effort to keep warm in winter. We describe our experiences in more detail in Chapters 4, 5 and 6 in the hope that they will be an encouragement to you. We believe that you stand to gain as much as we have and we want you to share the enjoyment that self-building has brought us.

Don't make the mistake of thinking that because we were both architects we were at any special advantage. We found our architectural education was seriously incomplete until we had built our own houses. It was sometimes as much a handicap as a benefit to have a professional qualification. There was so much to unlearn.

As well as housing ourselves, we have worked in different capacities with people building their own houses in Lewisham in south-east London. We have seen groups of people from the local housing-list design and build their own houses and get the same satisfaction, pleasure and excitement and enjoy the same benefits that we did.

We all learned by doing. Once started, we realized we were capable of more than we had believed possible. In overcoming our lack of practical building experience and the gaps in our skills and knowledge, we have surprised ourselves, and so will you. We have seen people grow in confidence as they acquired new skills, overcame difficulties and finally moved into the new houses they had created. They became proficient not only in building—learning how to handle a power saw accurately and safely, for instance—but also in how to negotiate with a local authority and how to organize a group effectively.

We also found that we were not in it all on our own. 'Self-build' is something of a misnomer: a self-builder soon finds that family, friends and the community around are warmly interested and quite prepared to become involved in the work. We found, and you surely will, that you soon become part of a network of people with various skills, all prepared to help one another. Mutual aid is the mode. It takes the place of relying on the established providers. Everybody helps one another and nobody takes control over decisions you want to make for yourself.

Chapter 2

Sources of inspiration

We introduce two of the people, Walter Segal and Christopher Alexander, who have, in their different but complementary ways, opened our eyes to the life-enriching potential of self-build.

Walter Segal

An inspirational character whose name will keep cropping up in this book is Walter Segal. Walter died in November 1985 but he continues to be the decisive influence on our thinking—you might almost say he is our hero.

He left Switzerland while still a young man and eventually settled in Britain, partly because he liked our tradition of improvising ways round difficulties. He was certainly as unorthodox and eccentric as any Englishman. He spent much of his career running a fairly conventional practice—designing office buildings, factories and homes for wealthy clients, and letting and supervising building contracts—but working in an unorthodox way. He employed no staff or specialist consultants. He felt he could be responsible for engineering and quantity surveying services, for instance, only if he did the work himself. He thus avoided the traps that certain architects fall into of designing a building form and then asking an engineer to find a way of holding it up—often finding the answer ungainly and expensive, or designing regardless of the amounts and quality of the various materials going into a building and then asking a quantity surveyor to measure it and tell them what it will cost. All too often the quantity surveyor's advice on how to save the inevitably excessive costs of the architects's proposals drastically affects the design, and in extreme cases even controls it.

Walter did not like employing architectural assistants for whose work he would have to be responsible and who, in orthodox practice, are often relegated to the position of hack draughtsmen. He therefore devised a very simple method of drawing rapidly, doing everything himself freehand on foolscap sheets of paper. He didn't employ a secretary/typist, so wrote almost no letters (Brian treasures one from him, written on a visiting card!), doing all his business on the telephone between the hours of nine and ten.

He was a very nice man, and he expected, and usually got, the best out of people. He was unhappy with the conventional competitive tendering procedure, where contract documents are circulated to an approved list of contractors, any one of whom, giving the lowest price, gets the job. He would circulate the documents and receive bids, then interview all the tenderers and accept the one he *liked* best! Very

9/2/77

Dear Brian,
Herewith the 1/1250 & 1/500 scale layouts for the site of the Italian Villa. I hope they are self-explanatory. There are three parking bays and entrance is through the existing gate. However the entire site, except the 30ft strip is needed.
Kind regards Walter

A Segal business letter!

shrewd really, as the lowest tenderer is usually too low and is desperately trying to get out of the job without loss, cutting corners and battling with the architect all the way. A contractor and architect who get on well are at least a *possible* recipe for a successful building. But even this did not ensure a sweet result and he became increasingly inclined to design only for clients who were going to build for themselves without contractors. By the time we met him he was ready to go even further.

Walter was dissatisfied with serving only the needs of clients wealthy enough to commission and build their own private houses. He wanted to get a public housing authority to let him demonstrate how the timber-frame method he had devised could benefit people on council house waiting-lists without much hope of an offer ever coming up; people of ordinary skill and intelligence who had no land and no money in the bank to finance a building scheme.

In Chapter 7, where we recount the story of Walter Segal's success in applying his gifts to the pioneering Lewisham self-build project, and in Part Four, which is devoted to the practical method of timber-frame construction 'the Segal way', we shall try and convey to you the essential common sense of his approach to design and building.

John McKean, in his monograph '*Learning from Segal*'[1] spells out the six key characteristics of a typical Segal building: first, structurally clear and calculated; second, having conceptual clarity and openness—the pieces can all be read; third, designed for giving satisfaction both to those who made it and to the occupants; fourth, client satisfaction; fifth, compact and appropriate planning; sixth, built very cheaply with no waste in material or time.

The self-build schemes he did under the sponsorship of the London Borough of Lewisham are generally acknowledged to be the summit of his distinguished career. Yet their successful completion was followed by a long pause, and no other local authority came forward to take advantage of the innovative path that had been trodden. We discuss some of the obstacles that checked the hoped-for surge of activity in Chapter 26.

Even while the first scheme was in progress some people foresaw the need to spread the word to other intending self-builders in an organized, systematic way, but it was not until a year after his death that a group of us got together and set up the Walter Segal Self-Build Trust[2] to further this purpose. It has charitable status, as befits its educational aims, which are:

> to advise people who wish to build their own home. We also advise local authorities, housing associations and other institutions which may not be aware of the potential of self-build, but who are in the position to promote such schemes. This advice covers such matters as how to establish a scheme, the costs involved and the financial options available. As well as the Segal Method, we can also advise on other self-build methods where these are more appropriate.

As the Trust has grown, it has developed a link with the Centre for Alternative Technology[3] where Walter Segal, in company with a team from Lewisham, tutored a self-build course each year throughout the decade before he died. The CAT is represented on the Trust's management committee, and the training courses have been further developed, to the extent that in the summer of 1987 (International Year of Shelter for the Homeless) an entire demonstration house was built by participants in a ten day course and some subsequent weekends at the Centre for Alternative Technology. A programme of self-build courses continues. Both these organizations are at the service of readers who want to go further than we are able to take them in this book.

Christopher Alexander

There are two books that have been immensely useful to us. One carries the magical title *The Timeless Way of Building*,[4] which immediately stirs the imagination of anyone who has longed to build with the sureness of taste, judgement and good sense that seemed to come naturally to our ancestors. The other, *A Pattern Language*,[5] has a title that needs explanation for its significance to us to become clear, but flows on directly from the first book.

Christopher Alexander, the principal author of both, reminds us that 'there is one timeless way of building. It is thousands of years old, and the same today as it always has been.'[6] He surely touches a chord in all of us when he remarks, 'Whoever you are, you may have the dream of one day building a most beautiful house for your family, a garden, a fountain, a fish pond, a big room with soft light, flowers outside and the smell of new grass.'[7]

He is not referring to an idle dream but to the deep longing that gives birth to our will to build, and he shows us all a way, the 'timeless way', that is accessible to anyone to fulfil it. 'It is so powerful and fundamental that with its help you can make any building in the world as beautiful as any place you have ever seen.'[8] He makes explicit a definable sequence of activities which will generate a building that is alive. 'The power to make buildings beautiful lies in each of us already.'[9]

The difficulty for many of us is that, although 'we know what we like', we have not much exercised our minds as to why we like it. We may respond with pleasure to an old Cotswold village street or a great timber-framed barn or a tile-hung cottage sheltering beside a copse, but we cannot put our finger on precisely what quality it is that moves us so. Alexander says: 'It is easy to understand why people believe so firmly that there is no single, solid basis for the difference between good building and bad. It happens because the single central quality that makes the difference cannot be named.'[10] He devotes himself in *The Timeless Way of Building* to identifying that nameless quality, and at the end we realize that we knew all along what it was but were afraid to say so in case we seemed foolish in the eyes of the experts.

Having grasped the nature of the quality without a name, it is another thing to devise buildings that encompass it. Alexander and his team spent years researching and tabulating the universally recognized features that are common to the buildings we love. They uncovered a language of patterns from which all these places were assembled. For instance, of one pattern, *Light on Two Sides of Every Room (159)*, which he claims 'perhaps more than any other single pattern determines the success or failure of a room', he says: 'Almost everyone has some experience of a room filled with light, sun streaming in, perhaps yellow curtains, white wood, patches of sunlight on the floor, which the cat searches for—soft cushions where the light is, a garden full of flowers to look out onto.'[11]

The next book, *A Pattern Language*[12], identifies and lovingly describes 253 such patterns (to be going on with, as it were, because the language grows as you use it and more patterns suggest themselves—it is an open way of thinking, not a closed system). Every time Alexander identifies a place that lives and takes us to the patterns that went into it, we recognize it. 'If you can search your own experience,' he says, 'you can certainly remember a place like this—so beautiful it takes your breath away to think of it.'[13]

Alexander invites us to employ a pattern language whenever we aspire to build in the timeless way. 'You can use it to work with your neighbours, to improve your town and neighbourhood. You can use it to design a house, for yourself, with your family . . .'[14]

We commend these two books to you warmly as they are truly enlightening: *The Timeless Way* for insights; *A Pattern Language* as a practical reference book to help you make the right decisions in sequence, going from the general to the particular, from the broad sweep to the fine detail.

Chapter 3

Using a pattern language to design in the timeless way

Brian recalls the design process that he and Maureen went through for their house, Romilly, in Herefordshire.

The Alexander books are a wonderful way of ordering your thoughts and deepening your insights but are not a set of formulae that will of themselves produce design solutions. Having considered the appropriate patterns, you still do the actual designing yourself. Your decisions will come out of your own experience: Alexander helps you to recognize things you find you already know.

The books are delightful reading and easy to comprehend. The pattern language has the structure of a network—each pattern connects to the one around it. As you move through it you select appropriate patterns and develop them for your particular situation.

With the overall concept of the project delineated this way, construction can be embarked upon with confidence. Nothing is so rigidly defined that further development is precluded. As the building takes shape and work progresses, adjustments can be made by the self-builder to take into account unforeseen factors, in contrast with the orthodox architect-supervised building contract where the design is frozen at the moment the documents are signed. Any subsequent variation provides excuse for a practically unstaunchable financial haemorrhage. The job thought out in this Timeless Way can be a happy blend of careful planning and last minute improvisation. True freedom!

To demonstrate its application, this is a brief selection from the range of patterns that apply to our own self-built environment in Herefordshire. We give the name and number, and quote the core description of the patterns used (not necessarily in Alexander's sequence), and then describe how we expressed it.

Old Age Cottage (155)
Old people, especially when they are alone, face a terrible dilemma. On the one hand, there are inescapable forces pushing them towards independence: their children move away; the neighborhood changes; their friends and wives and husbands die. On the other hand, by the very nature of ageing, old people become dependent on simple conveniences, simple connections to the society about them . . . Build small cottages specifically for old people. Build some of them on the land of larger houses, for a grandparent . . .
In our particular circumstances this pattern took a slightly different form. The 'small cottage' was already occupied by my widowed mother, and became a suitable habitation for an old person by our building the family home adjoining it. It was to our mutual advantage. She would be looked after in her old age and be able to stay in her own well-loved, cosy surroundings instead of going into a 'home'. We would have a superb building site with space, sun, a view and fertile land.

Settled Work (156)
Give each person, especially as he grows old, the chance to set up a workplace of his own, within or very near his home. Make it a place that can grow slowly, perhaps in the beginning sustaining a weekend hobby and gradually becoming a complete, productive, and comfortable workshop.
I was retiring from my profession, Maureen was getting under way with her new post-housewife one of papermaker. We could create a home and workshop together.

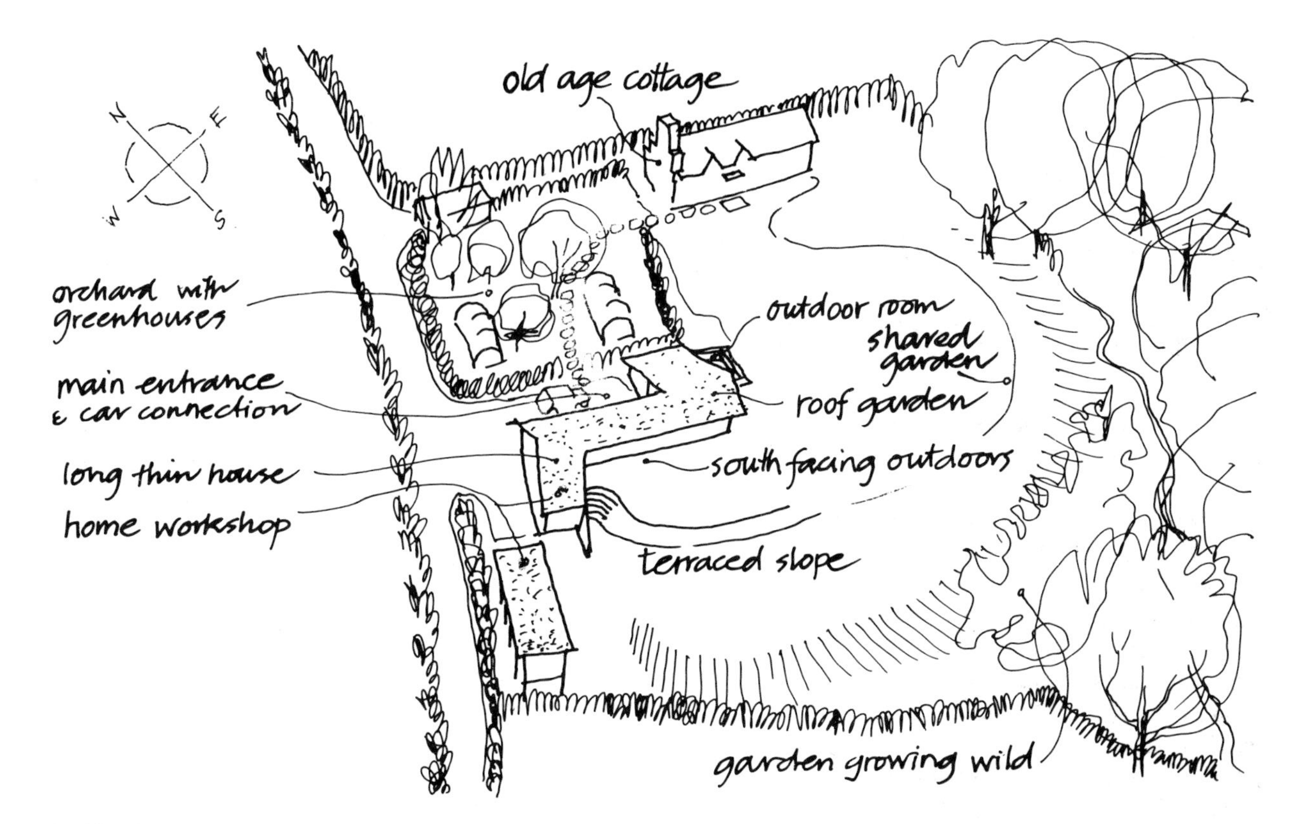

Building Complex, Romilly

Building Complex (95)

. . . Whenever possible translate your building program into a building complex, whose parts manifest the actual social facts of the situation. At low densities, a building complex may take the form of a collection of small buildings connected by arcades, paths, bridges, shared gardens, and walls . . .

The whole site, my mother's part and the portion ceded to us, would accommodate a complex of related buildings. The components would be her old cottage, garden shed and wooden garage in a much reduced and simplified garden; our small house, on a single storey and with wide doors for our own eventual old age; two workshops, one for home building and maintenance and one for papermaking; greenhouses, woodshed and garden-tool store; and eventually, a pavilion summerhouse, all in a productive garden.

Site Repair (104)

Buildings must always be built on those parts of the land which are in the worst condition, not the best . . . On no account place buildings in the places which are most beautiful. In fact do the opposite. Consider the site and its buildings as a single living ecosystem. Leave those areas that are the most precious, beautiful, comfortable, and healthy as they are, and build new structures in those parts of the site which are least pleasant now.

The site in my mother's garden was the dominating factor in approaching the design. It was indeed the scruffiest part of the land and certainly would benefit from Alexander-type 'repair'. It had been used as a poultry run and was overgrown but it had a splendid view from it. It was also blessed with an access road, electric power line, water main from the north; distant view, summer breezes, sun from the south, a tree belt to windward, a part falling away too steeply to use but with an intimate view to the east. A lot of time needs to be spent on site, soaking up its atmosphere and identifying its attributes. We had stayed often in my mother's cottage, but we camped out on the site a lot at this initial stage.

110: Main Entrance; 112: Entrance Transition; 113: Car Connection; 117: Sheltering Roof

South Facing Outdoors (105)
Always place the buildings to the north of the outdoor spaces that go with them, and keep the outdoor spaces to the south. Never leave a deep band of shade between the building and the sunny part of the outdoors.
We paid attention to our neighbours' cottages, which seemed so well suited to their location. They were long and thin, built parallel with the contours at the north edges of their plots, so that they formed a backdrop to their sunny gardens. We too decided to put the new house at the north edge of the undeveloped part of the garden my mother allocated to us, to leave the sunniest, most fertile, loveliest parts of the site to be gardened by her and us.

Long Thin House (109)
In small buildings, don't cluster all the rooms together around each other; instead string out the rooms one after another, so that distance between each room is as great as it can be. You can do this horizontally—so that the plan becomes a thin, long rectangle; or you can do it vertically—so that the building becomes a tall narrow tower . . .
We built long and thin, along the contour of the slope, running east-west, so that the rooms were shallow from front to back and were lit from windows on the south side, so that the main rooms would be sunny all day.

Main Entrance (110)
Placing the main entrance (or main entrances) is perhaps the single most important step you take during the evolution of a building plan...Place the main entrance of the building at a point where it can be seen immediately from the main avenues of approach and give it a bold, visible shape which stands out in front of the building.
The entrance is visible and welcoming to visitors, with a sheltered place by it where one can pause outdoors before being admitted, where one can actually feel a first relationship with the house by grasping a post, or reaching up to touch the sheltering roof edge.

Entrance Transition (112)
Make a transition space between the street and the front door. Bring the path which connects street and entrance through this transition space, and mark it with a change of light, a change of sound, a change of direction, a change of surface, a change of level, perhaps by gateways which make a change of enclosure, and above all with a change of view.
As our east-west house is end-on to the lane down from the main road, it was easy to make a courtyard on the north side with, first, the entrance to the workshop off it and, at the more private end, the house door. We found some nice stable paving bricks for a path along one side.

Car Connection (113)
Place the parking place for the car and the main entrance in such a relation to each other that the shortest route from the parked car into the house, both to the kitchen and to the living rooms, is always through the main entrance. Make the parking place for the car into an actual room which makes a positive and graceful place where the car stands, not just a gap in the terrain.
Like it or not, the car is almost essential equipment in the kind of rural environment we were coming to. Besides being the only way my mother can be transported to the shops, the hospital and to visit

162: North Face—a cascade which slopes down to the ground

118: Roof Garden

friends, Maureen needs a car to conduct her business and we would have many car-borne visitors. A little parking and turning space at the end of the house, with the courtyard, can hold up to five cars when Maureen has students. Usually it is just our car that stands conveniently close to the front door without masking it.

Sheltering Roof (117)
Slope the roof or make a vault of it, make its entire surface visible, and bring the eaves of the roof down low, as low as 6′0″ or 6′6″ at places like the entrance, where people pause . . . roof edges you can touch . . .
When approaching the house, it is good to have the feeling that it is a place of shelter. Even more than the walls, the enclosing roof emphasizes this character. The sweep of the roof, defining the shape of the house, and the overhanging eaves, throwing the rain clear, are very much symbolic of security and comfort. So our roof sweeps down over the front door, and Maureen has fashioned a little figure that peers down at the visitor as if it were saying, 'Hello and welcome.'

North Face (162)
Make the north face of the building a cascade which slopes down to the ground, so that the sun which normally casts a long shadow to the north strikes the ground immediately beside the building.
This car-space entrance-courtyard occupies the shady north side of the building, but needs sun let into it, so the roof slopes up gently from low eaves so as not to block out the sky.

Roof Garden (118)
Make parts of almost every roof system usable as roof gardens...
We were very conscious of the beauty of the landscape we were building in and wanted to intrude as little as possible. Our decision to build long and low below the skyline would help. We thought we must have natural stone walls, as our neighbours had, but we were left with a choice of roof to make.

In Chapter 4 we go into some detail as to how and why we used a sod roof, but at this early stage the decision to do it helped us to relate the home to the topography around. It would also provide us with a special bit of wild garden where flowers and grasses could grow untended—what the Friends of the Earth round here lovingly call 'unimproved pasture', because it does not get disturbed by cultivation and reseeding. There would be lovely views down the Wye valley from it when we wandered up there, but people in the valley wouldn't see much of us.

Intimacy Gradient (127)
Unless the spaces in a building are arranged in a sequence which corresponds to their degree of privateness, the visits made by strangers, friends, guests, clients, family, will always be a little awkward . . . Lay out the spaces of a building so that they create a sequence which begins at the entrance and the most public parts of the building, then leads into the slightly more private areas, and finally to the most private domains.
Some more precise form now began to develop in our minds for the interior spaces. Inside the front door is a semi-public space with tiled floor, coat pegs, a ledge

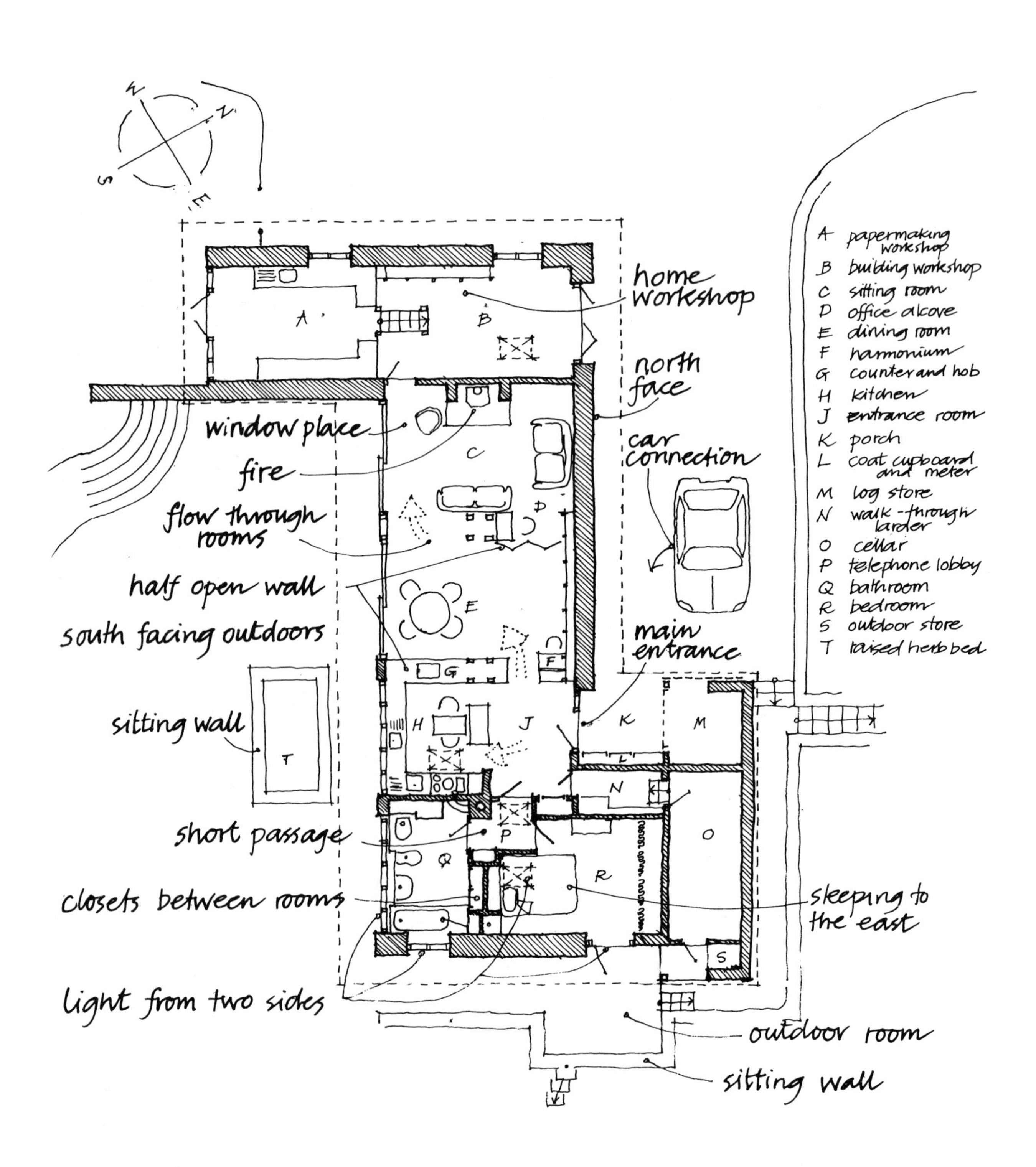
W
N
S
E
home workshop
north face
car connection
main entrance
window place
fire
flow through rooms
half open wall
south facing outdoors
sitting wall
short passage
closets between rooms
light from two sides
sleeping to the east
outdoor room
sitting wall
A papermaking workshop
B building workshop
C sitting room
D office alcove
E dining room
F harmonium
G counter and hob
H kitchen
J entrance room
K porch
L coat cupboard and meter
M log store
N walk-through larder
O cellar
P telephone lobby
Q bathroom
R bedroom
S outdoor store
T raised herb bed

Plan of Romilly noting several of the patterns used

to put things down, a glimpse into the house of things to come—either kitchenwards, if the call is informal, or to the dining space for more formality. Beyond that is the more intimate sitting-space around the fire, or in the other direction, through a buffer lobby, the lavatory and bathroom and the most private place, the bedroom.

Flow Through Rooms (131)
As far as possible, avoid the use of corridors and passages. Instead use public rooms and common rooms as rooms for movement and for gathering. To do this, place the common rooms to form a chain, or loop, so that it becomes possible to walk from room to room.
The impression of space is enhanced in our house by not compartmenting the rooms into separate boxes but allowing them to interconnect in the sequence of the 'intimacy gradient'.

Half-Open Wall (193)
Rooms which are too closed prevent the natural flow of social occasions, and the natural process of transition from one social moment to another. And rooms which are too open will not support the differentiation of events which social life requires...Adjust the walls, openings, and windows in each indoor space until you reach the right balance between open, flowing space and closed cell-like space. Do not take it for granted that each space is a room; nor, on the other hand, that all spaces must flow into each other. The right balance will always lie between these extremes: no one room entirely enclosed; and no space totally connected to another. Use combinations of columns, half-open walls, porches, indoor windows, sliding doors, low sills, french doors, sitting walls, and so on, to hit the right balance.
Our spaces are differentiated by structural frame posts, screens, counters, suspended shelves, but not usually by completely solid walls. The lighting at night is switched to allow the spaces to be articulated as pools of light.

Short Passages (132)
. . . Long sterile corridors set the scene for everything bad about modern architecture.
We managed to do without any corridors, but where a ventilated and sound-blocking lobby is required between bathroom, bedroom and the living rooms, we made it short, well lit and an interesting space in its own right. We put a telephone, some pictures and bookshelves in it.

Sleeping to the East (138)
Give those parts of the house where people sleep an eastern orientation, so that they wake up with the sun and light. This means, typically, that the sleeping area needs to be on the eastern side of the house; but it can also be on the western side provided there is a courtyard or a terrace to the east of it.
We wanted a special bedroom for ourselves, extremely private, compact in space but with lots of bookshelving and somewhere to hang our clothes. We would have glorious views from every other room in the house and thought there should be some place with a more cave-like, comforting atmosphere—and we would be in the room mostly at night anyway. We gave the bathroom the flood of sunshine and the view to encourage us out of bed in the morning, but the bedroom is just lit from the east, so the morning sun falls across the end of the bed. We are sheltered too from the buffeting of the prevailing west wind. A skylight over the bed gives a glimpse of the stars and illumination for daytime reading.

Bathing Room (144)
'The motions we call bathing are mere ablutions which formerly preceded the bath. The place where they are performed, though adequate for the routine, does not deserve to be called a bathroom.'—Bernard Rudofsky.

Concentrate the bathing room, toilets, showers, and basins of the house in a single tiled area. Locate this bathing room beside the couple's realm—with private access—in a position half-way between the private secluded parts of the house and the common areas; if possible, give it access to the outdoors; perhaps a tiny balcony or walled garden. Put in a large bath—large enough for at least two people to get completely immersed in water; an efficiency shower and basins for the actual business of cleaning; and two or three racks for huge towels—one by the door, one by the shower, one by the sink.
Alexander gives much significance to bathing—socially as well as personally important in his view—and we certainly enjoy bathing together. Our bathroom therefore needed to be spacious and beautifully lit and a congenial place to be naked in; quite the opposite situation of the all too common internal washing-and-defecating cupboard, artificially lit and ventilated, cramped and lined with cold surfaces. We instead

144: Bathing Room

gave the best, south-east corner of the house to it so we would have daylong sunshine from windows in both walls, cross-ventilation, a glimpse of the Brecon Beacons, enough room to move about from WC to bidet, from shower to deep bath. We lined it with wood, hung it with baskets of plants, stacked it with magazines—it was to be at least as nice as any part of the house.

Farmhouse Kitchen (139)
The isolated kitchen, separate from the family and considered as an efficient but unpleasant factory for food is a hangover from the days of servants; and from the more recent days when women willingly took over the servants' role . . . Make the kitchen bigger than usual, big enough to include the 'family room' space, and place it near the center of the commons, not so far back in the house as an ordinary kitchen. Make it large enough to hold a good big table and chairs, some soft and some hard, with counters and stove and sink around the edge of the room; and make it a bright and comfortable room.

The kitchen is in the middle of our house and much activity centres on it. It is near the entrance door, is surrounded by waist-high shelves and worktops, the walls are lined with open storage shelves and there is a central table. Cool stores on the north side open off it for food and drink.

Closets [Cupboards] between Rooms (198)
Mark all the rooms where you want closets. Then place the closets themselves on those interior walls which lie between two rooms and between rooms and passages where you need acoustic insulation. Place them so as to create transition spaces for the doors into the rooms. On no account put closets on exterior walls. It wastes the opportunity for good acoustic insulation and cuts off precious light.
All our rooms are separated from each other by thick walls containing built-in storage space; as well as being essential in its own right (Alexander suggests fifteen or twenty per cent of the house area for bulk storage), this serves to buffer sound.

The Fire (181)
Build the fire in a common space—perhaps in the kitchen —where it provides a natural focus for talk and dreams and thought. Adjust the location until it knits together the social spaces and rooms around it, giving them each a glimpse of the fire; and make a window or some other focus to sustain the place during the times when the fire is out.
The natural sequence in the long thin house is from kitchen to dining room to sitting room around the fire at the far end. There we have sitting space that can be grouped round the hearth or the window to the garden.

Light on Two Sides of Every Room (159)
When they have a choice, people will always gravitate to those rooms which have light on two sides, and leave the rooms which are lit only from one side unused and empty. This pattern, perhaps more than any other single pattern, determines the success or failure of a room. The arrangement of daylight in a room and the presence of windows on two sides, is fundamental.
In the rooms at the corners of our house, workshops and bathroom for instance, it was easy to put windows in adjoining walls. An advantage of a single storey

161: Sunny Place

and a low roof-pitch is that roof lights can allow light to fall in another direction than the windows, and we did this. Our experience of living in the house certainly corroborates Alexander's observation that 'this pattern alone is able to distinguish good rooms from unpleasant ones'. It is so nice in the parts of the house where we have exploited 'light from two sides' that I wish we had put two or three more roof lights in.

Window Place (180)
Everybody loves window seats, bay windows, and big windows with low sills and comfortable chairs drawn up to them.
The long south wall gives us the opportunity for low sills and sunny window places along it.

Sunny Place (161)
The area immediately outside the building, to the south—that angle between its walls and the earth where the sun falls—must be developed and made into a place which lets people bask in it . . . Inside a south-facing court, or garden, or yard, find the spot between the building and the outdoors which gets the best sun. Develop this spot as a special sunny place—make it the important outdoor room, a place to work in the sun, or a place for a swing and some special plants, a place to sunbathe . . .
The angle of the sitting-room window with the stone workshop wall makes just such a suntrap on our garden terrace. It falls short of the full requirement of the pattern by being exposed to the south-west

168: Connection to the Earth

163: Outdoor Room; 243: Sitting Wall

wind, though protected from the cooler and equally prevailing north-westerly.

Connection to the Earth (168)
Connect the building to the earth around it by building a series of paths and terraces and steps around the edge. Place them deliberately to make the boundary ambiguous —so that it is impossible to say exactly where the building stops and earth begins.
The paving stones of the terrace are laid with spaces between for mosses and plants; eventually they give way to lawn. The indoor space flows imperceptibly into the outdoors, connecting house, garden and distant view into a whole.

Terraced Slope (169)
On all land which slopes—in fields, in parks, in public gardens, even in the private gardens around a house—make a system of terraces and banks which follow the contour lines . . .
The building, cut into the bank, provides material for levelling a series of garden terraces down the slope to the south, so that when our eyes drop from the distant view, we can see our vegetables thriving. Screened by the bank is our row of compost heaps.

Garden Growing Wild (172)
A garden which grows true to its own laws is not a wilderness, yet not entirely artificial either . . . Grow grasses, mosses, bushes, flowers, and trees in a way which comes close to the way that they occur in nature: intermingled, without barriers between them, without bare earth, without formal flower beds, and with all the boundaries and edges made in rough stone and brick and wood which become part of the natural growth.
As well as the terraced slope, the garden takes its form from its location, sheltered by the woodland to the east, where, on the banks of a steep dingle, trees and plants grow naturally. High hedges break the wind from the north and the wing of workshops encloses it on the west boundary.

Outdoor Room (163)
Build a place outdoors which has so much enclosure round it, that it takes on the feeling of a room, even though it is open to the sky. To do this define it at the corners with columns, perhaps roof it partially with a trellis or a sliding canvas roof, and create 'walls' around it, with fences, sitting walls, screens, hedges, or the exterior walls of the building itself.
As well as our main terrace, which is sunny but very open, there is at the east end of the house a morning suntrap. There we have cut into the bank a semi-sunken outdoor room, paved with brick—a lovely place to breakfast and enjoy a slanting glimpse of the distant view past the wood, completely sheltered from the wind. Opening out from our bedroom, this complements its cave-like quality and gives us the opportunity to sleep outdoors in the summer. It makes the house larger without making the garden smaller!

Sitting Wall (243)
Surround any natural outdoor area, and make boundaries between outdoor areas with low walls, about sixteen inches high, and wide enough to sit on, at least twelve inches wide.
Our outdoor room is bounded by a sitting wall, popular at morning coffee time.

Home Workshop (157)
Make a place in the home, where substantial work can be done; not just a hobby, but a job. Change the zoning

157: The garage replaced with a Segal-method workshop

laws to encourage modest, quiet work operations to locate in neighbourhoods. Give the workshop perhaps a few hundred square feet; and locate it so it can be seen from the street and the owner can hang out a shingle.

The remaining part of our building complex is the home workshops. Where other people would need to devote space for bedrooms for their children or for visitors, we dedicated the equivalent space to workshops. Being at the extreme west end of the long thin house, they can be shut off at a narrow pass-door so that noise and smell do not enter the living quarters, and so that some separation of work and leisure can be attempted, if not totally achieved.

One end of the workshop is for Maureen's papermaking, the other (upgraded from the original appellation garage—it soon became obvious that it was too precious a space ever to put a car or even my motorbike in) is for general building and maintenance.

The requirements of a workshop are somewhat similar to those of a kitchen: a lot of wall space for open shelving; waist-high benches for mounting tools and placing workpieces; good light from two sides; everything to hand for the most common tasks. These needs we met, but we made it too small! So we built an extension, a freestanding timber-frame building that is a covered verandah, the construction of which is described in Chapter 4. Our plan is for Maureen to expand further by taking over the general workshop too, which will then be replaced during the rebuilding of the tumbledown garage belonging to the old cottage.

A Room of one's Own (141)
No one can be close to others, without also having frequent opportunities to be alone . . . Give each member of the family a room of his own, especially adults. A minimum room of one's own is an alcove with desk, shelves, and curtain. The maximum is a cottage . . . Place these rooms at the far ends of the intimacy gradient—far from common rooms.

A pattern we have not yet fully realized has to do with the privacy individuals need within a relationship. Maureen has her own domain as a working papermaker, and has even annexed an alcove in the living room for specialist books and a second desk for business correspondence. But my drawing board, files and writing materials get shifted about the house and the verandah, depending on the weather and the ebb and flow of visitors. So another outbuilding is planned to complete the complex, with the double role of writing room and guest wing—a summerhouse pavilion in the wooded part of the garden with a little wood-burning stove and a paraffin lamp.

Since the publication of the first edition of *The Self-Build Book* I have been reflecting on the question often put to me, 'Is your self-build house finished yet?' and how to reply positively. So I wrote a piece for *Resurgence* entitled 'Building like Gardening'. I pointed out that gardens are continuously worked on and are expected to change over time, become mature and beautiful but never be 'finished'. Why should buildings be regarded differently?

Homes, like gardens, do well to have care lavished on them continuously and gradually. Incremental growth to meet changed circumstances and timely renewal of what has deteriorated beyond usefulness set proper tasks for the self-help building owner.

Just such a point of growth was reached at Romilly when my mother, reaching the age of ninety-six, at last decided that she would be better looked after in the local residential home and gallantly handed her lovely cottage on to my daughter. She and her partner were ready to leave London where their small business barely paid their living expenses.

'Old Age Cottage' (155) was easily transformed into 'House for a Couple' (77) which the pattern says should provide 'a shared couple's realm with individual private worlds' . . . and have 'plenty of room for growth and change', but their needs for a workplace could by no means be satisfied in the garage shed, which was about to tumble down anyway. So we replaced it with a new Segal-method, timber framed, two-bay workshop; the larger half highly insulated and well-heated for their picture-framing enterprise, the smaller half to house my motorcycles, workbench and building tools.

This gave me the opportunity to undertake my own experiment in the technical development of the system Walter Segal had used so straightforwardly and economically at Lewisham (Chapter 7) but which no longer came up to the thermal insulation standards we now consider necessary. We did it without compromising too much the simplicity and ease of construction of his pioneer buildings. We used British-grown structural timber in fairly large sizes and this made room in the framework for plenty of Warmcell cellulose fill between the members. We wrapped up Walter's thin panelled wall in a thick woolly coat of Warmcell and fixed a rainshield of timber boarding outside that (the drawings on p.201 show the build-up of layers). We also put much thicker insulation under the turf roof.

The overall result is that it has proved feasible to keep an adequate working temperature indoors with only a tiny, low tariff night-store heater.

The exercise has increased my confidence, and I have been able to help other people do similar projects. But I haven't quite finished building for myself yet and have now to embark on that unrealized part of our building complex, 'A Room of One's Own' (141).

The pause in the programme while we did the workshops has given me time to rethink the longed for cabin many times over. A windfall gift from my son of secondhand timber has put a new slant on things. This time I have to design 'backwards' from the very sturdy and rather short baulks of Baltic Pine now stacked under the eaves of the new workshop.

I look forward to it as a welcome change from bookwriting!

What do the planners think about all this shed, workshop, studio, cabin building? Fortunately, I don't have to trouble them. In Chapter 20 you will come across the term 'permitted development'. In order to free local planning authorities from the impossible task of imposing planning control over every minor garden structure, a clause in the Town and Country Planning Development Order[1] permits, within certain limits (and you have to study the small print carefully), 'provision within the curtilage of a dwelling house, any building . . . required for a purpose incidental to the enjoyment of the dwelling house . . .' So provided you are the happy possessor of a dwelling with a moderately generous curtilage and stick to the rules you can build away to your heart's content.

Chapter 4

Building Romilly

Brian describes the experience of executing the house and workshop design, and acknowledges his unusual measure of good luck and the unlimited help given by friends and neighbours.

Our own house, Romilly, in Herefordshire has the thick stone walls and regular-sized bays characteristic of cottages in the locality. Early retirement gave me time to devote to it, and building in the country does not need to be a rush job. There is the garden to make at the same time and one has to work according to the season, and one waits for the right opportunity to get the material one particularly wants, or the craftsman to fashion it.

I was extremely fortunate to be able to employ the village stonemasons, who took time off their more urgent jobs of reroofing and repairing crumbling stone chimney stacks for local farmers to enjoy a rare bit of new stone-walling. Not that the stone was new—nothing at all suitable is quarried locally any more—but patient inquiry led me to derelict barns with dangerous gable walls, which for a very reasonable sum (as I was taking on the job of dismantling them) provided me with the thirty-five tons of lichen-covered rubble that I needed.

I collected it in my pick-up truck—a most desirable self-build tool—and helped lay it out in long rows of similar-sized stones on the ground. This helped the masons to find the next piece of the intricate jigsaw puzzle that is the key to a perfect plumb wall, with rough stones set in fine joints. It was the greatest pleasure, working with these two experienced countrymen. I learned so much from them, not only about building techniques, but about how the local economy worked and who could supply which resources locally. Materials, plant to hire and borrow, and skilled craftsmen in wood and metal were all available in the village, and I soon modified my intention to buy specialized items in London and bring them over.

Excavation of the site was done by a neighbouring farmer with a bulldozer in barter exchange for some design work I did for his farmhouse extension. The trenches (and septic tank) were excavated partly by another farmer's borrowed JCB (quite easy for an amateur to control, I was pleased and surprised to find) and partly by spade, pick and shovel. Hard and enjoyable work this, that gives one time to think, and get the real feeling of the site. (We embedded a 1976 bottle of elderflower champagne and a corn dolly in the foundations to propitiate the gods). Then the masons built a couple of courses of concrete blockwork around the perimeter to act as a permanent formwork to the floor-slab, and went back to their other jobs.

We got a big working party together of family, friends and relations to coincide with the deliveries of readymix concrete from the helpful local firm, who agreed to work through the weekend. This was quite an anxious operation, because it all had to be spread within an hour or two of delivery, and had to be levelled well enough to take the floor.

We got all the concrete blocks on site and then the masons came back and built the internal cross-walls and the inner skin of the external walls. They left projecting wall ties for securing the outer stonework, which would follow later. We now had the basis of a structure on which to put timber plates, purlins, beams and rafters for the roof.

I did a lot of thinking to get the roof slope right; to lie roughly parallel to the hillside; to be steep enough to give enough height at the apex for the water tanks to be inside, but shallow enough to allow a wide overhang to shelter the windows without shrouding the view. And, most importantly, not too steep for its turf covering to slip down eaveswards—and be comfortable for walking about on. These considerations produced an angle of fifteen degrees and an overhang of three feet.

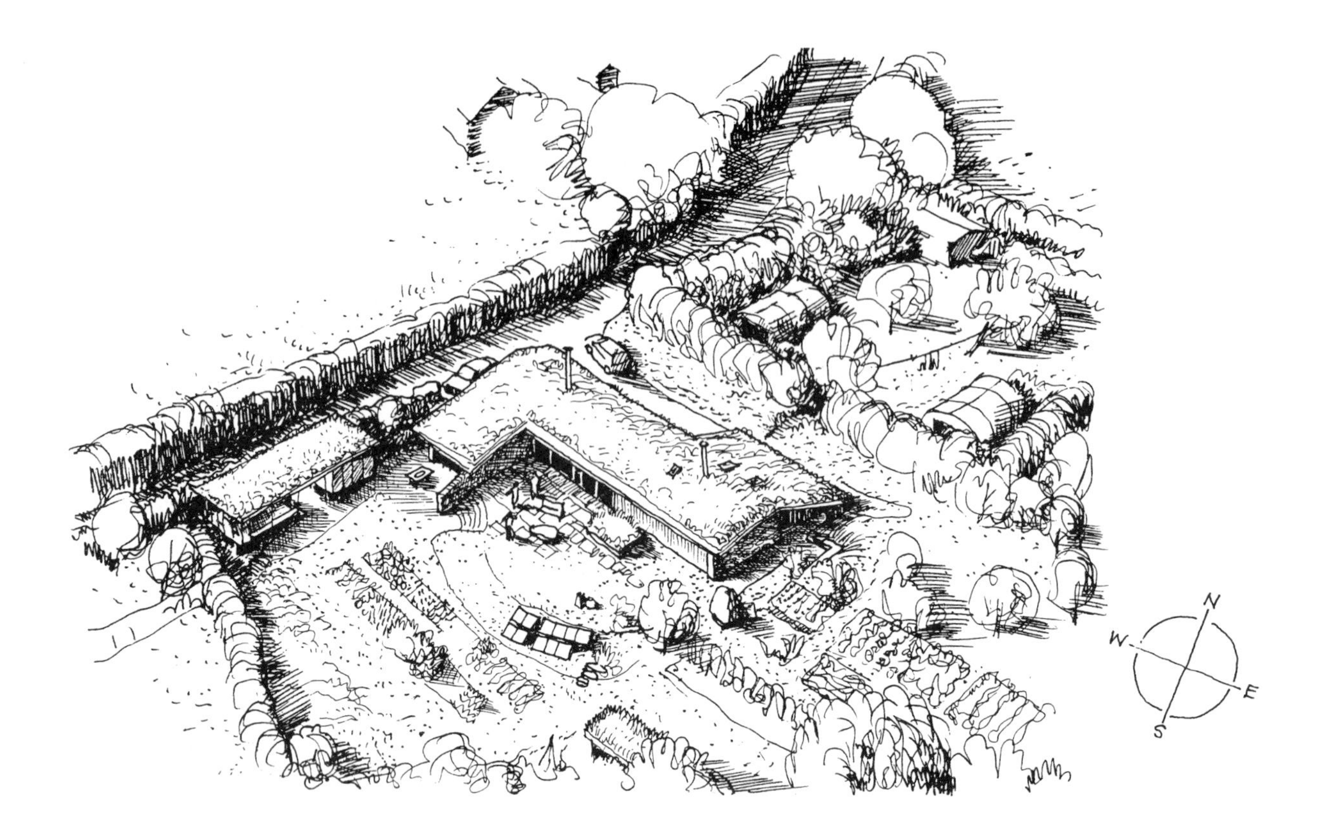

Aerial view of Romilly

Under this wide overhang, the masons could come back when they had time to spare and build the outer skin of stonework, in all weathers. They also had the advantage of firm insulating slabs, Styrodur expanded polystyrene, which I had pushed over the projecting wall ties, to build the rough core of the wall against—leaving only the outer face to find good stones for.

My other building tasks could go on independently, though I had to keep ahead of them with window and door subframes. These frames themselves were made to my design by the joiner in the next village.

The beams were heavy, but just manageable for me and my son on a weekend visit to lift into place, and Maureen and I could manage the rafters. This was the most daunting stage of the job; the house was nothing like a house yet, but there was so much done and so much in hand that there could be no possible thought of cutting our losses and turning back. Happily, Jon Broome and our friend Lil visited us frequently at this time and worked hard and long with us. Lil, being a carpenter's daughter, instructed me in handling a hammer with a sufficiently loose wrist action not to jar my arm nerves into agonized convulsions and how to carry screws on to the job stuck into a pot of industrial vaseline so they drive in easily. Jon reassured me about the design decisions made so far that I was hesitant about, and suggested brilliant little modifications. One needs help like this, and it is always forthcoming somehow.

It may be very sound advice to follow, that the house should be well finished before moving in. It is undoubtedly a complicating factor to have furniture and belongings in the way of the work. But waiting sounded too much like a counsel of perfection for us. We did it the wrong way, and moved into whatever part was weathertight. The roof was now fully boarded and the upstanding wooden kerbs

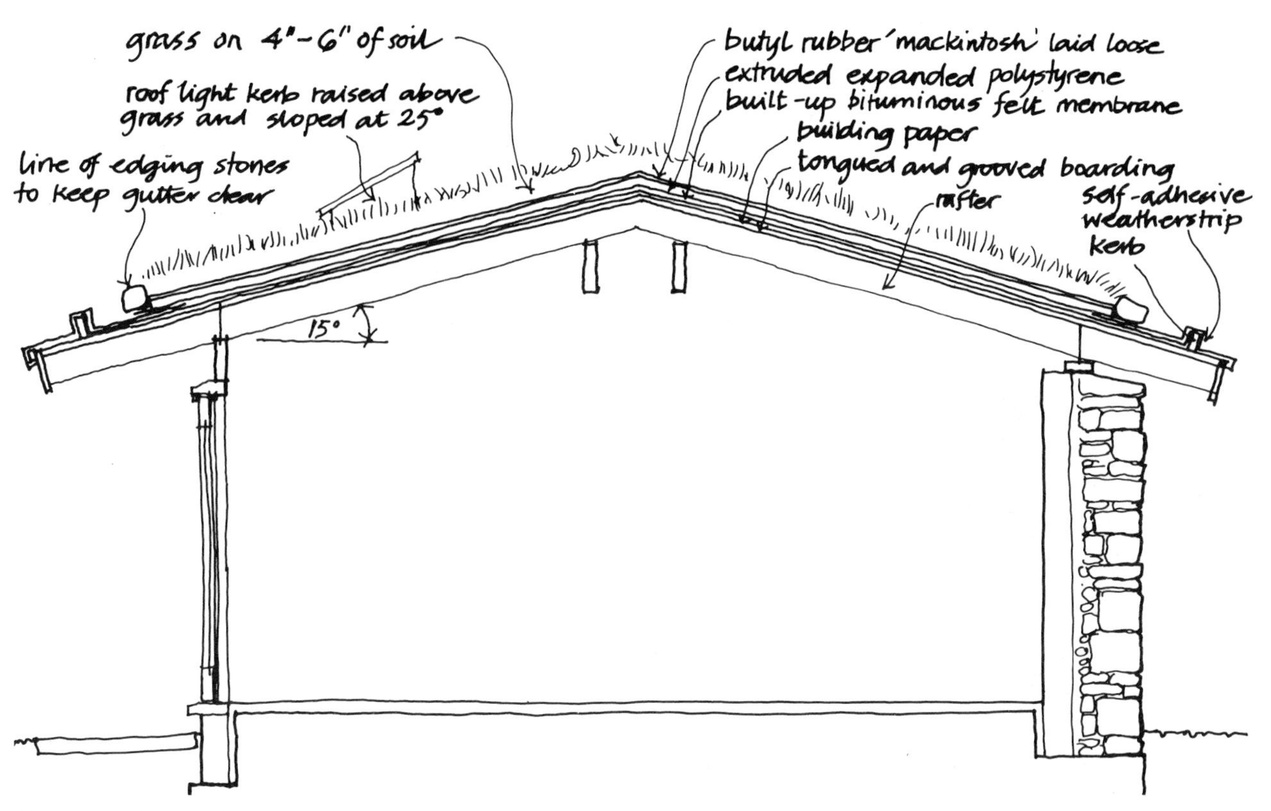

Section through sod roof

fixed on. These we set raking across the bottom of the roof slope to channel the water to outlets at the corners. We were now ready to set about making our unconventional grass roof.

The Romilly roof

This feature of our self-built house always excites comment. Some people are quite baffled by its unfamiliarity and ask, 'What's the idea?' Others relate it to the famous nursery story of the lazy man who grazed his cow on the roof while his wife was away shopping. He passed the cow's halter down the chimney and tied it to his foot while he had a nap. The cow fell off the roof and his wife returned to find him stuck half-way up the chimney.

The roof we put on our own house needed to be weatherproof, stable, good-looking, inexpensive and easy and safe to build. In our setting, the position on a beautiful hillside seemed to demand that it should be unobtrusive. Grass seemed obvious. We knew it would work—we had seen examples in Scandinavia and had put a turf-covering on the felt roof of our previous house as a way of arresting the decay it was suffering from exposure to the sun, wetting and drying and extremes of temperature. We liked what Ken Kern says about Norwegian sod roofs:

> There is no better example of how a native material can be utilized by an indigenous people to their practical advantage. In winter, the dead grass covering holds snow, providing additional insulation. In the spring, rain beats down the resilient, new growth grasses so that thatch-like they quickly shed excess water. In the summer, the full-grown grasses create air movement about their tall stems, effectively insulating the roof and even reflecting the sun's heat.[1]

Rather than add turf as an afterthought, we decided to design it as an integral element of the house this time.

Sod roof construction

First I made a deck of tongued and grooved deal boarding 20mm (¾″) thick (sold as 'economy grade' for flooring, where it would normally span 400mm (1′4″)). Even with the weight of wet turf and snow, the roof load is less than a floor load and the boards are able to span across my rafters spaced at two feet. These boards are about the only things in the house we fixed with nails—one through the middle of each board where it crosses the rafter. Six or seven rows of boarding are laid loose, with the tongues and grooves engaged, and cramped up with long, folding wedges made from scrap board-ends. However tightly the boards are wedged, they will subsequently shrink, hence the single nail in the middle to allow the movement to take place between tongue and groove. Then building paper is stapled down in a double layer, each sheet half lapping the next.

The waterproof membrane

The purpose of the paper is to provide a smooth sliding surface between the boards and the next layer, the bituminous roofing felt. It is important that this should not be nailed or stuck down to the boards. The wooden deck and the felt have differential movements under temperature, moisture changes and physical loading. If they are fixed together, the felt membrane eventually fails. This I learned from Walter Segal; it is the key to his success with flat roofs, and is where other people regularly fail.

Professional roofers usually bond the felt in hot bitumen, but to avoid the dangers of this I used the Shell Composite System of cold-setting bitumen products (Feltfix and Bituproof). We had to put our trust in the weather forecasters—there needs to be a dry spell of a day or two for the water-bound emulsion to dry off or it goes literally down the drain.

A single layer of base felt was lapped and bedded in Feltfix, which has to be very sticky to do its job and is spirit-based. We got well covered in it, but it stayed put. On top was brushed a thick, creamy coating of the emulsion called Bituproof No.3, using a broom and a squeegee. This dried in a day, when and if it didn't rain. A second coat went on, brushed at right angles to the first, embedded in which was a glass-fibre woven mesh called Marglas 193. Another day of fine, frost-free weather and a coat of Bituproof No.5 went on—it has more fibre in it. When that was dry, a last coat of No.5.

Insulation

While this was still tacky, we put down a layer of insulation, the closed-cell, non-absorbent, high-density, extruded, expanded polystyrene called Styrodur. Putting the insulation outside, on top of the waterproof membrane rather than underneath, is a recognized technique known as the 'upside down roof'. The membrane is impervious, so if it is outside the insulation, and therefore cold, and warm vapour from the room below reaches it, condensation will form. This will drip back into the insulation layer, rotting it, discolouring the ceiling and making the worried owner think the roof is leaking. If the membrane is kept warm, as in the upside down roof, condensation takes place on some other, more convenient cold surface instead, such as the window pane, where it is visible, easily mopped up and does little harm.

The pool liner

We were a little worried that the damp polystyrene might become broken up by the roots of the grass, so we added a belt and braces measure: an outer loose mackintosh of butyl rubber sheets thrown over the roof before the five or six inches of sods were finally put on. Where the bituminous roof is under the turf and protected from sudden changes in temperature and from the decaying effect of ultra-violet light, it will have a long life, but where it emerges at the edges, it needs protection or it will perish. A wonderful DIY material for protective flashings is a heavy aluminium foil with a mastic backing, which comes under many names. Flashband is one—we used Shell's version, called Self Adhesive Weatherstrip. It is easy to apply if tackled in small panels and on a warm day, but is rather expensive.

Do you mow it?

Everybody asks if we have to get a lawnmower up there; we don't, and we like the shaggy carpet of different grasses and a large variety of wild flowers.

The thin soil does limit the extent of growth. I cut it with a hook once a year.

Performance

We are very pleased with our roof. It has an indefinable comfortable feel about it—nice and solid. When the wind is howling outside, it is reassuring to know that all that weight up there is going to hold everything together. It performs its functions well, it looks good in our bit of countryside, helping the house to blend into the hillside. It is fun to go up on—children particularly love to play there and peer excitedly down the roof-lights at the doll's house interior. They don't seem prone to fall off the edge.

I am surprised more people don't have them.

Romilly—work inside

So, with walls and a roof, did we have the essentials of a house? No, we did not! Monumental as the task seemed to have been, and great as the satisfaction of topping out the structure was, one had to face the fact that the greater part of the work was still to be done.

The sheer amount of it did make me quail at times, but it was extremely interesting work and as each part of it was done there was the glow of achievement. The house got more and more comfortable and useful, but paradoxically the organization of work got more difficult as things began to compete in priority for attention. Towards the end there were always a dozen tasks to choose from and consequently progress overall seemed to be slower and morale could drop a little.

First to be tackled was carcassing—the installation of water and waste pipes, electricity cables, telephone and gas services, all to be done before finishes to walls, floors and ceilings were applied. This took careful thought, of course, as many of the service runs end up in the finishes. There is a practice in the building trade of dividing it into 'first fix' and 'second fix'.

As we were working with friends and relations and had to depend on when they could come, our divisions sometimes got a bit mixed up and sometimes I would be racing to put a bit of partition framing up before my electrician friend unreeled his cable along it. If I lost, he would stick a bit of conduit in the air and I would build round it later, which produced some hilarious bits of bodgery, but fortunately no permanent disasters.

Also a race, in our case, was the desirability of having windows and external doors before chill November. The local carpenter made up the frames to my drawings at the same time as the masons were forming the openings. This has the advantage that the frames are not built in as the work proceeds and subjected to damage—but how otherwise to get a good fit? The answer is to fix rough timber grounds to the openings and rebate the frames to fit against them later, with a cover fillet internally to cover the clearance gap. So, no cement smears on the frames and the work can go on independently: nobody gets held up.

Glazing can be quickly dealt with by specialists, but with due caution in handling large sheets of glass, which surprisingly are as whippy as cardboard, it can be done by yourself. We avoided putty, which needs skill to handle and is difficult to apply a finish to if the frames aren't being painted. We used DriGlaze tape, a wonderful invention. It is a self-adhesive (on one side), soft rubber strip, put on the rebate of the frame, against which the glass is placed, and on the fixing bead so that the glass floats in a rubber sandwich. Safe and weathertight. The rubber is compressed to three-quarters of its width before the beads are screwed home. It is ideal too for glazed doors, because of the cushioning effect.

Some self-builders I know take on plastering with no qualms, but it is not for me. I do not altogether like the bland, smooth surface it gives the interior, though of course it does take decoration well, after it has dried out. We decided to do without it altogether. The inner skin of the walls built by the stonemasons was formed of Lignacite, a concrete block incorporating sawdust in the mix that gives a good enough surface for emulsion paint, and they gave the joints a neat finish. The ceiling is the exposed rafters and boards of the roof deck. So, where to put electric cables normally embedded in the plaster? In most cases they look all right on the surface. Where that would not do I provided duct space, as we did for plumbing and waste pipes. We made the ducts of timber framing to match the dividing partitions and built-in cupboards. I put plasterboard between framing members to increase sound insulation and faced them with tongued and grooved boarding, fixed with screws and easily

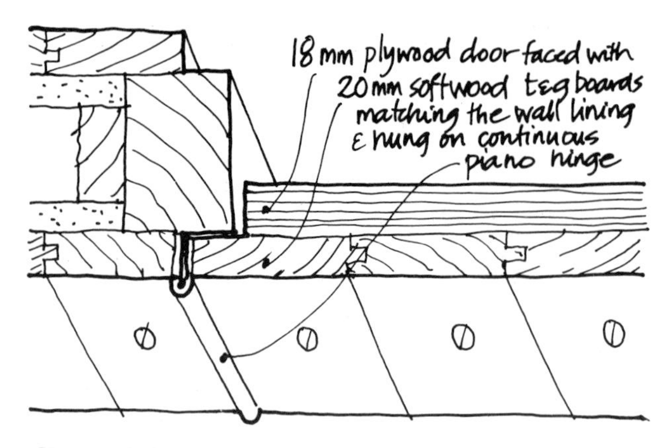

Home made doors

removed for access. I made all doors, cupboard doors, duct covers and wall surfaces the same, so that all presented a sheer surface with just wooden knobs and latches to show where the openings were. To brace the doors I screwed the boards to a 20 mm (¾″) plywood sheet. To avoid the intrusion of strap hinges traditional to boarded doors, I used continuous hinge—known as piano hinge—down the whole hanging side of the door. It is almost invisible and extremely strong, the load being distributed between a large number of small screws.

I liked exploring non-standard ways of doing things and the cumulative effect of such little deviations is to give the house a character that is out of the ordinary—not necessarily better, but distinctive.

Our roof construction of exposed rafters gives plenty of opportunity for hanging things up, such as a clothes airer in the bathroom, strings of onions in the storeroom, bookshelves in the living rooms, storage shelves in the workshops. Suspended shelves are very easy to make and stable for the amount of material that goes into them, and I commend them.

Because I am in the habit of coming in from the garden or the motorbike shed with filthy hands, it seemed a good idea to have 'elbow' taps. These were designed to enable surgeons to wash without touching the taps with their sterile hands; they are nudged on and off with the elbow. They are elegant-looking too.

Instead of prefabricated chipboard cabinets, which are quick and effective but dull, we designed our own kitchen benching, drainers and counter, with open shelves and the occasional drawer sliding on roller-bearing slides—perfectly sweet-running even when overloaded with cutlery and gadgets. The frames they were all built round are in the form of sturdy square wooden brackets bolted to the wall but kept well up off the tiled floor. Quarry tiles are difficult to fix to; the cantilevered construction allows us to get the floor wet and mop underneath everything.

Choice of floor finish is very important; one is in contact with it all the time. For the kitchen we chose quarry tiles, partly because they are robust and easy to swab down and partly for romantic reasons—they have a mature, cottagey look. They are fairly expensive, but a nearby tile works was running down its production and selling 'seconds' very cheaply, so we went over in the pick-up and bought thousands, selecting the best for the places that showed. Laying them well is quite difficult, but one soon gets into the way of it, tapping them into a mortar bed till it flows up the joints between them, then cleaning the surplus off the tile face. Sawdust is excellent for this—it soaks up the cement and is abrasive enough to clean the tile beautifully. Handfuls of dead grass are almost as good.

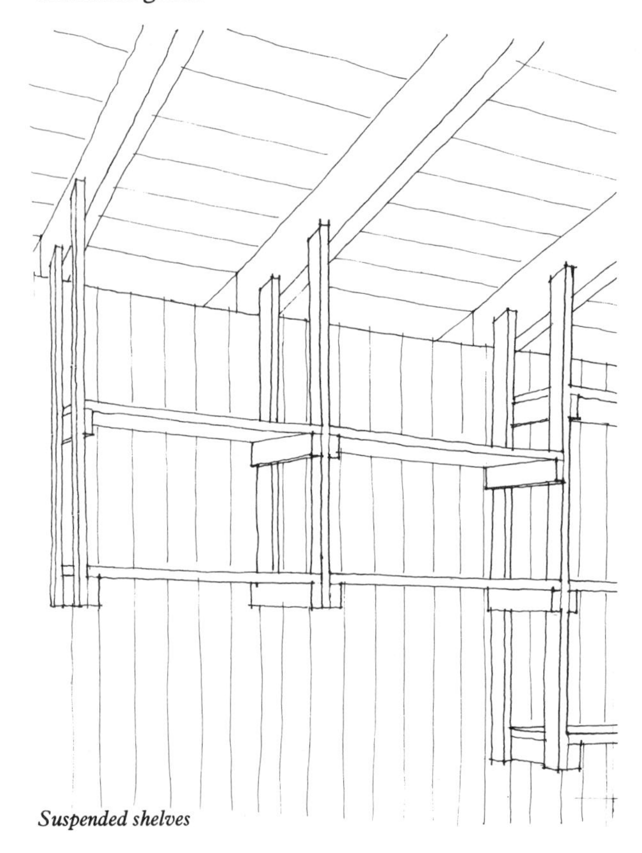

Suspended shelves

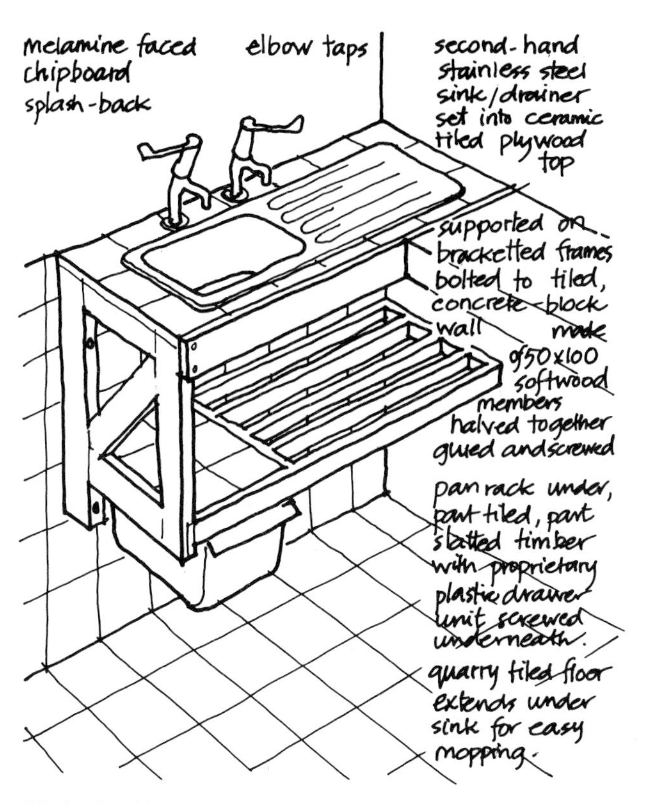

Kitchen benching

For the other floors we had a couple of strokes of good luck. Maureen is an avid reader of the advertisement pages of the local papers, and while we were still in Kent, years before we started building, she responded to a builder who had second-hand maple flooring for sale. A wealthy client had been abroad when the floor was laid and on his return, didn't like the colour! It had been stapled down properly through the concealed groove, so the top surface wasn't too badly damaged when it was ripped up, but it looked an awful mess all the same. Still, at £200 for enough to do a small house it couldn't be refused. We kept it for years in various sheds and greenhouses and five years later, we had to tackle the question of whether we could use it.

Another piece of luck, meanwhile, had been picking up a quantity of new Durabella flooring battens from a completed building site where the material had been over-ordered. The Durabella system is worth knowing about. It is a flooring-grade chipboard, water resistant, tongued and grooved all round and screwed down to timber battens that have glued to their underside a continuous pad of resilient foam plastic. You need only an ordinary tamped concrete subfloor to put it on—it must be perfectly level of course, but needn't be smooth, so you save a sand-cement screed. The battens are not fixed down to the concrete; the whole raft of boards and battens just sits there by gravity and has a nice, slightly resilient, springy feel. No fixings means that the damp-proof membrane doesn't get pierced.

We laid out the battens about a foot apart over the room and put the rough maple strips across them, fastening them down with special staples that are fired in by a clever machine. When it is struck by a heavy rubber mallet, it wedges the board up tight at the same time as the staple goes in. It was quite amazing how the twisted and ragged bits of wood nestled tight together and straightened themselves out as we worked. A day with a hired sanding machine removed the discoloration of years of casual storage and produced a blemish-free, smooth surface, which we sealed with a two-part synthetic lacquer by Sadolins. Fearful stuff to apply, this; even with all doors and windows open and masks on, the fumes could be stood only for a limited time. But on drying, the floor was perfect and has been a joy to us ever since.

We made two mistakes that you could avoid. One was due to my not thinking of floor insulation at an early enough stage. The gaps between the battens tempted me to put offcuts of expanded polystyrene sheets between them under the boards. I couldn't fill the spaces completely because there needs to be a circulation of air (and a slot at the skirting rather than a tight fit to the floor), and from time to time mice play happily in this space, making noisy rustling sounds with the insulation, which they love. We should have put our insulation under the concrete floor slab.

The other was the foolish idea of ordering enough lacquer, while it was cheap, to re-treat the floor later. This spare stock is now well over its shelf life and will probably be useless. Although prices have gone up since, I shall still have lost money.

A money winner, on the other hand, was the lucky acquisition of a set of panel radiators from a building contractor who had ordered the wrong size and so had them spare.

After using some in the house, there were enough

left over for me to make an array of solar collector panels on the south facing bank below the house. The radiators are enclosed in an insulated timber box with a glass front. About six square metres is sufficient surface to heat domestic hot water for a family for most of the summer months and often to make a useful contribution in the winter.

I followed the excellent instruction sheet issued by the Centre for Alternative Technology for the collectors[2] and designed my own pipework system and calorifier to work in conjunction with the orthodox boiler system. The main extra component is an additional indirect cylinder (with an extra long coil inserted) below the normal one and thermo-syphoning into it.

The arrangement diagram may at first sight look complicated but is really simple to follow. The system is entirely self-regulating; no heat can be accidentally lost from the panels which are below the level of the storage tank because it turns itself off when the sun goes in. Circulation can only occur when the sun heats the water at the lowest part of the system to a higher temperature than that at the top. There are no thermostats, electrically operated valves or pump. It is filled with automobile-type anti-freeze solution which prevents corrosion of the steel radiators and frost damage.

It has now operated for six years without any attention beyond maintaining the timber boxes and cleaning the glass.

It has been a great boon to us. Some people on first seeing it say, 'What a pity it only works in the summer when you don't really need it.' But we explain that this is just when we do need it, as we have lots of washing and bathing to do in hot weather precisely when we don't want to light the stove. This is pretty well always burning in the winter and automatically provides enough hot water (with sometimes a little solar pre-heating helping it out). In the half-and-half times, when the low and fitful sun only gets the water warm but not hot, it becomes practical to raise it a few degrees to bath temperature with the immersion heater. (To heat water from cold with electricity is too expensive to contemplate.) But the system does really come into its own in the summer. It is a special joy, after a day of full sunshine, to bask in a piping hot deep bath that comes absolutely free.

Why is it not more common? Many people who like the idea may be daunted by the difficulty of finding room in an existing house to fit in the extra plumbing, but self-building presents you with a great opportunity to avail yourself of the lasting benefit of a solar-heated water system.

The Romilly workshop

The search for space

It was not long after moving into the stone house that Maureen's papermaking overflowed it. What was needed was a sheltered outdoor room, a sort of verandah with protection from wind and rain but open along the leeward side. And she needed more storage space for raw materials.

The site influence on construction

There was no more room on the site to extend the main building along the contour. The answer was to strike out at right angles to the house and the slope and rise above the bank on stilts. Here was an ideal opportunity to learn from Walter Segal and design a similar structure to the Lewisham houses (described in Chapter 7).

A scale model

Unlike rooms in a house, Maureen's workshop didn't need wide spaces free of posts. With closer spacing I could use smaller timbers, and in fact built the whole structure with 100 × 50mm (4″ × 2″) posts, twin floor and roof beams, floor joists, rafters and diagonal braces—half the cost of the sturdy 200 × 50mm (8″ × 2″) that the houses are built of, but it is still far from flimsy at this smaller scale.

Workshop heights could be lower than domestic room heights too, so the whole building is a scaled-down model of a real Segal-method house. Such a building would make a very good rehearsal for anybody wanting to grasp the principles before embarking on the full-size enterprise.

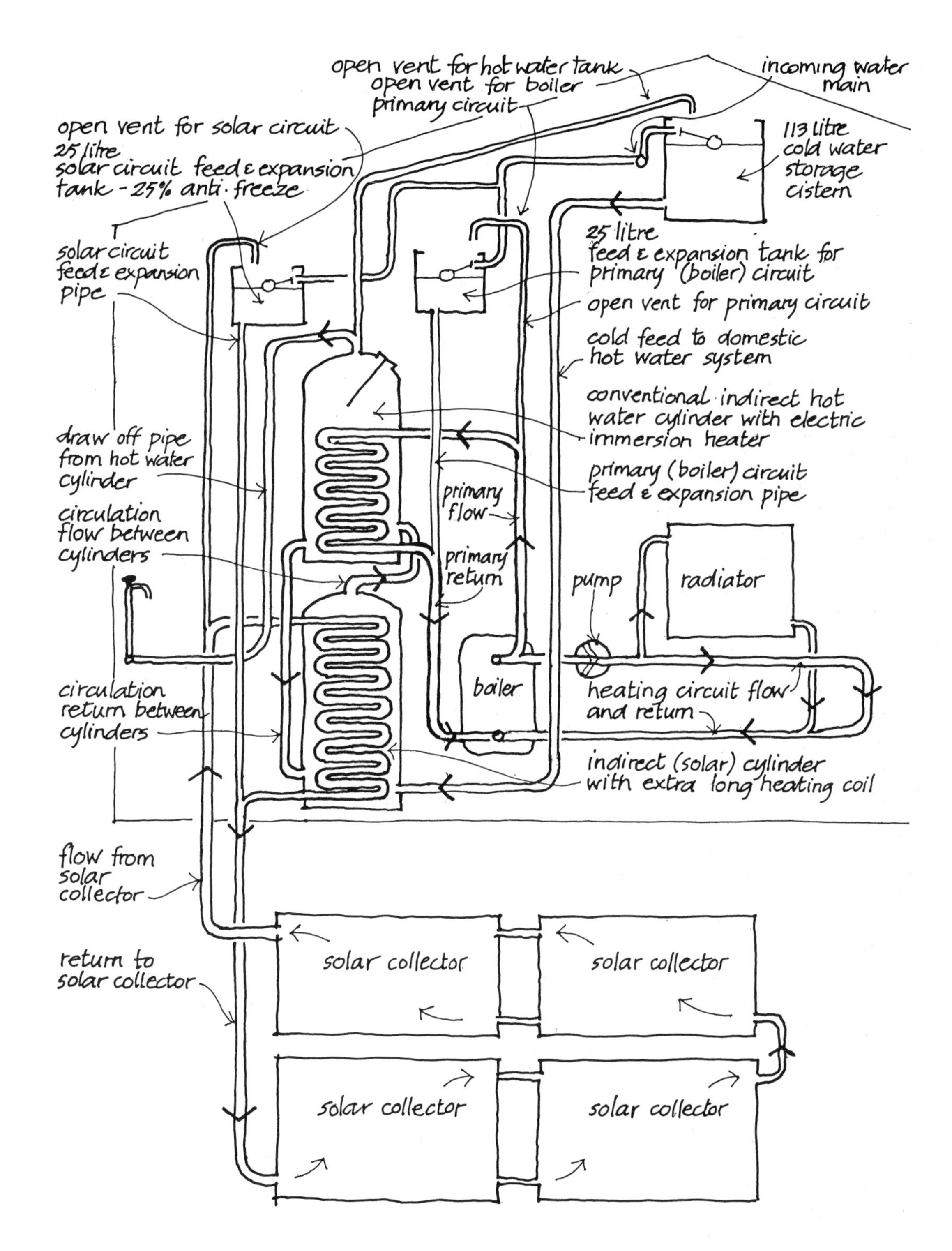

Plumbing diagram for solar collector

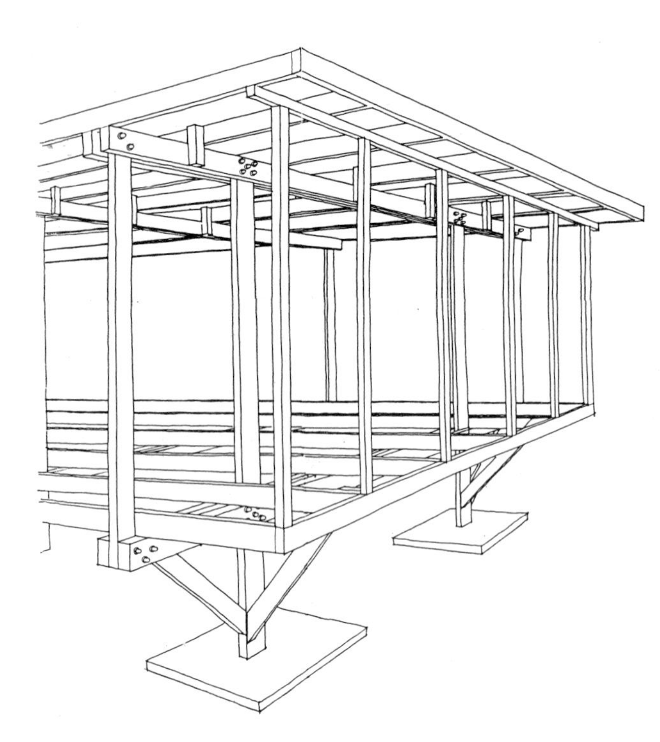

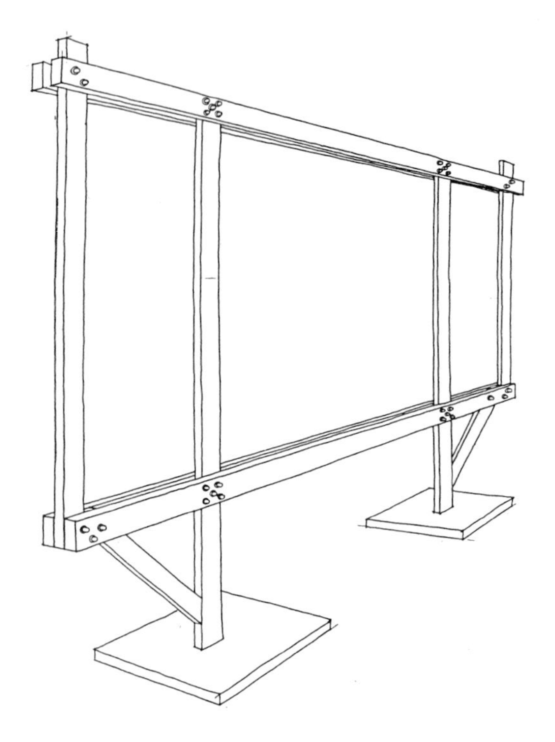

Sketches of the workshop structure and one of the frames

Opportunity to experiment

As a sort of demonstration model of a building, I could be less cautious with the workshop than when building the house. For instance, just to see what happens, I have completely avoided the use of timber treatments. These are extremely expensive nowadays, and are designed to be toxic to fungi and insects and are probably therefore toxic to us as well. Certainly they cause problems for bats. We have relied on good ventilation of the structure to avoid rot.

The building being so structurally simple and the fixings being all bolted or screwed, it will be very easy to monitor its condition and dismantle and repair any defect that occurs. Being a small building, I used small bolts made of ³⁄₈ inch Whitworth-screwed studding with washers and dome nuts at each end. This was a cheap substitute for the specially ordered, large-diameter galvanized bolts and nuts needed for a full-scale house. The studding was cut by hacksaw to the exact lengths required for the dome nuts to be tightened fully before becoming threadbound. A pattern of several bolts at each joint ensured that the bearing strength in shear of the softwood members was not exceeded.

Another innovation was to try a much simpler version of the turf roof. Being a shed, I felt it could do without belt and braces. This time the deck is of woodwool slabs; there is a sliding layer of building paper; the membrane is a single sheet of butyl rubber draped over and immediately weighed down with very cheaply obtained small offcuts of woodwool slabs. These have proved an excellent drainage layer for the sod covering, which I added as material came to hand. Otherwise, the design and method of construction closely follows the Walter Segal method described in detail in Part Four.

The clarity of this minimal building technique is very attractive, and I shall always remember the sheer enjoyment of putting up this little building. By doing so, I learned more of what Walter Segal has to teach us than any amount of academic study could have

Maureen's workshop

shown me. For a material cost of £2,500 in 1986 and some pleasant weekends of light work with friends and relations, we had a useful, flexible space—part summerhouse, part garden shed, part workshop.

In our case we added this building at a fairly late stage in our programme but I could suggest that you *start* your own self-build project with a little shelter on similar lines.

Chapter 5

Building 6 Segal Close

Jon describes his experience of building his own Segal house as part of the first Lewisham self-build scheme.

This is a tale of a remote dream come true largely by chance. Like other designers before me, I had always harboured some hazy notion of how good it would be to design and build for myself, to put into practice some of those ideas that I had developed over the years. I had been much impressed by the work of Walter Segal when a student, and had made a number of unsuccessful attempts over a period of six or seven years to get projects for houses, community buildings and a small office building built using Segal principles. Living as I do in London, the idea of buying a site and building a house was not realistic even on the wages of a professional local government officer. I did not have any financial resources and paid a cheap rent for a very nice, or so it seemed, little terraced Victorian two-bedroom house in Lewisham.

In 1978 I was fortunate to find myself recently returned to London from abroad without a job just at the time when the first Lewisham self-build scheme was turning into a real project. The pieces of the jigsaw were being agreed in principle after three years of negotiation. It was at this moment that Walter Segal was offered a visiting professorship in the United States. The details of the self-build houses needed to be completed for the final approvals and on Brian Richardson's suggestion Walter employed me for the six weeks he was away to do this. On his return he sought the opinion of Ken Atkins, the chair of the self-build group, who gave a favourable report. I lived close to the sites and had local knowledge, including having worked for Lewisham Council, and so Walter asked me to share the job with him from that time on. It was a wonderful opportunity for me to work with him, developing his ideas and putting them into practice. His was an approach that had been in my mind for a long time and had formed the basis of a paper that I had prepared in my spare time while working for Lewisham Council's architect's department. That paper was prompted by my frustration with the limits of what could be achieved within the council system. Nothing had come of it at that time, three years or so before, but Brian Richardson, who had read it, had put in much hard work with Walter Segal while I had been abroad.

I threw myself into the task of preparing the detailed drawings of the structures of the eight basic arrangements devised by Walter that were necessary to obtain the permissions to build. At long last the long-suffering self-builders were able to start to build and I spent many happy hours on site helping and advising. I was learning as much as the self-builders. I had the usual theoretical training and had worked at a drawing board for a few years. I did not, however, have a great deal of practical knowledge to offer about how actually to do things. I knew how buildings ought to end up but often not how to get there. You really have to think when asked by a person down a hole, up to their knees in mud, 'How do you build a manhole?' I had been gaily drawing these squares on plans for years without knowing what was actually involved in turning that symbol into a complicated construction of brick, concrete and clay pipe at a critical level with a metal access cover. I attended the group meetings and slowly learned how to work with groups of people, how to persuade and win people over and when to concede.

Some months into the project I was persuaded by my new friend Ken Atkins to put my name on the self-build waiting-list in the town hall because there was the possibility of a second scheme at some time in the future. A few months later one member of the first self-build group dropped out after starting to build

6 Segal Close

his house. He suffered a hernia—not on the self-build site, it turned out—and that, together with his preference for the pub rather than the building site, meant that he had fallen behind. He did not feel able to continue, so the people on the waiting-list were approached in turn. For various reasons, such as having been offered other housing or not feeling able to take up a building project at short notice, they declined the opportunity to take over the part-built house.

To my surprise and delight, my name came to the top of the list. The council decided that there was no conflict of interest in being both joint architect for the scheme and builder of one of the houses. I approached my sister Jenny to see if she would like to share the building and share the house when it was finished. She had been living in the north of England and her partner, Nick, was still there. They were planning to set up a household together in London and needed somewhere to live. I had always shared houses with people, because I liked it that way and enjoyed having

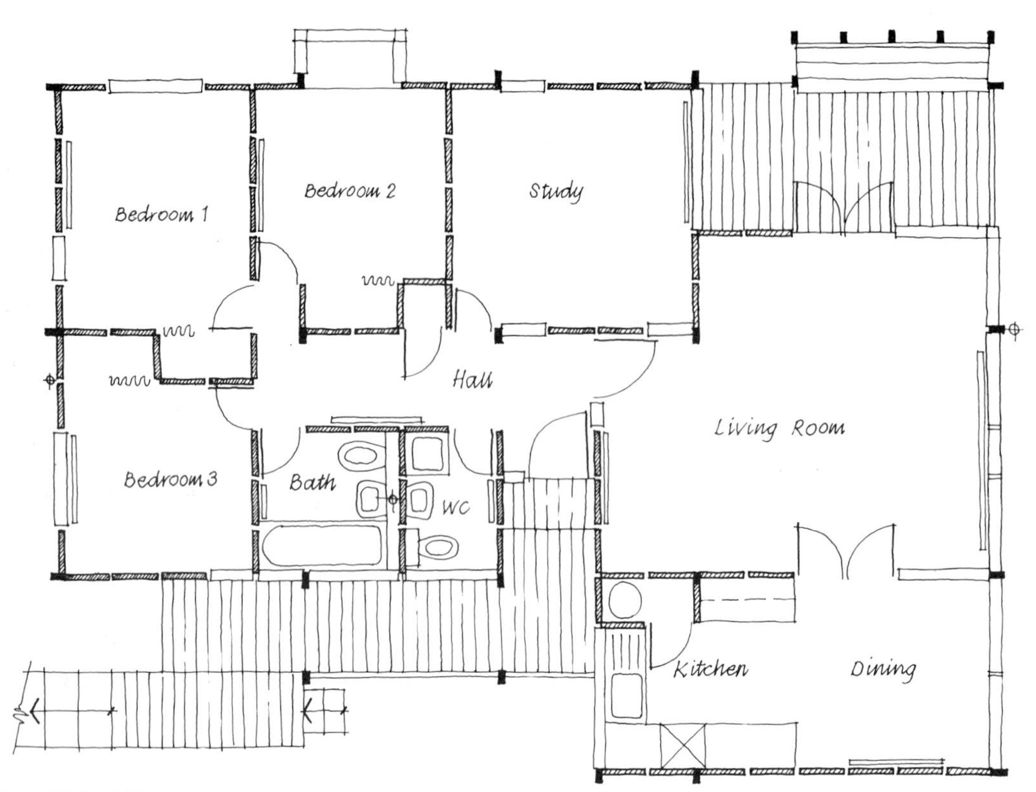

The layout plan of 6 Segal Close

other people around. And so it was that we three came to build together.

I find it really hard to imagine how we did it, looking back. I never seem to have any spare time nowadays, so how were we able to build for at least two full days a week for a year and half? The answer is a combination of single-minded determination, a desire to get it over with and the satisfaction and enjoyment of making something really useful with our own hands. Each of us had to plan a week's building, order the material needed and make sure we were putting in the time on site whilst doing a week's work and keeping another household ticking over, doing the shopping and so on. It meant that we did little other than build, work and worry for that time, relying on the satisfaction of hard work to keep us going. There were times when things did get too much—I think all self-builders experience them.

It was in some ways fortunate that we took over a house that had already been designed by someone else. I think that I would have been incapable of making the necessary choices if we had started from nothing; I would have made the process far too complicated. As it was, the roof and floor of what was originally planned as a three-bedroom bungalow was already built. Because we were building a Segal house, we did have the opportunity to amend the layout of the internal partitions and the position of openings in the outside walls. We decided to make the living room larger by omitting that part of the hallway which would have allowed one to reach the kitchen and dining area without going through the living room. I needed a place to work at home, but we also wanted two main bedrooms and a spare bedroom, so we arranged the rooms to turn what had been a three-bedroom house into what is in effect a four-bedroom house with one of these rooms as a study. The bedrooms are quite small, 3.2 m × 2.55 m (10′6″ × 8′3″) and 2.55 m × 2.55 m (8′3″ × 8′3″), but they are adequate as basic bedrooms. One of the hardest parts of the whole building process was making changes in the plumbing and wiring that had

The living room gives a really open feel

already been incorporated into the floor construction, so that walls that had radiators and electric sockets on them could be relocated. The location of services is the main limitation to the adaptability of houses of this kind.

We rearranged the size and position of windows from the original plans to take advantage of the panoramic views of the hills of Crystal Palace and to give lots of sunshine and light on two sides of almost every room. The effect is of a very light and airy interior. I had not realized how gloomy my little Victorian terraced house was by comparison. We decided to use Walter's original simple sliding-window design (without normal frames) in spite of the amount of careful work involved in making them. They fit into the overall look of the building so much better than any other type of window. We made one or two refinements that improved the airtightness of the design; by adding a clear plastic channel section that you buy in DIY shops for secondary glazing to the edge of the sliding pane, for instance. It was quite difficult to work out the shape, size and opening parts of the various windows so that we got ventilation in the right places—each room with a small window for ventilation in winter and with a large window to open and lean out of in summer, as well as low window sills in certain rooms and windows which were well proportioned when viewed from both the inside and the outside. Each elevation was quite different from the next—horizontal glazing facing the view on one side, small windows on the north and so on. We decided to have single glazing to save time and money in the first instance, with a view to installing double glazing later on. This has not happened yet and it is just one of the jobs on the list of things that should be done one day, along with fixing a towel rail in the bathroom and putting the house number on the door. It is in the nature of self-build that the process of building is never really complete.

We added a number of features to the house: a hanging seat on the verandah, a bay window and a window seat in one of the small bedrooms, to give it

Jon and the hanging seat on the verandah

more of a feeling of space and a view, internal windows between the study and living room and between the dining room and the living room, to increase the feeling of space while still enabling these rooms to be used separately without disturbing other people (Nick plays the flute). We built a walkway and wooden steps spanning over a bank leading down to the house from the access footpath. When you approach the house across this 'gangplank' you leave solid, earthbound feelings behind and enjoy the sensation of a building floating above the ground.

We had to work out a colour scheme right at the outset—one feature of building a Segal house is that once the roof is on and before the walls are built you start the decorating! We took over a pile of chestnut-brown-coloured external cladding sheets and brown stain rather badly applied to the exposed parts of the structural frame from the previous builder. We decided that we would have to apply black stain to the woodwork to give a good finish and a good colour scheme with the panel colour. Black is, however, not generally recommended for timber exposed to the sun as it gets very hot and can lead to shrinkage cracking. The window frames are picked out in a yellow pine colour to contrast with the other woodwork. Inside, natural timber is set against white ceilings and off-white walls in some rooms, with bright primary colours in others. We paid extra for tongued and grooved boarding to the ceilings in the hall, bathroom and shower room. This gives a good contrast between the feel of different spaces in the house and releases one from the straitjacket of the consistent floor and ceiling planes of most interiors. We kept the kit of basic white sanitary ware and kitchen fittings supplied through the bulk-ordering arrangements of the local authority that came with the house. Bright-yellow Formica splashbacks in the kitchen and around the bath and the shower, carpet in the living room, lino in the kitchen and red studded floor tiles in the bathroom and shower room (supplied and fitted by Ken Atkins, who is a floor layer by trade and who happened to have some over on a job—it is surprising how many people you know who have something to offer to a self-build project), with sanded and sealed softwood floorboards elsewhere, completed the internal finishes. The reaction of our neighbouring self-builders was one of astonishment at this colourful interior; they tended to go for a more traditional 'olde worlde' look, with dark-stained timber beams and white painted panels between.

We devised various interior fittings that make all the difference in giving a carefully thought-out and finished look to the house. The bathroom has the same tongued and grooved boarding used for the ducting to conceal the pipework and lavatory cistern. The mirror is removable to get access. Glass shelves are fitted at the end of the bath. We invented a way of using surplus timber to make shelf units that can be fixed to the vertical battens. It is very easy to fix anything in a house with a piece of wood on the wall every two feet. The front door is glazed and is set in a glazed screen, with the letter box cut through a piece of perspex. We fixed a ladder down from the verandah to the garden and up on to the roof for maintenance and to make it easy to pick the damsons off the top of the tree next to the house.

The actual building seemed endless. We would go home after a hard weekend's work and nothing looked as if it had changed. One expected things to take twice as long as anticipated, but it was a bit of a shock to find that many things seemed to take three times as long! Having cut mitres on window beads for what seemed like for ever, we worked out that the house required well over 1,000, each needing to be measured, marked, cut, adjusted and fixed! I don't remember feeling deprived of the things that I now do in my spare time, such as going to the cinema. I do remember a few evenings in the pub after a hard day's work, so tired that a silence would fall and glazed looks come into our eyes. This would be followed by collapsing into a hot bath, exhausted but satisfied. I became fitter than I have ever been and knew the hit parade by heart—which I do not now. We all learned a lot of new skills.

We had a couple of parties where friends were invited to work in return for some music, food and drink. They were a great success, everyone getting into a frenzy of activity. It was difficult trying to keep everyone supplied with tools and materials while ensuring that people knew what they were supposed to be doing. At the end of the day you really could see the difference, and we didn't have to undo much—except that all the light switches were fixed upside down, I remember! People came round when we moved in and remarked on their bit of handiwork. Friends had done things they never believed they could do—wire a house, for example. We decided to complete as much as possible before moving in because we anticipated that it would be difficult to get motivated to pick up a saw for a while afterwards. This has been largely the case, activity since having been limited to recoating all the woodwork with stain; and that was a job which in itself took three seasons to complete, not because it was a big project but because there seemed to be lots of other things to do.

Moving in was an odd feeling, having known the place intimately for months and months, every corner of it, but as a building site not a home. It seemed peculiar at first not having a pile of wood in that corner any longer, and not having sawdust in your hair or in great drifts on the floor. Everything was new and exciting. It was so bright and sunny. It was like being on holiday in a Californian beach house, and that feeling has not really worn off. We were moving two households into one house, which was made a lot easier by the spacious 'cellar' area you get below a Segal house. To our surprise the furniture fitted through the narrow 600 mm (2′0″) doors that are a feature of many of Walter Segal's designs. The exception was my plan chest, which had to have its top taken off. But then, not that many people possess items as bulky as this.

Jenny and Nick are both landscape architects and they planned the garden. It is small and on sticky clay soil. They did a marvellous job and it has grown into an exuberant jungle of exotic foliage that requires hardly any attention apart from cutting the lawn and hacking back the jungle from time to time. It was completed by a fence built of pieces of timber left over from the building. The design owes a lot to the architect Charles Rennie Mackintosh. Jenny and Nick have now moved back up north, leaving me in possession of the house and garden.

Chapter 6
Building again

Jon describes designing and building a second house.

After living happily at Segal Close for over thirteen years I am enjoying moving into another self-built house. There have been great changes over that period. My sister and Nick, now her husband, live in a seventeenth-century stone-built house in Yorkshire. Meanwhile, I had a relationship with Rona Nicholson who moved into Segal Close a week or two before our first child was born. The adjustment to sharing the house again and accommodating her belongings was closely followed by adjusting to sharing it with a new born baby.

It was to be hoped that Alex would not be the only one. We got to thinking that we would be wanting more space both inside and outside the house one day soon. Segal Close enjoys a magnificent view over other people's enormously long gardens. Although this gives the impression of great spaciousness, more like living in deep country than in the city, the house itself only has a very small garden. We approached the neighbours with a view to buying the end of their garden to put the extendibility of a Segal house into practice. They were not willing to negotiate. There followed a dispiriting and increasingly half-hearted search for a site to start a new house from scratch. We looked at overpriced scraps of land around South London. We were about to abandon the idea when we learned of a garden at the back of a friend's house half a mile away in Forest Hill which its owner was keen to sell. So started the new self-build project.

I am going to recount the intervening three years as a self-builder's diary and thereby hope to capture some of the feeling of what it is to build a house; to describe the strange blend of dreams and mundane work, how the process affects the different people involved and hopefully to bring to life that mysterious process by which idea is turned into reality through the exercise of skill and perseverance.

The dream

I had built but was not the designer of Segal Close. I had designed for other self-builders but had not had to carry out the designs myself. I wanted to experience the intimate relationship between house and occupant that could come from fulfilling my dream of designing and building my own home.

Rona had a slightly different vision which nevertheless amounted to much the same thing, which was to live in *our* house rather than *my* house, a house that we had created together. It was to have no timber battens on the walls inside and be set in a large garden.

May 1991: the site

We visited the site and were enchanted. It was a square of 30 metres set at one end of a block of back gardens between houses adjacent to an alley. Half the area was occupied by a large lawn surrounded by mature trees and borders planted with roses. One quarter was an orchard with a pond. The final quarter was an overgrown vegetable garden complete with greenhouse and shed. The area between the houses had been occupied with tennis courts during the twenties and thirties and the shed was the old pavilion. Rona fell in love with the place as I had done and we resolved to proceed.

June: the offer

We sought advice on the value of the plot. This is related to size, position and outlook but most importantly to the development potential of the land. Enquiries at the Town Hall suggested that the Local

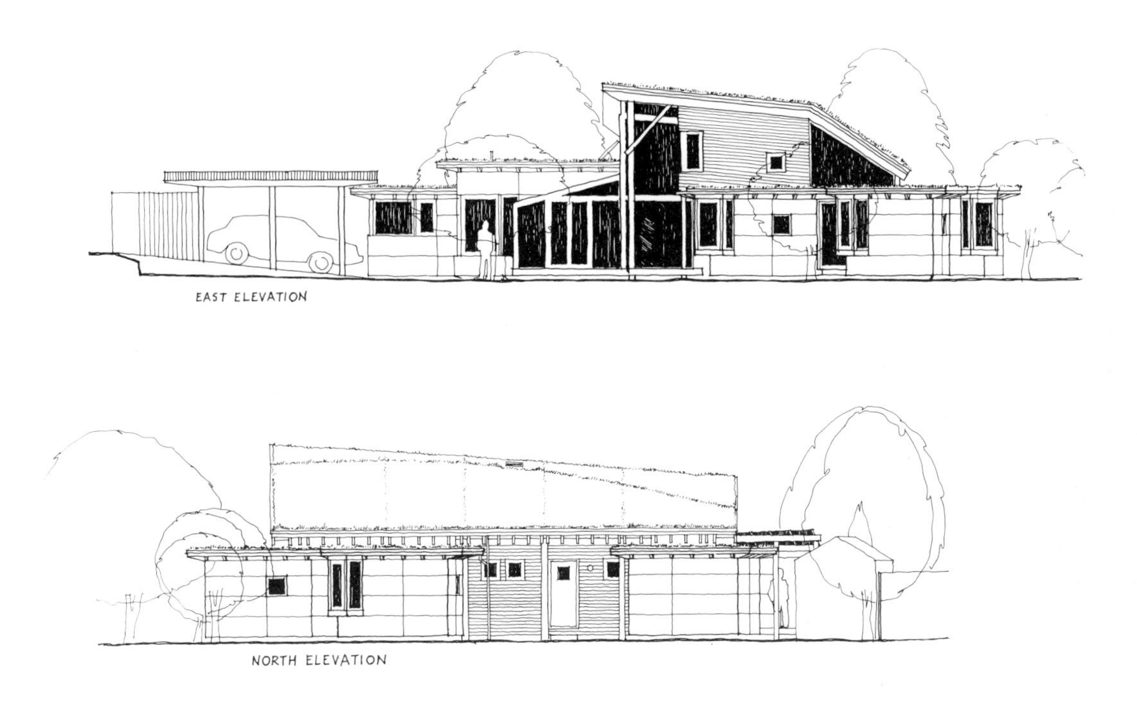

Elevations

Authority would only give planning permission for a single house on this parcel of land but it could be a large one. We wanted three children's bedrooms, a bedroom for us, a spare bedroom, a family room, a more formal living room and a study. That is a total of eight habitable rooms. The cost of land in a particular area can be expressed as the cost of land per habitable room. At the time the range in London was between five and seven thousand pounds per habitable room depending on location. We offered £48,000 for the site, which was £6,000 per habitable room for the house we wanted to build. It was a difficult judgement, wanting to offer enough to secure the land but not more than necessary. This was all made possible because when Rona moved into Segal Close she sold a house in Greenwich, the proceeds of which allowed us to offer cash for the site. Some back of envelope sums suggested that our savings would go a good way towards paying for the building—and so we decided to proceed. Much to our relief the offer was accepted.

July: the solicitors

A number of critical questions had to be answered before we would know if it was possible to build the house we wanted in the garden. Would the planning authority consider that the neighbour's privacy would be unreasonably reduced, would we be able to use the alley for vehicular access, would we be able to connect to the sewer in the main road? We instructed a solicitor to produce a draft contract of sale which would be conditional on our obtaining planning permission. The usual process of searches revealed some curious covenants but no problems with obtaining water, gas and electricity on the site. Vehicular access was negotiated with the local authority and it was decided to route the drainage through the vendor's adjoining garden, past the side of his house and into the main road. Preliminary discussions with the planners confirmed that they would recommend to the planning committee that permission would be given for a single

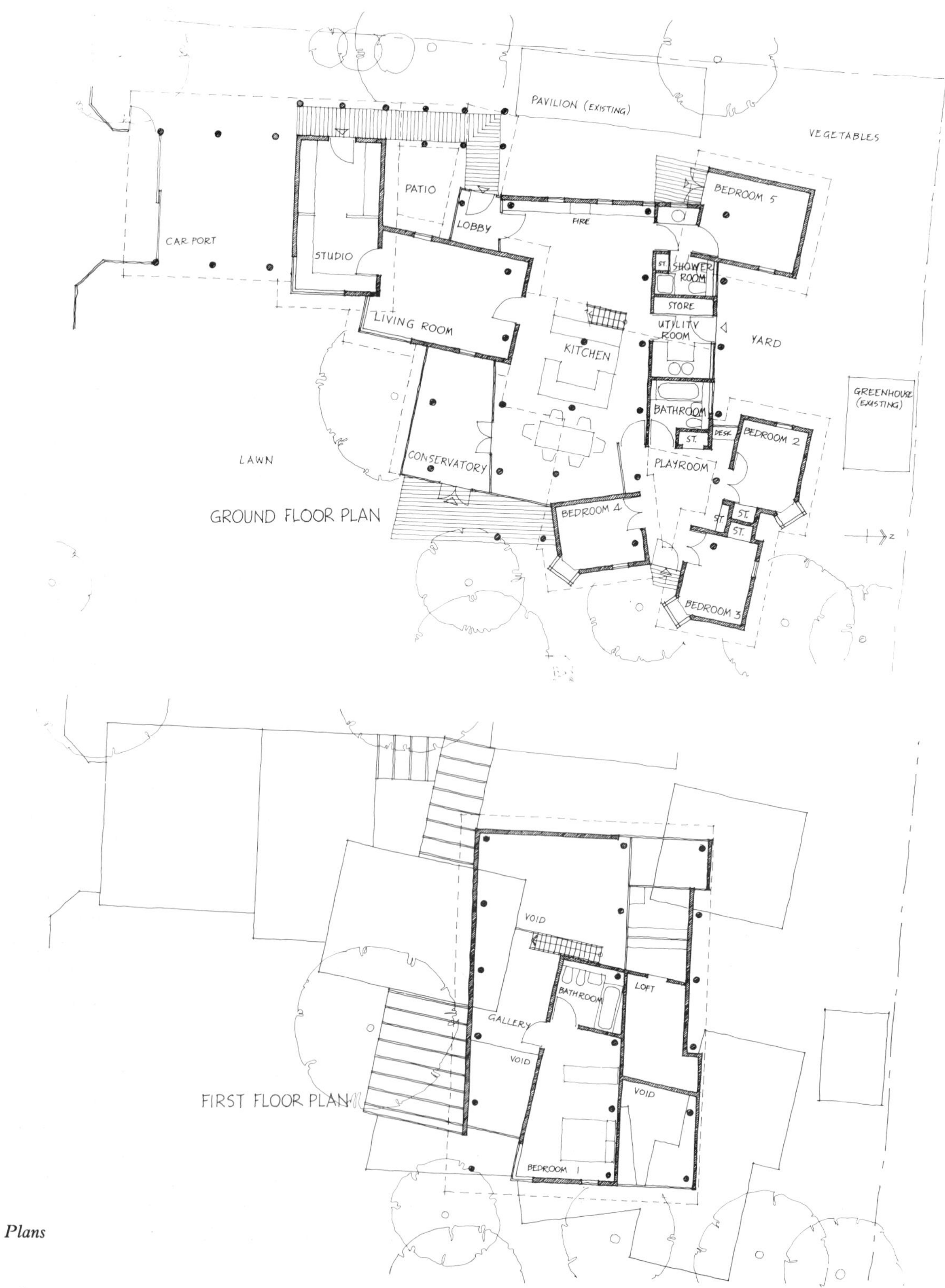
PAVILION (EXISTING)
VEGETABLES
PATIO
BEDROOM 5
LOBBY
FIRE
CAR PORT
STUDIO
SHOWER ROOM
ST.
STORE
LIVING ROOM
UTILITY ROOM
YARD
KITCHEN
GREENHOUSE (EXISTING)
BATHROOM
ST.
DESK
BEDROOM 2
LAWN
CONSERVATORY
PLAYROOM
GROUND FLOOR PLAN
BEDROOM 4
ST.
ST.
ST.
BEDROOM 3
VOID
BATHROOM
LOFT
GALLERY
VOID
FIRST FLOOR PLAN
VOID
BEDROOM 1

Plans

house on the site. They were familiar with the idea of self-build and timber frame construction following their experience with the earlier Lewisham self-build projects. Meanwhile, the vendor's solicitor was being disquietingly unhelpful.

August: the design

We had to assume that all this could be sorted out and get on with developing a design ready to make a planning application. We were then immediately confronted with one of the biggest and most difficult questions to answer in the whole enterprise: what sort of house did we want? Anything was possible. For my part it was important that it was to be a reflection of my architectural ideas and philosophy. It would be an opportunity to extend the tradition of timber frame self-build which has been developed by Architype, following on from the work of Walter Segal. Architype had been developing the Segal method to incorporate an ecological approach to building and this would have to be a central idea. I was also eager to explore the forms and spaces that could be built in timber. The plan developed relatively quickly with a big family room at the heart. This contained the kitchen as well as family living space and the eating area. A children's area contained the three childrens' bedrooms opening off a play area. A separate adult's wing contained a 'posh' living room and a study at the extremity shielded from noise and disruption. We would have a bedroom and bathroom upstairs with a gallery overlooking the family home, which could be used for doing things whilst still being in touch with the hurly burly below. A guest room for visitors or live-in nanny was to be in a separate wing with a side entrance and its own shower room and kitchenette. Rona works from home so the study was near the entrance with a separate door, so that her business visitors could come and go without having to enter the residential part of the house. The family room was to look out over the garden with a way out on to a verandah through a conservatory. The front door was to lead into a glazed lobby off the family room, to reduce heat loss and provide a transition space as you enter. The play room had a separate garden entrance. A utility room opened off the kitchen with a door to the back yard and there was to be a large storage loft upstairs.

The house was to be sited on that quadrant of the site now occupied by a now overgrown vegetable patch. The plan was devised to be built around the existing trees on the site. The form of the building centres on a rectangular roof over the family room supported on tree trunk columns. The single storey wings are in the form of 'boxes' inserted under the edge of the roof. These boxes are clad where they appear inside in the same materials as on the outside. This together with the views out into the garden and the tree trunk columns echoing the trees outside are intended to reduce the distinction between the inside and the outside. The main rectangle is clad in timber boarding with contrasting panel cladding to the single storey boxes. The various roofs are covered in turf.

As the layout was developed it was necessary to check the practical aspects of the design: would the structure work, what would the implications be on energy consumption, what roof pitch is possible with a grass roof? We sought advice on these aspects. The design period was one of very intensive work in the evenings.

After about a month we had a design drawn up and submitted for planning permission. The pressure to make a planning application and determine whether we would be able to proceed with acquiring the site limited second thoughts during the design stage.

November: planning permission

It was necessary for the conditional contracts for the purchase of the site to have been exchanged before the planning meeting—to guard against us obtaining permission and thus making the site far more valuable without any obligation on the vendor to sell to us at the agreed price. An element of drama was brought to the proceedings because when the day of the meeting came the vendor's solicitor had not exchanged in spite of promises to do so. It ended up with me going to the meeting from work still without confirmation that the exchange had taken place. I had to phone home to check if the deed had been done whilst I was on the train. If not I was to ask the chair of the committee to use his discretion and withdraw the item. As it was the vendor's solicitor was contacted at home by the vendor and instructed to exchange forthwith. Our solicitor was able to relay this information in time for us to proceed at the meeting. The committee granted permission to cries of outrage from the objectors. So we had the planning permission necessary for the purchase of the site to proceed.

January 1992: the model

The design was developed in the evenings over the next few weeks and I built a card model at a scale of 1 to 20. One of the most characteristic features of the design was introduced at this stage, the curve to the main roof. It arose out of the need to reduce the height of the roof generally whilst still keeping enough headroom in the upstairs bathroom. I was very pleased to adopt this more organic approach although later it did prove to be relatively complicated to build. In deference to the neighbours, it was also decided to retain the existing shed on the site instead of building a new one attached to the house. This was a good decision which saved time and money, was very useful during the construction and retained a link with the history of the area. The model proved very useful in visualizing the spaces in the design and the finished house is very similar although a number of elements were changed in the course of construction.

May: building regulations

The next few months saw intensive evening work developing the design, specifying materials and preparing drawings and calculations for the Building Regulations submission. I calculated the structure as we normally do at Architype for timber frame self-build projects with the exception of the structure to the main roof. This was beyond my capabilities on two counts; firstly the calculation of timber poles is not covered in any of the British codes—although it is in New Zealand and Australia where this form of construction is relatively common (their data was obtained and used for this house). Secondly, the structure of braced columns is what is termed an indeterminate structure. This means that it is not capable of analysis using straightforward calculation and has to be handled by a computer so I commissioned a professional structural engineer to carry out this part of the analysis. It consisted of around one hundred pages of computer data which with my forty pages or so of beam and column calculations formed the structural submission.

The Building Regulations submission was completed by a specification which included calculations of heat losses, window areas and condensation prediction prepared by the company that supplies the recycled newspaper insulation which was to be used.

The amendments to the design at this stage—the roof form, the shape of the conservatory, the angle of one of the bedrooms to fit around the trees better and the retention of the existing shed—required that a revised planning application be lodged.

Whilst this information was being considered by the local authority, site clearance was completed and a start was made on the foundations with the agreement of the Building Inspector.

June: start on site

This was a hectic period organizing tools, insurance, temporary telephone, water and electricity, a programme for the work, schedules for first deliveries of timber, materials for temporary gates and fencing, storage arrangements for materials, a bank account for payments and all the other organizational matters necessary. I negotiated with a carpenter, Steve Archbutt, one of the self-builders from the second Lewisham self-build, to work on the job. I started working four days a week at Architype, three days on site.

We created a new access with temporary gates and ramp formed of hardcore and repaired the shed. This caused Steve to suffer the first and only bad industrial accident on the job: he sawed the tip of his finger off and had to be rushed to the hospital by one of the neighbours. Fortunately it healed remarkably quickly without permanent damage. Relations with some of the other neighbours got off to a bad start with complaints about lorries coming to the site with deliveries but others proved very helpful after the earlier objections and one neighbour provided lunch every day for us.

July: foundations

I spent more late nights preparing a setting out drawing for the foundations. The setting out was quite complex given the irregular shape of the building. Setting out with pegs and string was not helped by our Alex, now a year and a half, pulling the string lines down.

I arranged for a JCB to remove the topsoil from the area to be built on and I was reminded of the destructive power of such a machine working in a restricted area up against existing trees. I organized the hire of a post hole auger to drill out the foundation holes. This was a small hydraulic machine which can drill a two foot diameter hole up to four feet deep which is then filled with concrete to form the foundation. The sixty or so foundations took about two weeks to dig using this machine with much of the time taken removing the spoil by wheelbarrow. Some of the foundations were required to be more than two foot in diameter and these were undercut by hand with a spade.

The concrete was made on site with a mixer and it took just over a week of hard physical graft to fill all the sixty holes. I hired a labourer friend of Steve's to work with him on the foundations on days when I was not on site. At weekends we had a great deal of help from a few friends who worked at clearing, digging and setting out. The foundations were completed by setting a paving slab on a mortar bed where there were to be timber posts. These slabs were levelled-in using a dumpy level.

August: drains and services

The trenches were hand-dug in a period of high rainfall and the site was reduced to a sea of mud. Plastic drainage was used which is very easy to lay. Ducts were laid for the electric main and telephone and a plastic water main was laid in. Ballast was laid over the whole area of the building to prevent weed growth under the house. This also provided a good base for building on.

September: timber delivered

The timber delivery was delayed because I found it difficult, working in the evenings, to produce the schedule in time. The order in the end coincided with the summer shut-down at the sawmill in the Welsh borders. The timber was ordered in one lot for the whole job and is British-grown Douglas Fir. Locally grown timber was specified to reduce the energy embodied in the construction—imported timber requires more fuel to bring it from Canada or Scandinavia—and to encourage the use of UK grown timber for structural purposes. This adds value to the timber and encourages replanting and proper management of British woodlands. The timber was of very good quality with hardly any pieces with splits or twists that rendered it unusable. Much of the material was also in long lengths (over seven metres) and large sections, 63 × 250mm. Douglas Fir is a moderately durable species of timber and so no timber treatment has been used on the job. The timber was kiln dried and the poles were specified to be debarked by hand with the maximum 10% taper in the length.

The timber comprised two large articulated lorry loads. They looked large from the ground, the first and largest load took six men four hours of very hard work to unload and get into the specially erected scaffolding rack. This stored the timber under cover until the main roof was on in December when the rack was dismantled and the scaffolding returned off hire. It enabled the different types of timber for columns, beams, joists, cladding to be racked separately for easy access. I sorted the timber and found, inevitably, that some pieces had been wrongly delivered. I arranged for them to be changed.

October: framing

Another stint of working late in the evenings was needed to produce drawings showing the key dimensions for making the frames. I also produced detailed drawings for the fabrication of the galvanized steel plates that were necessary to bolt the beams to the posts. The six main frames were made flat on the ground and lifted up into position. Each frame consisted of three tree trunk columns with double beams at the roof and floor levels bolted together with braces at roof and floor levels. The tree trunk columns had to be notched for the roof and floor beams. Because the poles were irregular we stuck stringlines for a reference point on each log. We devised a gauge to ensure that the seatings cut into the logs for the beams were parallel and the correct distance apart. We made one or two mistakes, drilling the wrong size hole for bolts for instance, and they took time to overcome. We suffered our first burglary; the shed was broken into at night and all the power tools were taken. I replaced the tools and bought a secondhand tool vault. The people at the tool shop said they had workmen in every day who had lost the lot. The insurance covered the loss.

The frames were raised into position with the help of friends on three Saturday mornings. This required much phoning around to assemble the required number of people at the same time. This was a moment of great uncertainty, would they be too heavy, would there be a danger if they fell? It took twelve people to lift each frame into position using pushers bolted to the roof beam fixing plates at the top. We had a heavy rope belayed around a tree in case of emergency but as it was they went up in time-honoured fashion without any difficulty. The last two frames had to be carried in vertically in two parts because there was not enough room to raise them whole.

Once the frames were up we adjusted them for position and verticality and positioned the DPC at the base. One refinement over Walter's original detail was to use bitumen coated waterproofing membrane cut to the shape of each post in addition to the normal lead detail. The bitumen is squeezed into the end grain of the timber and gives a better seal at this vulnerable position. Also where the posts were external to the completed building and therefore subject to regular wetting, I inserted a 50mm upstand between the post and the foundation so that the post was not actually standing in the wet. This upstand was made out of a 50mm slice of large diameter plastic drainpipe filled with a strong mortar.

Notching the tree trunk columns

November: beams and joists

I spent more late evenings doing a dimensioned roof layout showing the position of the rooflight and noggings which would support the top of partitions later. Double beams were fixed between frames at roof and floor levels; a heavy and difficult job when at roof level. They had to be lifted 6m by one person at each end on a ladder, offered into position, cramped and marked for length with the correct angle for the cuts. They were lifted back down for cutting, then back up, cramped into position, drilled for bolts using metal plates as a template and bolted in position. A lot of time was spent manoeuvering ladders into a position for drilling that was comfortable. We did not use any scaffolding during the job but constructed working platforms when necessary. This was the time of my nastiest moment on the job; the drill jammed in the on position and rather than be thrown from the ladder I let go and it wrapped the cable around my arm. I managed to turn it off but it was a frightening and painful moment, twenty feet in the air and not helped by being very cold and raining at the time. The roof joists were heavy to get into position being 50 × 250mm in section and 6m long. The ends of the joists to the rear slope of the roof needed a very deep notch to sit in the joist hangars and I had to check the shear load. I employed Steve's friend to help him when I was not on site during the week for this part of the work and I got help from friends at the week end. The beams and joists at floor level were easy after this. I had the first day off, not feeling well, with flu.

December: Roof

We fixed fascias using a string line to get a straight line on a structure which is all at angles. I had to swop material around to get enough depth to the fascia at the

Topping out

front. The Warmcell recycled newspaper insulation was delivered in compressed bales, with a blowing machine on free loan. I carried out a trial section in the main roof to check the viability of blowing in insulation from below once the roof deck was fixed. In this way the insulation would not get wet. It appeared to be possible so I proceeded to fix the roof.

We fixed netting (used normally for preventing debris from falling from scaffolding on demolition sites) over joists to retain the top of the insulation in the roof. Then we fixed 50 × 50mm spacers which form a ventilation gap under the roof. We stapled PVC ventilation angles to the edges of the roof to keep insects out of the ventilation voids. We subsequently discovered wasps passing through these vent angles and so abandoned this product for the ventilation voids in the external walls in favour of much cheaper aluminium woven insect mesh. We fixed the roof boarding. Fitting the boards to the curved roof felt more like boat-building than carpentry.

I took time off work but had to go in a couple of times to deal with crises—business not good so we had to make some redundancies. We fixed the single-ply PVC roofing. This system retains the original Segal idea of a loose-lay membrane but with the added refinement of a washer detail that retains the ability of the membrane to move horizontally whilst keeping it anchored to the roof against uplift from the wind. This is achieved by screwing PVC discs to the roof at 600 centres with metal washers. The roof membrane is then glued to these discs. We used these discs on the main roof as we estimated that it would be some time before the roof was ballasted with soil during which time the roof would be vulnerable. The PVC material comes in a roll which is seamed together with a hot air gun. The membrane is laid over an underlay and welded to a metal angle at the edges. There are a number of advantages over Segal's earlier felt roofs; the process is self-buildable and the membrane is very flexible, which is a particular advantage in this case where the membrane has to accommodate the curved shape of the roof. The rooflight is a triple-glazed dome with insulated upstand to which the roof membrane is bonded. This is a very simple and effective detail. Roofing started during a very wet spell. The hot air welding technique worked well in the damp but gluing the membrane to the discs did not. I needed to dry everything out as far as possible with the hot air gun. Roofing ended during a very cold spell. Working on the roof was tricky when it was icy as the PVC was incredibly slippery. We lost control of a roll of roofing which went over the edge and took an hour to sort out. I had my other nasty moment—I was working alone and slid down the roof out of control. Fortunately I was able to stop against the upstand at the edge.

There was no work on Christmas Day. We had a topping out ceremony on the 28th which was attended by sixty-five friends who came back to Segal Close for drinks. The house is not big enough. It felt like a big moment and the end to a good year's work.

January 1993: Floors to wings—New baby

The floor structure of single-storey wings was set out on blocks and fixed together. Setting out the angled shape proved more straightforward than anticipated. Our new baby was born in the kitchen at Segal Close. The floor structures were supported on concrete stools cast in plastic buckets obtained from the flower stall in Catford. The procedure was to level the structure on blocks, jack it up and slip the concrete stool under on a thick mortar bed. The jack was then let down so that the floor structure sank down on to the blocks squeezing mortar out until the concrete stool was set at the right level. The temporary blocks were removed when the mortar had gone off. This procedure worked very well. Rona and the new baby were admitted to hospital for ten days when she was three days old. I worked on site with Alex in his pushchair which worked for a few hours at a time. It was enough to keep Steve going.

February: Walls and roofs to wings

The stud wall panels were nailed together using the floor structure as a base and erected in turn. Work proceeded quickly. Problems occurred as usual as soon as a more complicated part of the construction was encountered, in this case the framing for the bay windows. It proved very difficult to set out the structure in mid-air. This was to cause some problems later, having to adjust subsequent work to take account of the discrepancies in the setting out and level of the framing. Steve phoned to say that there was a problem, so I looked in on site on my way to work for the first time. It was a mistake in cutting the taper to the overhang to the living-room roof—nobody will ever notice.

Blowing Warmcell into the flat roof

The sequence for the smaller flat roofs on the wings was to put in the insulation from above rather than below as on the main roof. The procedure was as follows: stapled vapour-control paper to underside of roof joists, nailed battens—which will eventually support ceiling boards—to the underside of the joists, blew in insulation from above, fixed the edge fillets, metal angle and finally the waterproof roof covering. We had hoped to be able to do all that for a small roof in a day but this proved impossible. We were able to get the deck fixed in a day at which point temporary plastic sheeting protected the insulation until the next day.

Stud walls: panels being erected

I misjudged the fine line between wasting money by ordering too much and wasting time by having to order more later. I had to re-order the vent angle but was able to obtain some surplus metal angle from the nearby self-build job under construction. From now on we were working under cover.

March: External walls

This was a period of intensive decision-making on site; should we have a full-height window slot at both ends of the west wall, how was the conservatory roof going to be supported on the front wall to the kitchen, would 16mm diameter steel studding be strong enough to support the front wall to the kitchen?—and a myriad of other questions.

We fixed 50 × 50mm battens to the external walls. These formed the outer layer of a double layer wall construction which gave a cavity of 125mm to be filled with insulation avoiding cold bridging at the studs. It also meant that the wall was constructed of smaller sections of timber which were cheaper and easier to handle. Work proceeded rapidly but, as usual, time was needed to puzzle out what to do at awkward junctions.

We nailed bitumen-impregnated softboard to the outside of the external walls. The permeable softboard formed the outside layer of the breathing wall construction. The boards are tongued and grooved on all four edges which meant that they could be fixed without having noggings under all the edges. This saved a lot of time.

We carried soil up on to the roof. I had sought the advice of an ecological landscape designer who advised that the topsoil that I had set aside from the foundation excavations was far too rich to support vegetation on a roof. The drought-resistant species that can survive through a long hot summer need a very impoverished soil; too rich a soil encourages the growth of lush green species that die as soon as it gets dry. We ordered chalk and ballast which were mixed with the sieved topsoil in varying proportions to give

Outside walls clad with bitumen-impregnated softboard

different soil types on the different roofs of the house. The back slope of the main roof is very steep, about 30 degrees, and this has a framework of battens laid on the roof which supports turfs from the garden laid upside down. The root mat in the turf prevents the soil being washed away and the battens stop the turfs sliding down the roof. The soil mixture was carried up ladders on to the roof in buckets and spread to a thickness of 80 mm on top of an underlay to protect the roof membrane. There are 50 tons of soil on the roof! We had help from friends at the weekend and I hired in Steve's friend during the week to help with the labouring again. The roofs were seeded with grass seed mixtures with some wild flower seeds mixed in, the species being selected to suit the particular soil type on that roof. There is no moisture-retaining layer or drainage layer in the build-up of the roof so the vegetation experiences extreme conditions from waterlogged to completely dessicated. It is anticipated that the vegetation will take three seasons to become established. The effect is very colourful in the spring with the wild flowers but the roofs dry out completely in summer. The growth has been insufficient in some places to stabilize the soil on steep slopes and erosion-prevention measures have had to be implemented. The soil that had washed away has been replaced, reseeded and covered with a biodegradable jute fabric through which the plants can grow. This has been successful in preventing further trouble and the jute will rot away in time when the vegetation is established. At this point Steve decided that he was going to find his fortune working in Germany. It was a sad loss as we had worked well together. Steve was tremendously energetic with a great drive to get things done. He recommended a carpenter friend, Noel Gaskell, who lives locally. Noel visited the site and it was agreed that he will take the job.

April: Roof fascias

We took Easter week off for a family holiday in Wales.

On our return we screwed roof fascias and cappings in place. It took a surprising amount of time; sanding, painting, drying, measuring, marking, cutting, drilling, notching and fixing.

May: Insulate main roof

The Warmcell insulation to the main roof was blown in from below. First vapour control paper was stapled to the underside of the joists, the insulation was blown in and battens were nailed to the underside of the joists ready to take the ceiling boards. Temporary staging was built over the large voids to gain access to the ceiling.

June: First fix plumbing

I spent more late nights working out the plumbing. The headroom in the loft was very tight for a gravity fed hot water system. However, it looked as if it was possible and so this was the basis for the design of the pipework. I had to fix the size of all the windows so that the heat losses could be calculated. Then the radiators could be sized so that the pipework could be installed in the right place. This meant having a complicated discussion with Rona about ventilation. She wanted a small opening window in each room for ventilation at night in addition to the main opening window. The house was designed with a minimum of opening windows and relied on a controlled ventilation system. Because of the layout of the house this could not be a passive system but had to rely on a fan drawing air out of the rooms that are a source of moisture such as kitchen and bathrooms. The rate of extract is controlled by a humidity-sensitive extract register in the ceiling. Fresh air is drawn from outside the building through humidity-controlled inlets in each of the habitable rooms in the house. The inlets admit more air if the air in the room is moist, if the room is occupied, for instance. A compromise was reached whereby small openable vents would be provided in each of the ground floor bedrooms. This enabled work to proceed with the plumbing. The wall construction was modified to accept the additional vents and they were manufactured along with the rest of the windows. They did look rather unnecessary, however, when they actually arrived on site so the wall construction was modified back again. This was fortunately one of the very few bits of work that had to be done again throughout the job.

The pipework in the house is a plastic push-fit system which is absolutely marvellous; quick to make the joints, many fewer fittings necessary because the pipe is flexible and the system is foolproof (of all the hundreds of joints in the building only two dribbled slightly when the system was tested and they were both on the conventional plumbing fittings. Not one of the plastic joints leaked a drop, which is very impressive.) I had to have a week off with a virus infection and felt exhausted after. I never used to be sick like this. I also had to spend a weekend at the office to meet a deadline on a project. Maybe I was just trying to do too much. I needed to get the window details sorted out so that I could order the windows which were now overdue—more late nights. I constructed the bay window structure and did other preparatory work for the windows so that I could measure the openings and order the windows. We had been a year on the site. Progress had been good generally, but we had never recouped the time it took to sort out the foundations at the beginning and there had been cumulative delays since. These totalled around two months at this stage.

July: First fix wiring

I spent yet more late nights working out the electrics. I had a terrible time trying to decide on light fittings; good fittings are almost inevitably imported and far too expensive. I checked to see if it would be necessary to increase the size of the cables because they are running in insulation which might cause them to heat up. The advice was that it would not be a problem.

August: Flooring

Bitumen-impregnated fibreboard was nailed up under the ground floor joists to support the insulation. This was one of the worst jobs so far, lying on our backs in the restricted space under the building. Also at this time the drainage connections were brought into the building.

We started flooring out of a sequence as the intention was to have had the windows fixed and the building weathertight. I decided to go ahead without windows and put up plastic sheeting for protection.

I had purchased a quantity of second-hand oak from vinegar vats from a factory that had closed down in London Bridge. The vats were kept full of water when the factory closed to stop them from collapsing so the material had a very high moisture content. I decided to go ahead anyway. The floor is now drying out and gaps are opening up but the oak looks absolutely wonderful when polished and sealed. We had a working party with friends which was a good day and good work was done. We had a week off in Scotland.

September: Plasterboard walls

I decided which make of radiators to use so that we could incorporate fixing noggings in the wall construction. I chose radiators with inset headers so that the pipes come into radiators unseen from behind. They look good in the finished job.

Plasterboard was fixed with screws using a battery powered screwdriver. Tapered edge boards were used which will have flush taped joints with filler. This technique took a while to perfect, in fact I never really perfected it, but it gave a flush finish to the walls and was probably quicker and certainly cheaper than the classic Segal battened wall.

I carried out a trial section of wall insulation. This was injected into the cavity in the wall from the outside through holes drilled in the fibreboard. I think on reflection that this was not the best method. Much better would have been either to staple scrim, lightweight fabric, over the inside of the studwork and inject insulation through holes in this, so that you can see exactly what is going on before fixing the plasterboard or to use the wet spray method where the insulation is mixed with a little water and is sprayed into position rather than blown. This latter method has the additional advantage of sealing the building better against accidental air leakage. I was working on a chapter in a book on Housing and the Environment in my spare time!

October: Windows

There were over fifty of them, some of which were large and high up. The largest and highest was six foot wide, seven foot high, double glazed and twenty foot above the floor to the top. It weighed about 500 pounds and four of us could just about stagger around with it. The question was how to get it up into position. The method adopted was to construct a kind of giant's ladder below where the window was to be fixed. This ladder could accommodate five people who lifted the window one rung at a time. The top of the window slid against battens fixed either side and a second set of battens prevented the window from toppling backwards. The window was in place within minutes.

The tolerance gap between the window frame and the structure was stuffed with mineral wool insulation from the outside and sealed with mastic from the inside, against the normal practice in Britain of sealing around windows on the outside which has the danger of trapping condensation inside the joint.

November: Plasterboard ceilings

I completed various plumbing and wiring bits and pieces in preparation for fixing ceilings. I filled the first floor under the bedroom with Warmcell which also acts as good sound insulation. I fixed ventilation ducts to bathroom and kitchen.

December: Fibreboard ceilings

Slotted fibreboard panels were glued to battens in the single-storey wings. The idea was that the slots would conceal the position of joints, particularly where the panels have to be cut and fitted around the tree trunk columns, but in fact it was very difficult to get the panel joints neat. These fibreboard panels were one of my least good ideas; the panels are very easily damaged and do not look good with the poor joints. I decided not to fix these panels in the kitchen and wasted a couple of hundred pounds-worth of material.

I also worked at painting windows, filling window fixings with timber pellets and hanging external doors and fitting ironmongery so that the building was secure for the first time.

We had been on site eighteen months now. The original programme showed us moving in at this time. It felt as if progress had slowed down and a lot of time seemed to be going in fiddling around with many things at once without ever being able to finish one complete stage in the job. I felt very weary now but we had to keep plugging away as there was still a lot to do. I prepared a revised programme for completion which showed another five months work. In fact it was to be another eight months before we moved in. One of the implications of the extended construction period was that the original estimate for paying for labour had more than doubled. We decided to raise a second mortgage on Segal Close.

January 1994: Timber ceilings

The ceiling to the main roof is square edge Douglas Fir boards nailed to battens with a gap between. This allows the boards to follow the shape of the curved roof without cutting. There was a great deal of planning to do and Rona and I spent a day visiting kitchen, bathroom and lighting showrooms looking at fittings.

February: Stairs

It became clear that the original design for the staircase would obstruct the space and view in the big kitchen. A rethink was initiated, standing in the space to see the effect of different arrangements. This ability to change your mind as you go along is one of the important advantages of building your own house. We decided to make the stair treads of Douglas Fir plywood which has a lovely colour and bold grain pattern. It is manufactured in Canada using a glue which does not suffer from the undesirable property of giving off toxic gas into the interior of a building, unlike the more common urea-formaldehyde glues.

We cleared out junk from Segal Close prior to putting the house on the market. We were somewhat apprehensive about how difficult it was going to be to sell what is a very comfortable house in a good position but nevertheless unusual in design at a time of recession in the housing market. As it was, the first estate agent we placed it with produced nothing. Our first free advertisement in the architectural press produced a number of architects who were only interested to see inside a Segal house. We were just beginning to get anxious after about three months and we were about to place the house with another local agent but we decided to take out another free ad first. This produced exactly what we had hoped for, someone, an architect, who walked in through the door and said 'I have been looking for an interesting house to buy for four years and this is it'. He offered the asking price, the building society surveyor valued the house at the same figure and the sale went through in about two months. This gave us a date to work towards.

March: Plasterboard joints

We found it difficult to get a very good finish at first but we soon improved with practice. The random orbit sander was magic for sanding down the joints but it makes an astonishing quantity of really fine dust.

April: Second fix plumbing

We fixed the bathroom, the boiler (which is a high efficiency gas condensing model) and the hot water cylinder. This has two primary coils, one from the boiler, the other from the projected solar panels to provide pre-heat. After seeking advice I decided to install a mains pressure hot water system although this meant adapting some of the pipework that had been installed already. This part of the work found me back at the plumbers merchants practically daily, sometimes more than once in a day, for bits.

May: Second fix electrics

We ran a heavy armoured cable under the building from the meter in the shed to the intake position in the utility room. I hired an electrician to terminate this cable at each end, which was beyond me. While he was at it he installed the consumer unit. He made the whole operation look very easy. we switched on and . . . no bang, it worked! A small milestone, power in the house.

We brought the water pipe into the building. Another small milestone, water in the house, and not too many leaks! We were due to be moving in now according to the revised programme but it was to be another three months before we were able to do so.

June: Wall insulation

I injected Warmcell insulation into the wall cavity from the outside through holes in the fibreboard. I took some boards down to check that the cavity was fully filled. There were some places where voids remained, particularly at the corners of the building so I drilled more holes and topped up the insulation. I injected Warmcell into the internal walls as well to provide sound insulation.

The heating system has two zones each of which is controlled by an electronic gadget which learns the thermal behaviour of the building over the period of about a week. The controller then optimizes the start up time of the boiler to achieve the set temperature at the time required thus obtaining maximum efficiency from the system. This takes into account the effect of the outside temperature.

We brought the gas into the building from the meter in the shed. My friend Terry, one of the original Lewisham self-builders, popped by one Saturday morning and tested the gas insulation for leaks. There were none, again! I was getting quite used to things working first time by now. This was the moment to fire up the heating system . . . again it worked first time!

Another of the Lewisham self-builders came in for a day to plaster the kitchen ceiling. The process of plastering made a great deal of mess and confirmed my prejudice against wet construction.

The staircase

July: Floor finish

The second-hand oak flooring was very uneven in thickness and so we used a planer to achieve a relatively even surface for the floor sander. Even so, the powerful, hired-in belt sanding machine found the oak heavy going, because it was so hard. The first time we used the machine it streaked across the room out of control and crashed through the plasterboard because the fixing for the handle was not secured properly. It was a time-consuming operation sanding with successively finer grades of abrasive and applying two coats of sealer. The result, however, is wonderful with a deep, rich colour and interesting grain pattern.

The kitchen units had been made by a friend who is a local cabinet maker, using the same Douglas Fir plywood that we used for the stair treads and a formaldehyde-free grade of MDF (Medium Density Fibreboard). They were beautifully made and look very well with a contrasting colour and grain pattern to the oak flooring. The internal decorations were under way using organic paints and sealer and the house was full of the delicious scent of citrus.

August: Moving in

The pace became pretty hectic at this time and the site diary has no entries for the last three weeks or so because of the lack of time. I ended up working one or two late nights on site till midnight. The cooker was fitted and the skirtings and architraves were sanded, sealed and fitted. Rona visited the site the day before we were due to move and had serious doubts about whether it would be possible, given the state of the place. Help from a friend equipped with a broom, hoover and damp cloth turned the place from a building site to a new home. The move itself passed off remarkably smoothly and we were installed amidst a jumble of belongings by lunchtime. A good many bottles of champagne were consumed that evening. The house has proved to be very comfortable and the children love the space for them to run around in. There were one or two drawbacks to living in an unfinished house; there were no doors, which visitors

The house approaching completion

found a bit disconcerting in the toilet. There was still a great deal of work to do before the house could be thought of as finished. We decided to keep Noel on and he worked till December fixing cladding and weatherboarding to the outside and hanging doors.

Which brings me to the time of writing. I am now taking a bit of a breather until the good weather before embarking on the walkway to the front door, the entrance lobby, conservatory, verandahs, upstairs bathroom, shower room, kitchenette, shower room and carport. And then there is the garden of course. We have to wait until we can save yet more money; it is a balancing act how long we can wait to reclaim the VAT on the job, which comes to a few thousand pounds; we will not be able to reclaim VAT on any material bought after that.

The project has proved more ambitious than I imagined; it has cost more than the back of envelope budget because of the extra labour we have had to employ and because we chose to go for the best and not cut corners. It has taken longer than envisaged largely because of the complexity of the design, a far cry from the simple, light constructions that Walter Segal designed. It was not a good idea to have a new baby in the middle of the process but we have survived and can now look forward to many years in our new house—I shan't be building a third.

Part Two

Others have done it

Some examples of pioneering self-build projects showing the wide range of people involved and the variety of ideas and techniques that are possible. To reassure you that you are not venturing out into entirely unknown terrritory, we describe some recent self-build achievements.

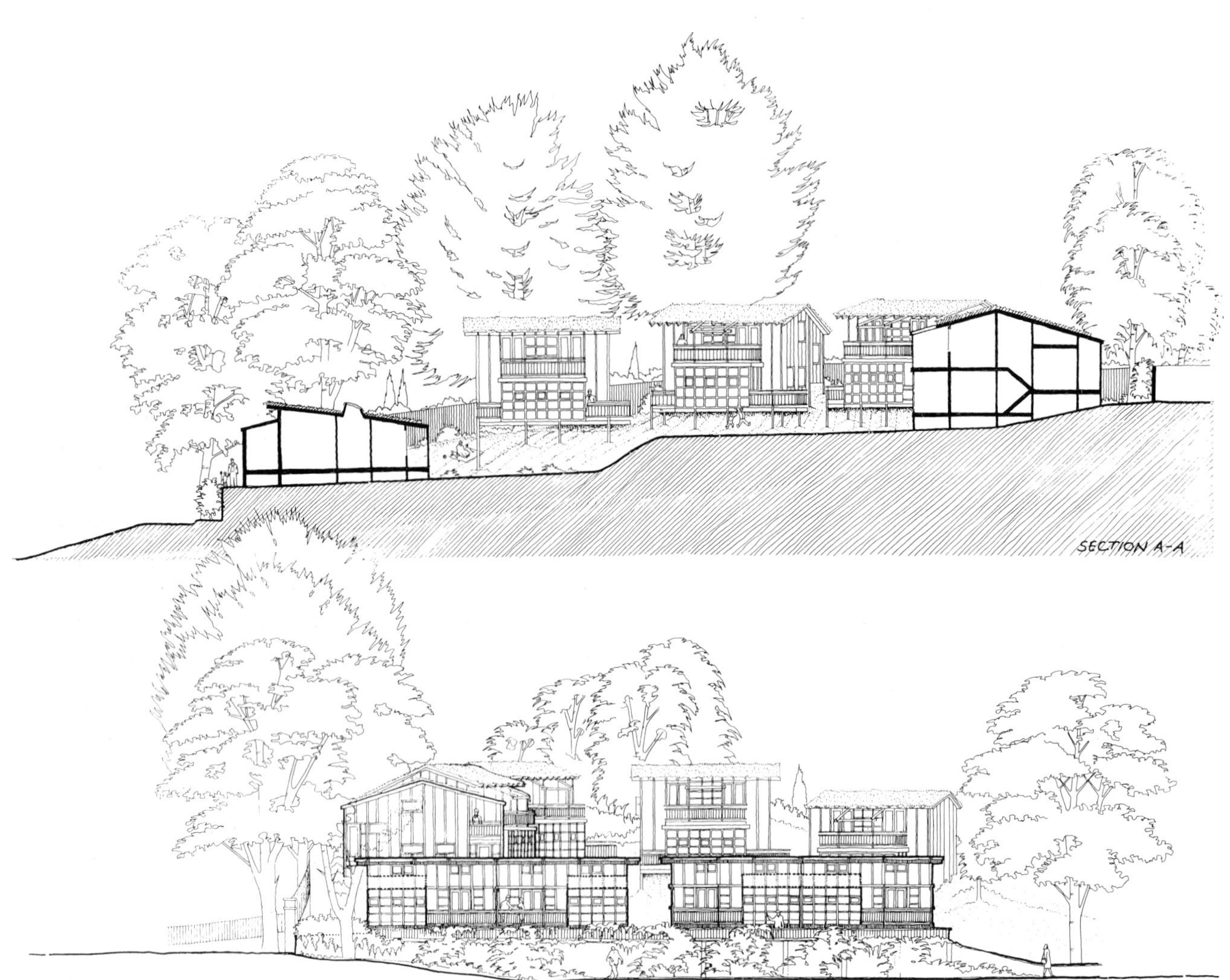

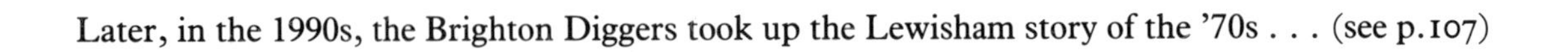
Later, in the 1990s, the Brighton Diggers took up the Lewisham story of the '70s . . . (see p.107)

Chapter 7

The Lewisham story

London local authority makes self-build accessible to anybody

In the late 1970s Lewisham became the first council to fund a self-build project using the Segal method of building. People in housing need, mostly without building skills, were enabled to design and build their own houses. This is the story of how it happened.

Dreamers awake

The Lewisham story is one of dreams come true. In the mid-1970s a number of people had dreams which coincided, and on wakening, these dreams proved to be realizable. Colin Ward was dreaming about introducing a new factor into housing, dweller control, and was writing and lecturing and dropping words into influential ears. Walter Segal was dreaming of using his architectural genius towards 'self-help house building, based on mutual help among members of a friendly society on leasehold land; that is foremost in my mind,' as he wrote to Brian in 1974.

We were both being paid to dream together, in the Borough Architect's Department. Our imaginations were fruitlessly occupied with trying to devise forms of council housing that really suited tenants and wore the proper architectural expression that modest but really cared-about houses should have.

We think councillors on the housing committee were probably dreaming that their architect's department would come up with something other than the run-of-the-mill, narrow-fronted terrace house, which although able to scrape through the cost yardstick at estimate time, always exceeded cost budgets at final account. Such was standard fare in the 1970s, by which time we had at least learned the sad lesson of the system-built tower block and were trying to fight off the Department of the Environment's (DoE) preferred but ungainly, and unpopular, walk-up maisonette block.

The Borough Housing Officer was burdened with an endless waiting-list, too many unhappy tenants and a huge stock of defective houses. She was dreaming of tenants who didn't keep wanting to move and didn't continually insist on their dwellings being repaired and maintained.

The people of Lewisham, waiting for the trickle of council lettings to wear away at the waiting-list until they came to the top, or waiting for transfers from their currently unsuitable council flats, were dreaming of a chance to do something for themselves.

Even the Minister of Housing, then Reg Freeson, dreamed of something better happening in the housing field as a result of his term of office, and invited the local authorities to suggest innovative schemes to him that might break the pattern.

These dreams and dreamers began to converge in 1974 when Colin Ward, having addressed a National Housing Conference presided over by Reg Freeson, spoke to Walter Segal. It was probably at the Architectural Association, during one of the meetings of the Dweller Control Housing Group, which included also John Turner and others interested in tenant control, the London squatters movement, Third World housing, etc. (Jon also used to go to these meetings, but Brian's report of this meeting is from a letter Colin wrote to him in May 1974—the first letter on his self-build file!)

Walter told Colin about his co-operative self-build aspirations and his search for a local authority to take on the idea. One of the reasons Colin gave him for thinking of Lewisham was the presence on the Housing Committee of Nicholas Taylor, the enlightened architectural journalist and author of *Village in the City*.[1] He had turned from journalism to local politics

Lewisham phase one: one of the small sites, Elstree Hill

as a way of getting his ideas into action. Another reason was Brian's established position there as Assistant Borough Architect responsible for housing. Colin had for long revolved in anarchist circles with Brian and he knew that Brian shared his philosophy of 'Anarchy in Action'[2]—that is, that anarchy is not just an unobtainable Utopia waiting for the day everyone turns good, but the best way of organizing social affairs now.

When Brian heard Colin talk of Walter Segal as an anarchistic architect, he was all agog; and at a party at Colin's house Walter and Brian met. Brian was at once entranced by his bold, radical, yet utterly common-sense ideas. The spark that led to the birth of the Lewisham Self-Build Housing Association was struck that night.

The Borough Architect, though intrigued, was not convinced that there was any mileage in Brian's new enthusiasm, but he allowed him to develop the idea for committee consideration, provided that it did not impinge on his working hours!

Getting the proposal accepted

Reports prepared by the full-time officers of the council did not get considered by committees unless they were requested, but we knew Nicholas Taylor (and his colleague Ron Pepper, who chaired the Housing Committee) well enough to 'seed' the invitation without difficulty. Nick asked the Borough Architect to look at alternative methods of housing such as setting up a co-operative self-build housing society.

In the Borough Architect's name, therefore, we came up with a tentative proposal for possible forms of a council-sponsored self-build group, using Walter Segal's system of design 'that is capable of being constructed by people without particular building trade skills and which produces structures of very low cost', to quote the committee report.

The committee in October 1974 did not dismiss the idea out of hand, but had doubts and reservations—hooks on which we could hang a more considered proposal. We had to come up with answers to questions of the durability, sturdiness, fire-resistance and insurability of the timber-frame construction, so further reports followed, which dealt with these questions and others about forms of tenure, methods of organization and finance, and proposing specific sites.

All this meant enlisting the help of other council departments. The Housing Officer, Borough Treasurer, Borough Solicitor and Borough Valuer unselfishly spent a lot of time on our brainchild. Needless to say, this was all in addition to their already full workloads, and we were pleasantly surprised how, by degrees, they all became keen on the idea.

On the question of sites, we were on a good wicket because at a time of boom Lewisham Borough Council had been buying every site on the market (and by Compulsory Purchase Order many that were not!) and had somewhat overdone it. As the boom tapered off, government cost controls became ever stricter and it was soon the case that only pretty large sites, fairly square-shaped and level, with good bearing soils (foundations on shrinkable London clay could be very expensive) and without trees to preserve were viable for council-house building within the cost yardstick imposed.

The council found itself with several 'unbuildable' pieces of land that were small, sloping, soft-soiled and tree-covered. Walter Segal loved them! His design approach turned all these handicaps into advantages. A year spent successfully persuading the Lewisham councillors and officers that self-build was desirable and possible did not mean, however, that the issue was settled. Now we had to test the water and find out if we were right in supposing that enough people would actually want to build their own council houses. Our colleague Jonathan Street, a public relations officer of the borough and editor of the free newspaper *Outlook*[3] that went to every local household, put an article about the scheme on the centre spread of the issue which went out in May 1976. It included an invitation to a meeting at the town hall on 17 July 1976. Over one hundred people came and most were interested enough in what they learned to leave their names. On the spot a dozen of them formed a steering group to carry the idea forward. From this point on, their enthusiasm was the driving force that carried what had now become a serious issue from stage to laborious stage.

The steering group and the council officers worked out a questionnaire. The replies clearly articulated the shape a self-build group would take. A strong majority opinion emerged on the ownership aspect. No one was prepared to build a wholly council-owned house for themselves to rent. At least some stake in the equity was essential. Few wanted to build a co-ownership scheme where they would be tenants of their own co-op. Most preferred an equity sharing scheme, with the council owning the freehold of the land (a council stipulation) but the ownership of the house being shared, the shares being adjustable over time to give the self-builder the opportunity finally to own the whole as he or she became able to afford it.

The option of a small ownership/large rental proportion was asked for as a way of admitting to the scheme people who had few savings and a low income. The scheme, they said, should be open to all on the council's lists without qualification as to income, age, sex or skill. And finally, the lucky few for whom sites were available should be chosen from the many aspirants by ballot.

Another public meeting in September 1976 confirmed the joint steering committee/officers' report on the result of the questionnaire and on 1 December 1976 the Mayor of Lewisham drew the self-builders' names. The first fourteen were to be the pilot group, the rest were in reserve. Subsequent recruits were added to the end of the list in order of application.

Our first approach to the DoE, the government department controlling finance and therefore to a large extent housing form and content, received a positive response. The minister had asked for imaginative, innovative proposals; we had come up with one. 'Good, just get on with it and tell us about your experience in due course', was the gist of what a senior official told us.

We savoured our delight only for a few days, before learning that this enlightened lady (coincidentally called Mrs Segal) had retired. Her successor was a great deal more cautious. Every stage of the planning and costing of the scheme had to satisfy his scrutineers. We likened them to the Devil (St Mark: 5,9): 'My name is Legion: for we are many.'

The hurdles put in our way seemed endless. The job of the DoE regional architect was to ensure that all schemes conformed to design standards. (They were erroneously called Parker-Morris, after the committee which laid down flexible minimum standards which, however, came to be enshrined as rigid maxima and minima.) As the Walter Segal houses were to be individually designed, with Walter Segal's help, by their occupants, who might decide to have large or small bedrooms, more or less storage space, etc., this was difficult to resolve.

Cost control was extremely rigid. There was an elaborate system called the Housing Cost Yardstick (since replaced) and although Segal houses were intrinsically economical even without the free labour factor, it was difficult with all the unknown quantities inherent in such a flexible scheme to produce financial calculations that were satisfactory to them.

Also, by historical accident, the high price paid by the borough for some of the sites suggested to the yardstick rulers that they should be developed with multi-storey flats (an impossibility given their characteristics, and not what the council wanted anyway), rather than the detached houses that Walter proposed and which he well knew the people wanted. We had to persuade them that if we didn't build on them this way, we couldn't build on them at all.

The financial rules also required that competitive tenders be obtained by the Borough Architect for all building contracts, and we were going to build these houses for the cost of materials plus an estimate made by a quantity surveyor of the value of the labour put in by the self-builders—economical and reasonable but non-competitive and different from the norm. Again, an exception had to be pleaded for.

As well as the DoE-imposed design standards and cost limits, the scheme had to meet the myriad other constructional and planning controls, made more time-consuming in our case because of the unorthodox construction method and variety of plans involved. The District Surveyors who administered the London Building Act in our borough demanded particularly full details of the timber-frame construction since the original purpose of the act had been to prevent a repeat of the Great Fire of London.

This posed a problem, as the houses were not yet individually designed—how could they be when the self-builders who would occupy them could not yet be formally constituted? But how could they constitute themselves as a self-build housing association and commit themselves to the project until the technical, legal and financial framework was properly established? However, as an act of faith, the houses were designed in some detail during this period of flux. Sketching on squared paper, families drew up their ideas of the houses they wanted. Walter examined these, then prepared a range of house plans, single- and two-storey, with two, three or four bedrooms, each with a shower room as well as a bathroom and each with a verandah. The group received the plans, made their basic choice and then, at individual visits to Walter's studio at Highgate, discussed with him the particular features and amendments they wanted to incorporate. Walter then drew up a scheme for each self-builder.

Meanwhile, the official negotiations went on like an endless dance.

The nightmare

Everybody at one time or another was waiting on someone else in a great circle: self-builders, architect, quantity surveyor, Borough Architect, DoE officials, Borough Engineer and Surveyor, GLC Fire Prevention, District Surveyor, Fire Brigade, Borough Valuer, Borough Solicitor, Borough Treasurer, Council Committee . . .

This was a pretty wretched period for everybody and it went on for a long time—long enough for Brian's beard to turn grey. Walter was disappointed that he couldn't get the buildings started; so were all the Lewisham Council officers, who were putting in lots of time for no apparent result; and most of all, so were the self-builders and their steering group. How could they decide whether to hold on to their faith in the scheme and stick out for what looked like jam tomorrow, or to accept whatever other inferior housing opportunities might come their way today?

Meanwhile, important decisions had to be made about jobs, what schools the children should go to, whether to book holidays, buy furniture or save up. How long could they put up with the overcrowding, or living with relations, or putting up with hostile neighbours or whatever, while waiting for the project to start?

There were some useful things to be done while waiting. Richard Gant, in charge of the Churchdown Adult Education Institute at Downham, was one who had been excited by the *Outlook*[4] article, seeing the possibility of using his educational resource in the community to some direct purpose, and he offered his workshops and lecturers to the group.

A series of classes was held, some with Walter explaining the structural principles and with practical exercises in making typical joints for the timber frame. Other lecturers explained plumbing, electrical wiring and so on. There were talks on legal and organizational aspects by Council Solicitors and Valuers. Useful as they were and as much as they helped to fill the waiting time, these occupations were no substitute for definite progress.

The meetings between officials went on and on, round and round, but in a gradually upward spiral, baffling the self-build group, whose own meetings at the Cranbrook pub, where a sympathetic landlord let them spend endless evenings in an upstairs room, became more and more heated and desperate, as Brian often found when going along to report supposed progress to them.

Ken Atkins, who from being reluctantly pressed into the steering group had become the vociferous chairman, became so frustrated that he entered into the DIY spirit and, having found out what unanswered piece of correspondence was currently blocking progress, would go up to the GLC offices in Vauxhall Bridge Road, site of much of the delay, tour around the building until he found the right door, then

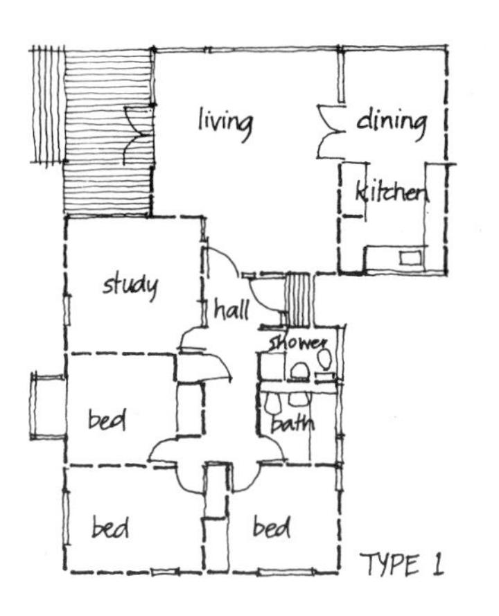

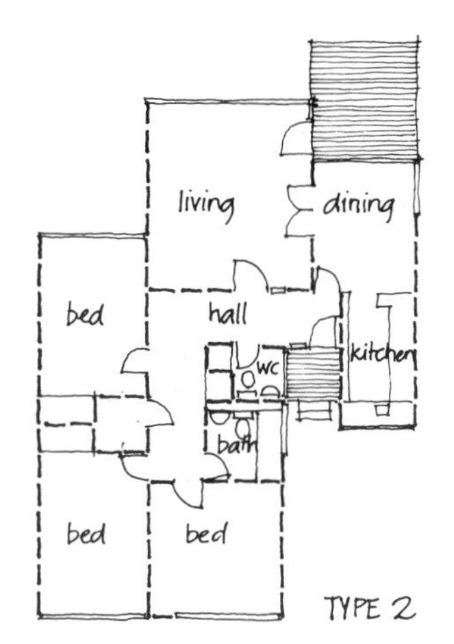

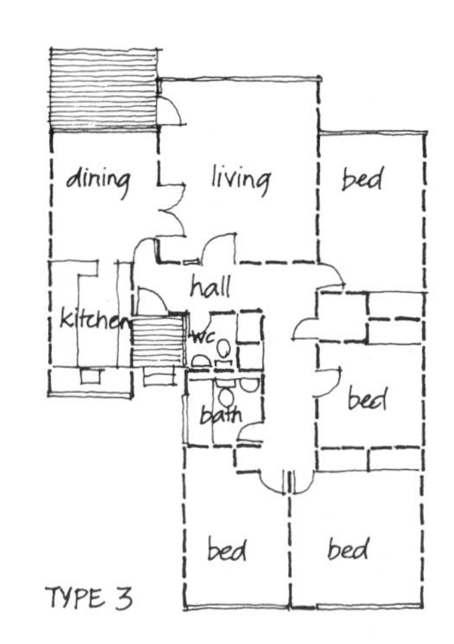

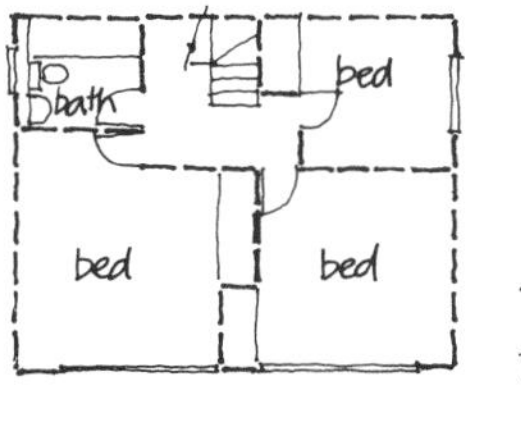

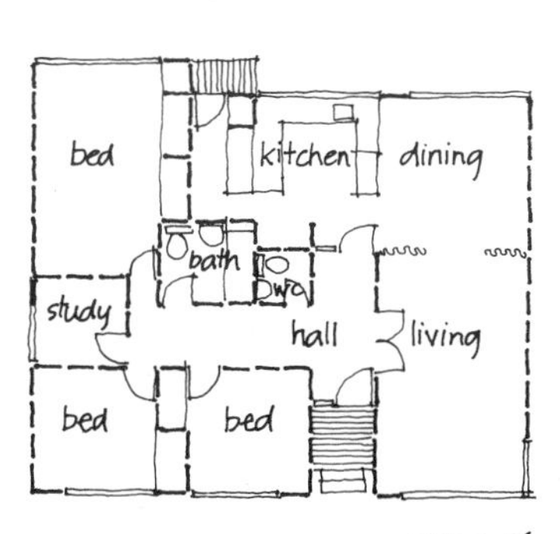

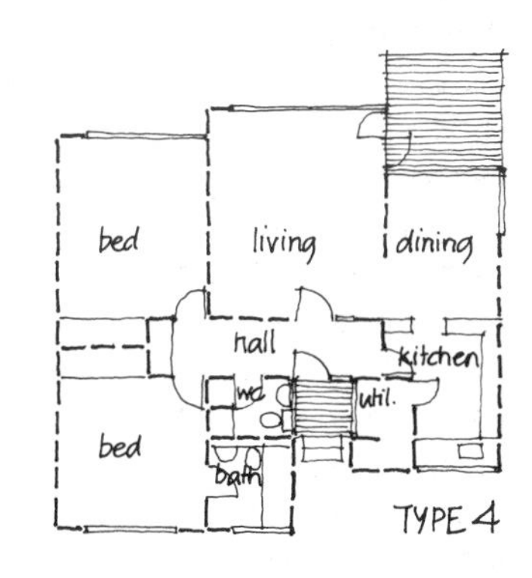

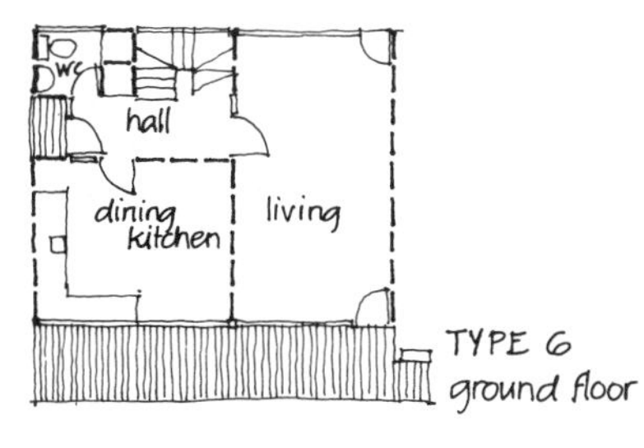

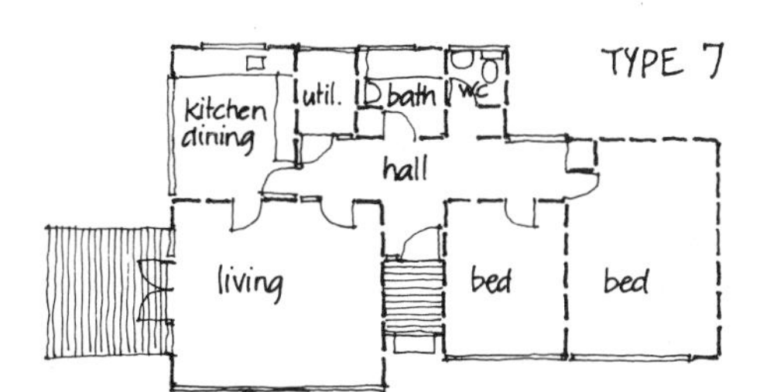

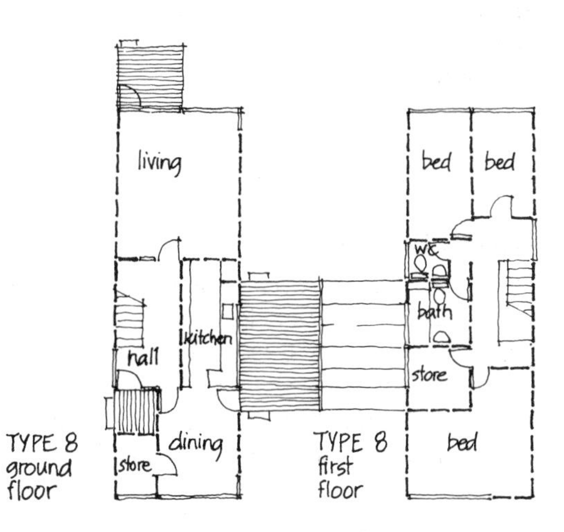

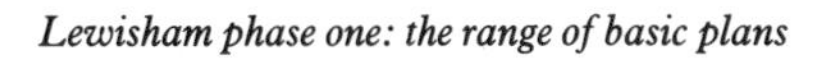
Lewisham phase one: the range of basic plans

tackle the person on the spot and explain politely but insistently the reason prompt action was required. It was an effective technique.

During their sessions at the Churchdown Institute and the Cranbrook, the self-builders devised and registered their form of housing association. They studied the 'Model Rules' book[5] issued by the National Federation of Housing Associations for the Self-Build Groups and rejected it.

The whole concept around which these rules were framed, that of a well-disciplined, tightly organized group of mainly skilled building tradesmen working in sequence on a whole group of houses, all exactly comparable with one another, was inappropriate to them. The ideas that working hours should be logged and penalties applied for non-attendance for whatever reason, even illness, that women and children should be barred from site and relegated to tea-making and administration, were an anathema.

The Lewisham self-builders decided on one rule: there should be no rules, beyond the regular payment of a nominal subscription that would build up a fund for buying some essential tools, electric hand-saws and the like, and some headed notepaper.

They left it at that to see how they went, and never added any more. Their faith and patience were eventually rewarded. The persistence of the council working party and Walter Segal (now joined as co-architect by Jon Broome, who had in the meantime left LBC, spent two years working in Trinidad and returned at just the right moment to help Walter turn out the flood of drawings and documents now required) succeeded in reaching a point where DoE approval to let a contract was a reasonable expectation.

The awakening

In March 1979, not far short of five years after Brian's file opened with Colin's letter, and before all the official approvals came through, Housing Committee chairman Ron Pepper (who had to authorize and bear responsibility for every council housing scheme) said, 'Let's not wait any longer for confirmation and just get on with it.'[6] So the lead house, appropriately Ken Atkins's at Elstree Hill, got off the ground at last (the below-ground works, requiring only labour and not costly materials, had been going on prior to approval, on all sites).

Ken and Pat's two-storey house

The scheme on site—the 'lead' house

Walter had predicted at the outset that the building would be the easy part, and so it proved. Ken and Pat set the pattern of work. Having prepared their site and made the main frames they invited the rest of the group for the communal 'barn raising' effort, which only took half a day. Pat was out shopping and couldn't believe her eyes when she returned to see the whole volume of her house materialized as if by magic. In the next nine months during evenings and weekends they finished their house.

The floodgates opened

Ken and Pat did make the going seem straightforward if not actually easy, and as the official permissions for the other sites trickled through, the rest of the group followed suit with great gusto. Jon, as site architect living nearby and constantly on call, was kept very busy, chasing deliveries and supervising the setting

A family affair

Lewisham phase one: a bungalow

out of work; Walter was frequently on site too, and was an unflappable inspiration. The technical performance of the self-builders amazed everybody charged with supervising the job, and probably themselves most of all.

Each household was responsible for building its own dwelling from start to finish. This was made possible because each person was able to carry out all the operations necessary using the simple Segal technique of timber construction. The group would co-operate on the shared parts such as roads and drains. People did work together in many cases because there are real benefits to working with others, sharing tools and knowledge and having company. The essential point is that they did so out of choice, not necessity. There was a balance between individual and co-operative working, and it was this that played such a part in keeping relations on site good. It liberated the group from the more restrictive organizational procedures used on most self-build sites. Women and children were encouraged to take part so that the whole enterprise was a family affair, leading to an atmosphere more like a family outing than a building site.

Lewisham phase one: a pair of two-storey houses

It was a happy time, seeing the results of so much planning and waiting around taking rapid form as really nice houses emerged, houses that pleased everyone: self-builders, councillors and their officers and consultants alike. The most worthwhile result, though, seemed to be the flowering of talent and initiative among the self-builders; they seemed to be growing with their houses.

The first Lewisham Self-Build Housing Association built fourteen two-, three- and four-bedroom houses and bungalows on four small sites.

Finance and tenure arrangements

The financial arrangements for phase one of the Lewisham scheme were devised so that people with low incomes and no capital could build. They were based on a shared ownership arrangement where the self-builder buys a ninety-nine-year lease for part of the equity and pays a portion of the standard council rent for the balance of the equity. The cost of the lease is reduced by a sum that represents the value of the self-builder's labour in building the house on the council's behalf. The self-builder assumes full responsibility for maintenance, even though the council may own part of the dwelling. This was important from the local authority's point of view, overburdened as they were with maintenance problems. The 'labour allowance' was fixed as a proportion of the cost of the materials and it reduced the cost of the lease to a more affordable level. The self-builders were granted a guaranteed council mortgage. They had the option to increase the proportion of the equity that they had on a lease in ten per cent portions. Many of the first self-builders in Lewisham now hold all the equity on a lease, because it is currently almost as cheap to buy the lease as to pay rent: rents have risen steeply, whereas the cost of the lease was fixed at the outset. The council financed the scheme as it would its normal housing developments.

The second Lewisham self-build project

The first project was such a success that the council decided to proceed with a second scheme. The site that was identified shared many characteristics with the sites that formed the first scheme. It was steeply sloping and wooded and had a subsoil of shrinkable London clay. The whole project of thirteen houses could, however, be accommodated on this single site, unlike the first scheme. A number of lessons had been learned from the first project and as a result the second phase was changed in a number of ways.

Due to the delays experienced on the first phase prior to building, it was decided to work up proposals in advance of inviting people to join a self-build group. A basic frame for a two-storey house was designed and a layout worked out that satisfied the basic planning restrictions on overlooking, building lines and so forth; it avoided the existing sewer running across the site while offering good orientation and aspect to the individual houses. A great deal of individual choice was still possible because of the inherent flexibility of the building method. People were able to decide how they wanted to divide up the internal space in the houses, whether they wanted a separate kitchen or dining room, or whether they wanted two, three or four bedrooms, for example. It also allowed for optional extensions to be added on to the basic house type. It was agreed with the planning authority that additional planning permission would not be required provided that these extensions were no more than ten per cent of the floor area, or in other words about $9m^2$ (100 square feet). These extensions were formed by cantilevering out from the basic structural frame in different positions on the front or back of the house. The structural frame was identical in all other respects for each house. In this way the houses are all quite different from one another although the self-builders were not involved in the scheme until much later in the process.

The scheme was again financed by Lewisham Council, but this time more flexibility was possible in the equity sharing arrangements. People were able to set an equity share that suited their particular financial needs, starting with a minimum holding of around twenty per cent which represented the value of their labour input.

The results of the dream

All in all twenty-seven houses were completed in the two Lewisham self-build schemes, providing detached houses with gardens for local people in housing need. Of the self-builders, half had been council tenants and half were on the housing waiting-list. As well as the most obvious benefit of living in improved housing conditions, there were many other consequences for the self-builders. The changes in people's lives have been far-reaching.

One family lived in a 'concrete jungle' of slab blocks on an estate where noise from the neighbours was a continual problem, where the kids vandalized the lifts —in short, they suffered from all the drawbacks of high-rise living. They now live, to their great delight, in their own three-bedroom house set in its garden with trees all around. The father's outlook has been transformed by coming into contact with all sorts of different kinds of people with different attitudes. His horizons have been widened to include all manner of

Lewisham phase two: Honor Oak Park

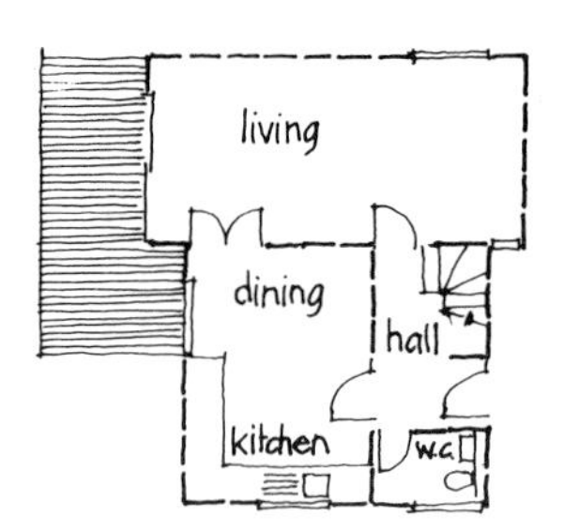

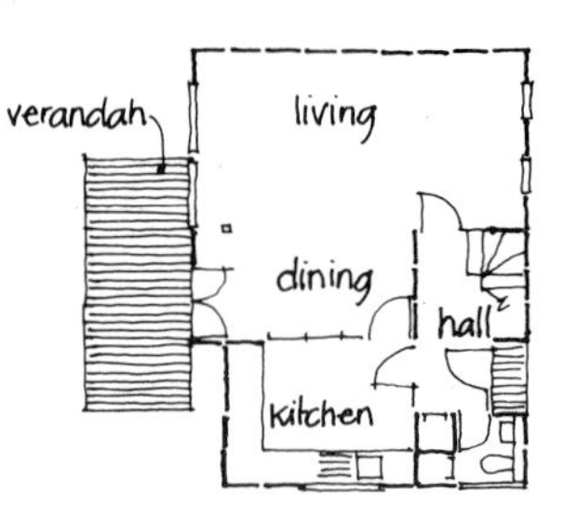

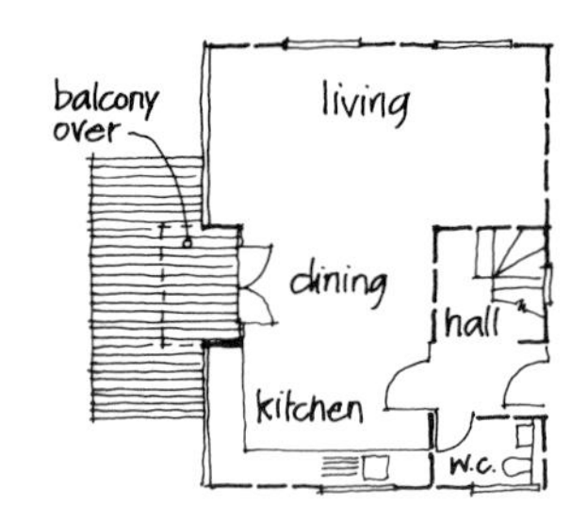

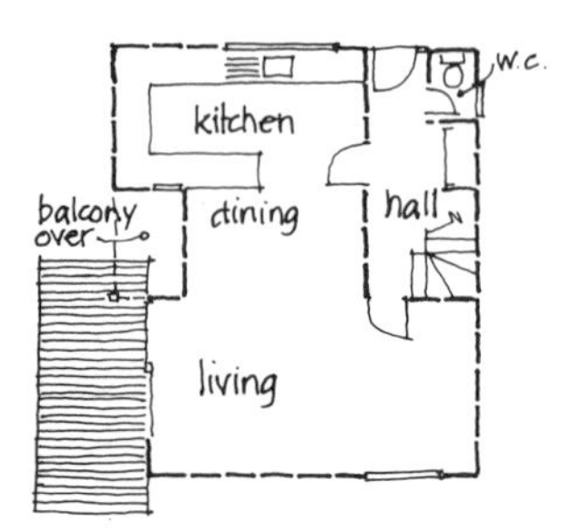

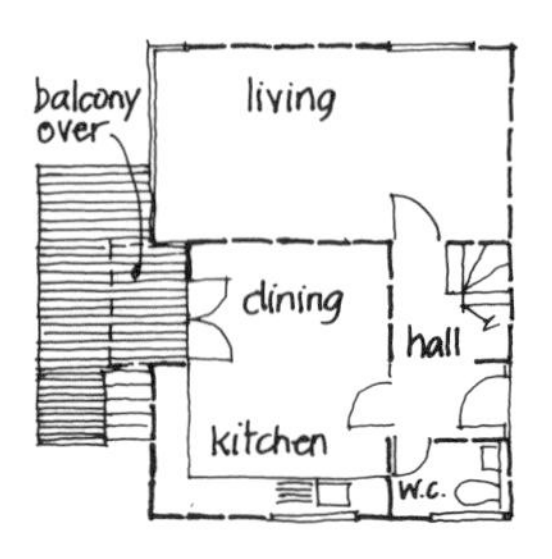

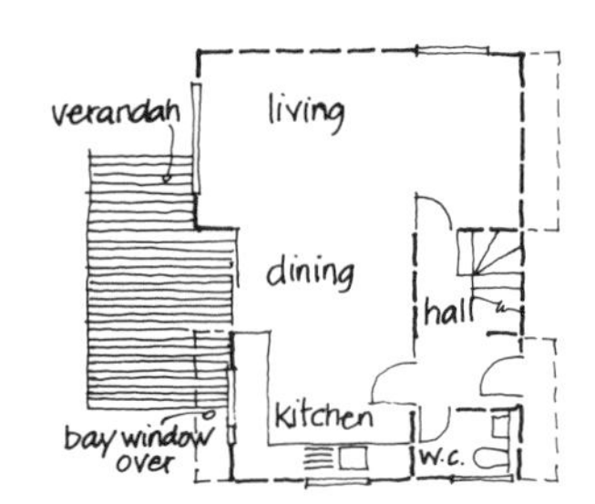

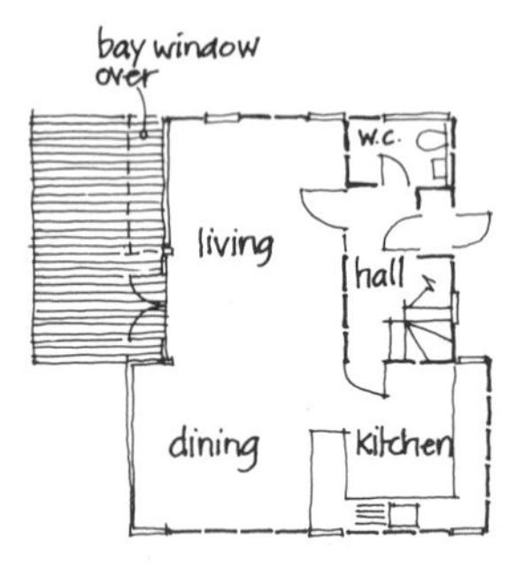

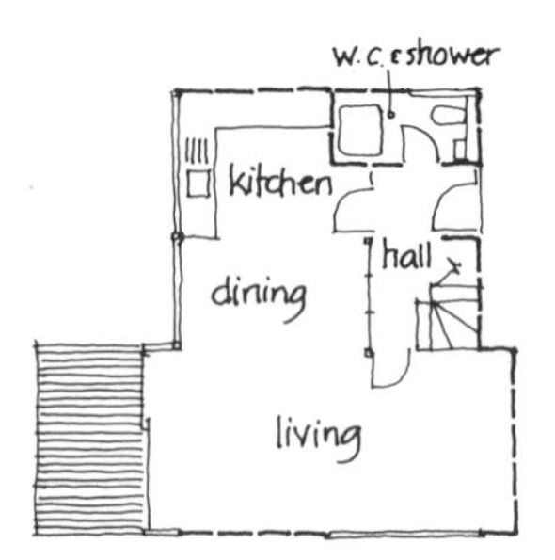

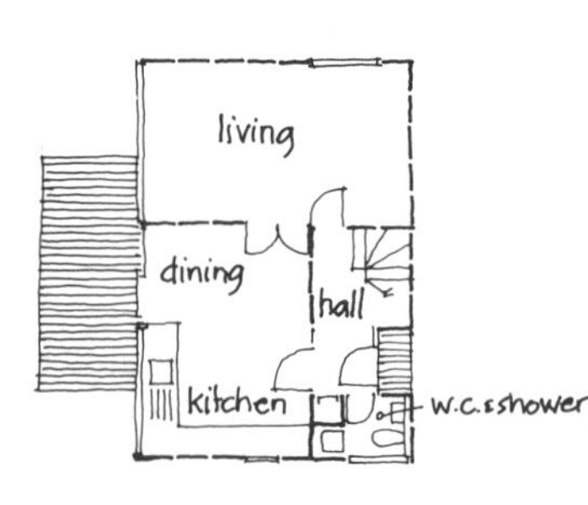

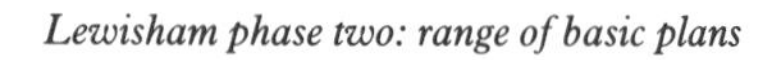

Lewisham phase two: range of basic plans

Lewisham phase two: houses among trees

Lewisham phase two: a completed interior

political and social issues and he now teaches other people how to build; he has given lectures to the assembled housing ministers of Europe at a conference in Switzerland and has travelled to Germany, Holland, USA and Italy on housing matters.

Then there is the couple in their late fifties who built a two-bedroom bungalow for themselves to retire to while having to care for a very infirm aged parent who needed a great deal of help with her daily routine of getting up, eating, going to the toilet and so forth. They would never have had the remotest chance of joining most self-build groups. As it was, they did take longer than anyone else to build, but this was not a problem because the group was organized so that each builder was working independently.

The whole experience was educational, in the widest sense of education for life, for the people who took part: forming a group from the random selection of very different personalities that came together and developing an effective way of reaching decisions that would be acceptable, negotiating with the many authorities involved, working alongside other people and so on. One of the members was heard to remark when he finished, 'Nobody will ever be able to tell me that I cannot do something—anything—ever again.' This person is now in business in his own right as a builder, building among other things Segal-type timber buildings.

Another advantage for the Lewisham self-builders is the ease and economy with which it is possible to build on and to change the internal arrangements of the houses. This is likely to ensure that these buildings have a long, useful life (which is more than can be said for much of the housing built recently, some

Lewisham phase two: building around themselves

of which has had to be demolished after just fifteen years). A stock of adaptable houses was identified as a 'national necessity' in the Parker-Morris report, published in 1961, and it remains so to this day.

For example, two of the houses have since been extended. One family had another baby and built on an extra bedroom. It took about three weekends to make the shell of the new extension weatherproof, after which it was a matter of the finishing and decorating. The cost was about £1,200 in 1986. They were able at the same time to take out the wall of the living room and re-erect it in a new position 4 feet further out, thus increasing the size of the room at very little cost. Another family found that their dining area, whilst adequate for normal family use, could not accommodate a large group of people for a special occasion. They too built on, turning part of their verandah into a new dining room, complete with bay window with a window seat. The modular nature of the building enabled them to reposition the old kitchen window and the patio doors out on to the verandah to form part of the enclosure for the new extension. This too took three weekends or so to get enclosed and cost around £1,500 in materials, including some rather expensive but beautiful boarding for the ceiling.

It is also significant that only four of the fourteen phase-one homes built fourteen years ago, and one of the thirteen phase-two houses, have been sold. The other twenty-two homes are still occupied by the people who designed and built them, not because they could not move if they wanted to for lack of buyers, but because they are satisfied with what they have got. This situation is in marked contrast to many self-build schemes where a high proportion of the houses are sold within the first few years by people trading-up in the housing market. The Zenzele scheme described in the next chapter is an example of this. The Lewisham self-build projects are radical: the notion that you can offer people from the housing waiting-list or existing council tenants the land, finance and technical resources to design and build their own houses without expecting them

Lewisham phase two: 'great satisfaction and self-confidence'

to have had any previous experience of building is a far-reaching idea now (even with all the recent talk of participation and community architecture). It was certainly a bold step to have taken in the mid-1970s. Nobody knew if it would work. We believed that ordinary people had the skill, persistence and energy to succeed. That belief was well founded as it turned out and these projects have been a great success.

There are now groups of delightful, light and airy, comfortable houses built with great economy on small sites scattered throughout the borough that would otherwise remain unused; houses designed and lived in by contented people who are part of a community with shared aims and experience. As long-established communities have broken up for one reason or another, the community spirit has largely disappeared from London, but Lewisham self-builders will put up the flags and bunting when a new member is born into the community. This shows one way that a sense of belonging can be re-created in the places we live. What is most significant, however, is something that is not immediately obvious: the great satisfaction and the self-confidence that people have gained.

Since these projects were devised the whole local government context has changed radically, with a drastic cutback in the resources available for housing. For this reason it is unlikely that local authorities will wish to use their scarce financial resources for direct funding of projects of this kind in the future, preferring to spend them on other housing priorities.

Lewisham 1995: Fusions Jameen

The recent story

Lewisham Council's support for self-build continues, however, by making sites available. There are three projects under construction at the time of writing, all being developed by South London Family Housing Association. Fusions Jameen are a black housing co-operative who are building eight bungalows on two sites close to Segal Close in Brockley and rather similar to the 'classic' Segal bungalow design. The co-operative are now developing another site in Downham with thirteen 1, 2 and 3 bedroom houses with grass roofs. Greenstreet Housing Co-operative are developing ten two-storey houses on a site in New Cross. The site is in a conservation area and the houses are designed to fit into the surrounding Victorian street pattern. The idea is that extensions are possible into the roof space. These houses are all designed to a high level of thermal efficiency and incorporate green design thinking. This will bring the total to over fifty Segal method self-build houses in the Borough.

Chapter 8
Zenzele

Single, unemployed young people from St Paul's, Bristol, build for themselves

This pioneering scheme deserves wide acclaim. It convincingly refutes the false notion that young people are unable to improve their lot by their own efforts. Given only a modicum of support, and the opportunity, they did it in Bristol.

An organization new to us, the Bristol Self-Build Development Agency, booked a place at a day seminar we held during the period of the Walter Segal exhibition at the Festival Hall.

The man who came, and paid close attention to the advocacy of the Walter Segal method we were giving out, was Chris Gordon, the Agency's co-ordinator and a key member of the Zenzele Self-Build Housing Association, about which we had heard quite a lot.

Jon was excited to hear about the Agency and booked a day with him in July 1988 when we could come to Bristol and find out more. It turned out to be Carnival day in St Paul's; the crowded streets didn't make finding our way to Chris's office any easier. Although there were plenty of police about, diverting traffic, Upper York Street did not appear on their A-Z street maps so we had to wander about till we found it. Chris was faithfully there, missing his first Carnival ever (it has been going longer than Notting Hill, and in spite of the association in the public mind of St Paul's with the great riot, is always trouble-free). But he grinned and said for him Carnival really got under way about two the next morning, when the parties were in full swing. Meanwhile he devoted his attention to us wholeheartedly, saying with evident truth that there was nothing he liked better than to talk about self-build.

Chris worked at the time in a small, bright, workmanlike office, where he was funded by the local Inner City Task Force, which got its money from central government. It was part of Task Force remit to encourage self-build as well as to create employment opportunities and so on . . .

He had just been reading a newly published book —a collection of *New Society* pieces called *Grassroots Initiatives*, which had an account of Zenzele. Short, but the best yet, said Chris. He said he would like to write a book about it one day, saying what it was *really* like. It was a success story, and a two-page spread allows space only for the good things that came out of it. It wasn't as easy as that, said Chris.

This chapter does not pretend to be anything like the full story, but will go a little beyond the *New Society* article and will just hint at what lies behind the laconic reports that record the triumphant success of the Zenzele project.

The official 'completion report of the Bristol pilot project for the unemployed—twelve self-built flats erected by Zenzele Self-Build Housing Association Ltd, a member of the National Federation of Housing Associations', for instance. Prepared by the project managers, I.E. Symonds and Partners, Quantity Surveyors, in conjunction with other advisers, it is pretty terse stuff.[1] The objectives are set out:

> a) To provide incentives and work experience for the young unemployed people in the scheme; the skills then acquired should increase their job prospects.
> b) To enable the young people to acquire their own accommodation, which they could take pride

Zenzele, Bristol

in building themselves and which would provide some equity for the future.
c) To demonstrate that a home of your own is a possibility for even the most disadvantaged members of our society.
d) To test these objectives by assisting in setting up and completing a pilot scheme with a local group.

Well, they did carry out a pilot scheme, and realized all those objectives. Work started in 1984. During 1984 the building society granted mortgage facilities and the flats were completed in fourteen months, at a cost of sixty-five per cent of their mortgage value and within the budget. Wonderful. One would think that such common-sense ideas successfully translated into action must have meant that the project was an entirely enjoyable experience.

Not really, said Chris. Setting up a group was fascinating, but not keeping it going while the bureaucratic procedures were gone through. And how was it set up? Why named Zenzele? And what procedures were troublesome? The idea germinated in the mind of a Bristol citizen named Stella Clarke. Not involved in local party politics, apparently, but more in local affairs, including the university, she is a magistrate and a difficult person to say no to. One of the local enterprises she was associated with was Project Full Employ, and the project co-ordinator was an African woman called Tana. She put the self-build idea round the young people of St Paul's, who had problems getting employment and accommodation, and the first person she spoke to was Chris Gordon. Tana eventually married and returned to Africa, but her suggestion of the Zulu word for 'build-it-yourself'—Zenzele—has stuck as the name of the scheme.

News of the project was passed on by personal contact and a group of twelve gathered—half black and half white young men as it happened, because there were no racial connotations in the concept of the project, only social and economic.

It was a wonderful idea and there was great enthusiasm but the negotiations for approval to the scheme, for land and for finance, took ages and ages —just as at Lewisham and Lightmoor (see Chapter 8). The self-build group quickly learned the skills necessary for organizing their own business—registration, account keeping, training and so on. But the mechanisms for making things happen seemed to be out of reach of the young men; all they could do was wait for the outside advisers to pull it off. The personnel of the group changed radically during this terrible period. Many hopeful people worked hard at establishing the group, but just couldn't hang on and hang on, and left without seeing anything come of their efforts; a bitter experience. Chris is still in touch with many of them and some regret not holding on longer, because the rewards of the persistent final twelve—by now eleven black men and one white man —were great.

A site had to be located. The group identified one, outside St Paul's in Eastville, that was owned by the Bristol City Council. It had quite a low historical cost, and at that time the council was permitted to sell at that. No site investigation could be afforded so a risk was taken and work proceeded without it. Indeed, old cellars were found, and necessitated some deep foundation. Hard work, but fortunately not too great a financial setback.

The support group of advisers was under the leadership of Stella Clarke, who became honorary management consultant to the self-builders, and included Phil Barnes (a local authority financial adviser who became honorary financial consultant), architects, quantity surveyors (who also provided project management) and the manager of a local building society. With solicitors and the local branch of NatWest bank, a financial deal was worked out, using Housing Corporation funds for development finance. Above all, the DHSS had to agree to support any remaining unemployed members' mortgage repayments.

All this took time and demanded the co-operation of a lot of different people in different organizations. When the financial picture emerged, the architects had to develop a brief with the self-build group, arrive at a budget and design within it. Chris Gordon is full of admiration for their efficiency in achieving this by no means easy task.

Then there was the job of assembling a sufficient number of skills to carry out the work. The job, unlike a Segal scheme, was not designed to by-pass the conventional trades that are wet, heavy and require special skills—bricklaying, roof tiling, plastering and reinforced concrete work. Rather, the undertaking and mastering of these skills was an intentional part of the scheme to increase the job opportunities of the participants.

So, after all the waiting and negotiating, to the straightforward hard work of building. In the foundations, everybody tried their hand at block laying, the ones who could cope going on to develop their ability, the ones who were hopeless taking on the labouring work. And so through the other stages of the work and the other trades.

The person to whom much praise is due was the working foreman, John Meehan, who took technical charge of the whole operation on site. Curiously, he is not named in the official completion report quoted above. He had been selected by the self-builders from a shortlist compiled by the project managers and paid a salary out of the scheme budget. According to Chris he was a model foreman, always on site from 8 a.m. to 5.30 p.m., always ready with a plan of work, authoritative in all his judgements about the building process and the quality of work. At the end he said it was the most challenging thing he had ever done. Like the others, he benefited from the experience and went on to be site agent for a major contractor.

The City Council's Building Inspector too was helpful, not creating difficulties but seeing that everything was properly done.

The bills of quantities were not of the usual type, a vehicle for obtaining competitive tenders and of no use for any other purpose than comparison with the final account. These bills were comprehensive and accurate and were used, like Walter Segal's own schedules, to order materials. Zenzele opened trade accounts and monthly payments were made from the money made available by the Housing Corporation. Materials were called up by the foreman in accordance with his plan of work.

All the work was done by members of the group (though Chris confessed that friends and relations did occasionally join in, although they were not formally recognized because of insurance complications) with the small exception of some plastering.

Zenzele: pride in building for themselves

There was a plasterer in the group (also an electronics engineer, a jeweller and a law student!) but the craft was difficult to teach others and there was just too much for one man, so the external render was subcontracted. Chris Gordon's brother was the carpenter in the group and did some lovely work in the glazed fire-stop screens among other things. Chris himself had worked for a roofer and tiled the entire roof.

At the outset, everybody entered into a written agreement to work a minimum of twenty hours each week if in employment, but with a commitment to work as much more as they could. In the event, many worked two or three times the minimum, their partners keeping them supplied with food and drink so that there were no interruptions; the meals were consumed on site! Chris says that without all kinds of support from the womenfolk the men couldn't have done it.

When various members of the group got offered employment during progress, they took it and were rescheduled to the lesser rate of working, though it was still a heavy one. This happened to so many of them that the programme was extended a couple of months. This agreement was freely honoured by all members of the group but one. There was no fixed system of penalties imposed, but after repeated warnings and by decision of the whole group, he was simply expelled and replaced.

When the whole block was finished and habitable, bar decorations and internal partition details, the final twelve took occupation and raised their mortgages. They were in well-built flats, spacious for their purpose of accommodating single people, but soon the members were family-building and many of them were able to sell profitably enough to move into a family house. When we went round in July 1988, only two of the constructors were still in residence.

By the end, eleven of the twelve were in paid employment. Some set up their own businesses, notably a building group called Zenzele Construction, which Chris joined. They had a battered old van that was a familiar sight round St Paul's, and Chris was constantly waylaid by admirers of the Zenzele self-build who wanted a go themselves now they saw it was possible.

So with the help of Stella Clarke again, and a lot of work put in by Judy Dugdale from the regional National Federation of Housing Associations office, Chris left the building firm to the other partners and took up full-time work with a new organization, the Bristol Self-Build Development Agency. He has surrounded himself by a very powerful support committee. On it he has Stella Clarke, again in the chair, the National Federation of Housing Associations, two regional Housing Corporation representatives, the Bristol Churches Housing Association, the Bristol and West Housing Association and its related but separate body, the Bristol and West Building Society, a self-build group chair and the Director of Housing for Bristol City Council.

Such a group, serviced by Chris Gordon, a driving enthusiast for self-build, must succeed in breaking down the obstacles that so nearly wrecked his original Zenzele scheme. He already has three more embryo groups in formation and dearly wants to inaugurate others, including one catering especially for Bristol's Asian families and one for the Vietnamese. But still the wheels grind slow, and the search for and acquisition of affordable sites are the sticking points.

Bristol City Council still has considerable undeveloped land assets, much of it derelict for a long time. But land values in Bristol, as elsewhere, seem to bear no relation to the character and condition of the site, and market prices have gone sky-high.

Let us hope that government will encourage local authorities to dispose of land at *reasonable* cost to laudable local enterprises like the Bristol Self-Builders.

Bristol after Zenzele

The initial burst of enthusiasm generated by the success of the Zenzele scheme led to two groups being established in Bristol in 1988. Both experienced delays and work did not start on site until 1991. Both schemes also suffered problems during construction. The Bankole project of 7 two-bedroom houses for shared ownership was completed in early 1993 by a group of unemployed people. The La Maison group also containing unemployed people, built 14 one-bedroom flats. Work started on site in November 1991 but there were problems with the foundations caused by existing mineworkings and this work was subcontracted out. There were also problems within the group and construction took longer than anticipated, so the scheme was not completed until Spring 1994. It proved impossible to raise the necessary mortgages on completion and the flats all had to be rented.

Meanwhile, one of the council housing officers was keen to demonstrate what a Segal approach could offer in the Bristol context. A research budget of £40,000 was earmarked for a demonstration house on the site of a prefab that had burnt down in the 60's. Council tenants in the Broomhill district were invited to a meeting in early 1992. Interested people were interviewed and an unemployed young couple with two children, living in a flat in a tower block unsuitable for children, were selected to build a three-bedroom bungalow. This they did in just ten months with the help of a carpenter/trainer who was on-site two days a week. The project was a success and the council organized an open day to interest housing associations, building societies and the elected council members in self-build in the Bristol area.

Arising out of this initiative, Bristol City Council resolved to promote a group self-build scheme. A meeting of people from the council transfer list was held in late 1993 to establish a group to build a Segal-method project of eight bungalows. The group is mainly young families, with one single parent family, and all are without regular employment. Construction, started in November 1994, is progressing well although the site supervisor had to leave for personal reasons. The group is now being assisted part time by the clerk of works for Bristol Churches Housing Association, the development agent for the project.

Bristol City Council is sponsoring two other self-build projects which are at the feasibility stage in 1995.

Chapter 9

The Lightmoor Project

A vision of a place for a community to live and work is realized in Telford

A wider view of self-build which involves developing economic and community life as well as providing domestic accommodation. A modest start inspired by a grand idea—grass-roots self-builders acting as catalysts for social change.

The scenario

'The Lightmoor Project began as a "scenario" written for the nine working parties brought together by the Town and Country Planning Association (TCPA) and the Rowntree Trust to decide what might be the 1980s equivalent of the Garden Cities which ushered in the 20th Century,'[1] explains Tony Gibson in one of the stream of leaflets emanating from the TCPA advocating the self-help approach to neighbourhood development.

One of the members of the working party was Martin White, Telford Development Corporation's (TDC) Chief Planner. As a result, TDC earmarked 250 acres of third-rate agricultural land it had no plans to deal with, for the new community. There was going to be a different way of reaching Ebenezer Howard's goal of a harmonious garden city: 'The starting point would be the prospective residents, spelling out what they wanted, reinforced by expert advice but working from the bottom up; rather than the professional deciding what would be good for everyone, working from the top down.'[2]

I [Brian] was lucky enough to catch Tony Gibson lecturing at a WEA (Workers' Education Association) one-day school at the Canon Frome Community in Herefordshire. It was in November 1985 and plans were well advanced (after the usual prolonged negotiations) for a spring 1986 start on the Town and Country Planning Association's own Utopian dream at Telford—the Lightmoor Project.

Tony described to us the common objectives and attributes that he identified as unifying a 'combustible' mixture of people in a community:

to be on good terms with those around;
to be asked to do things;
to have the door on the latch;
to have the satisfaction of getting something done;
to contribute to the enhancement of the neighbourhood.

I warmed then to his exposition of the proper relationship of the people to the professionals. He said professional knowledge is important and not to be discounted, but without involvement of people from the start, the professionals can become a problem to themselves and everyone else. He spent a lot of time devising ways of decision-making that involved everyone and not only the fifteen per cent of the population who are what he calls wordbrokers—the talkers.

The project

So pioneer-minded people were invited to work out their own plans for the land. Margaret Wilkinson, TCPA's representative in Telford, emphasizes that Lightmoor is all about people, not about houses alone. They would be able to thrive because opportunities to make a living would be built in to the neighbourhood, within easy reach of home.

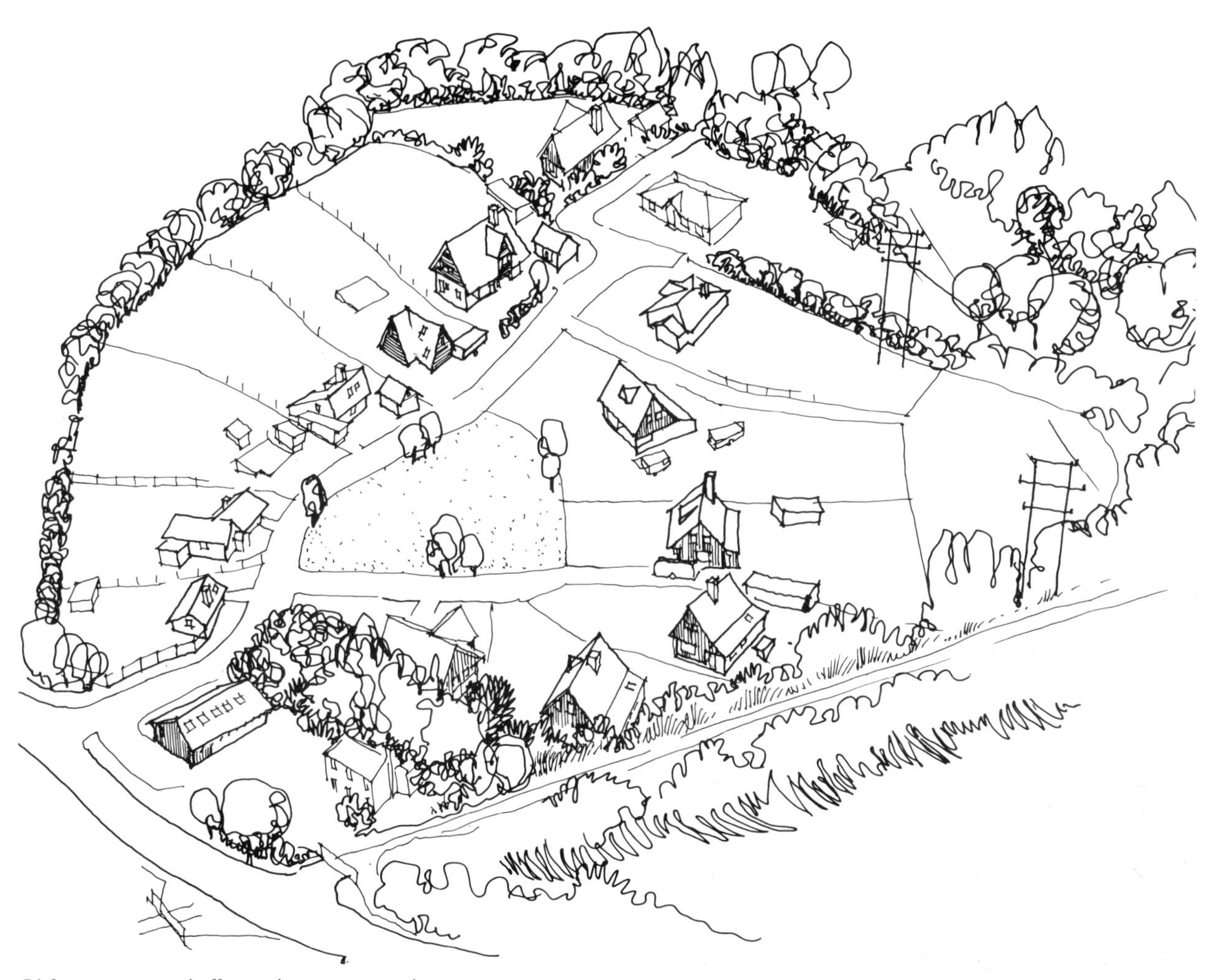

Lightmoor: an organically growing new community

Margaret explained in a letter to Brian that 'the Telford Development Corporation obtained a Section 7.1 approval under the New Towns Act from the Department of the Environment for the establishment of a new community at Lightmoor, which permits mixed use of the site. Each individual plan for house and/or workshop has to be approved under Section 3.2 of the Act . . .

The crucial element that distinguishes this procedure from normal planning control is that the 7.1 approval allows residents the opportunity to set up their own home-based enterprises either in a workshop sited on their plot or incorporated in their home designs. This coupled with the half-acre plot size, on which there is room to keep livestock, which could provide the family with eggs, milk, cheese, is meant to enable residents to opt, if they so wished, for a belt-and-braces economy . . .'[3]

Consistent with the idea of an organically growing 'bottom-up' community, Lightmoor started small. The first phase covered some 22½ acres, including a meadow, pit mounds and wet land. This site was of poor quality, and consequently, being considered undevelopable, was low-priced. (It was ironic that later on this valuation was threatened by the very success of the self-builders in improving the land and demonstrating its usefulness.)

There were fourteen families in the first phase (Margaret was working with a group of prospective phase two residents) and they were formed into a body, the Lightmoor New Community Ltd, a non-profit-making company. The three local authorities

—Development Corporation, District Council and County Council—and the TCPA are represented in the company, with the residents in the majority.

This Community Company holds the head lease from the TDC and issues subleases to self-builders—a legal framework that has caused much frustration and expense in its negotiation. The first-phase residents have their own residents' association, at which all matters affecting the community are discussed. Any problems are taken by the residents' representatives to the Council of Management of the company.

Before submitting his or her plans for detailed planning consent (the Section 3.2 mentioned earlier), each self-builder had to have them approved by the Council of Management of the company. The financial arrangements for the company and its legal and financial relationships with the self-builders are complicated and specific to the particular situation. Broadly, phase one has had two types of financial backing. First, the Nationwide Building Society has provided Lightmoor New Community Ltd with a group mortgage covering the land costs, the infrastructure of the whole site and individual house costs for the first nine self-builders. The next group of five are building as the Lightmoor New Community Housing Association and they have covered their building costs by way of a Housing Corporation loan, which will in turn be converted into a mortgage with the Halifax Building Society on completion. The special feature of this scheme, worked out by Coventry Churches Housing Association, is that the mortgage valuation is based on the completed house rather than its construction costs, and this reflects the 'sweat equity' of the self-builders. The difference between the value and the cost of 'sweat equity' is invested in a high-interest account and is used over a period of eight to ten years to keep mortgage repayments lower than they would otherwise be. The repayments to the Halifax Building Society for this period are handled by the Coventry Churches Housing Association, which charges a small management fee—they call the package a 'management' mortgage. The advantage is that it constitutes a group low-start mortgage without the need for an initial cash deposit. Individual self-builders can buy themselves out of the arrangement at any time by paying off the outstanding mortgage debt.

It was nearly three years after meeting Tony Gibson that I [Brian] actually got to seeing Lightmoor for myself. My first visit came out of my contact with the Green Wood Trust, who among other things advise Telford Development Corporation about the management of the woodlands they own in and around the Ironbridge Gorge. A founder member of the Trust and teacher of woodwork is Gerwyn Lewis, who reminded me that we had met years before when he came to the Centre for Alternative Technology on one of the weekend courses, where a team of us from Lewisham expounded on the Walter Segal approach to timber construction. Gerwyn was sufficiently enthused to take on building his own house and found an opportunity by joining the Lightmoor project.

So one day I made a foray to Telford, where I always get lost in the new town's road system, to discover where Lightmoor was. I found the entrance to the site, where there was a large timber-frame building, and asked my way from a man there who told me that he was Simon Harper, a furniture maker and chairman of the Lightmoor New Community Housing Association. He had his business in the Youth Workshop, which stands in a half-acre area leased by TDC to the TCPA.

It was a self-build project, with timber portal frames made on site and economically founded on short timber piles put in by a tractor-mounted fencepost driver. It was well insulated, with 200mm (8″) of Rockwool donated by the manufacturer, and had been designed by architects Catterall Morris Jaboor, about whom I was going to hear more later.

Simon directed me down the new site road and, having passed a couple of brick-built houses, was caught sight of by Gerwyn Lewis, who, although I was uninvited, made me welcome and showed me his handsome timber-frame house under construction. It was designed by local architect Robin Heath, working in conjunction with Richard Jaboor of Catterall Morris Jaboor.

Because the plan form of Gerwyn's home is complex, and principal divisions had to be structural walls, Robin Heath and Gerwyn chose conventional stud framing, rather than post and beam. The light weight allowed relatively cheap strip footings compared with the raft foundations recommended by the ground survey consultants for the brick and block houses on the site, because of possible mining subsidence problems.

Next to the house, Gerwyn had previously built a well-equipped joinery workshop, which would eventually be the house's garage. The 'garage' could be built ahead of the other permissions because it was carefully kept below the 31m² (334 square feet) size limit and was thus not subject to building control. Attached to it was an ingenious solar-heated store for kiln-drying timber.

I was interested, and there was a lot more to see, but night was falling. Later in the summer I would arrange another visit, and come back with Jon.

The visit

On arrival, we started off at Brenda Cooksey's house, No. 1, an immaculate brick-faced, concrete-tiled bungalow of orthodox appearance externally. Inside the front door (that was, as predicted, on the latch) it looked like a real self-build palace under construction. Like so many of us, Brenda and Dave had moved in from the caravan as soon as it got more comfortable indoors. They were in the long process of finishing off around themselves, and were not too bothered about when it would all be done. Meanwhile, life has to be lived.

Brenda told us that she was a 'Green' smallholder who had found living with her goats and other creatures in a council house pretty intolerable and needed her own place with plenty of space. But such enterprises do not produce much money and making ends meet involved husband Dave continuing with his self-employed bricklaying job. Even income from this was variable and unpredictable and had necessitated spending five days a week in digs in London for a while to pick up sufficiently well-paid work. Maintaining regular outgoings on the property constituted a continuing worry.

In fact Dave and Brenda and their (also brick-built) neighbours the Browns were the only survivors of the original respondents to Tony Gibson's invitation to come and pioneer a self-help community in Lightmoor. Brian asked Brenda how she came to hear about it. Tony Gibson's leaflets were spread far and wide, and in addition there were local advertisements and notices. A Parish Hamlet Committee member in Telford showed one to Brenda, who at once saw it as a route to solving her problem of space for the goats.

It seems that many people who were drawn into the scheme at this stage, attracted by the idea of creating their own living and working community, with a variety of small-scale enterprises mutually aiding one another, had to abandon it because of financial pressures. At that time, no system of financing the company could be found such that it could admit members on very low or non-existent incomes. Mortgage money was available to the company from the Nationwide Building Society on completion of stages of the work, but the way the sums added up meant that only those with fairly substantial incomes or a previous home to sell could prove that they would be able to tackle the repayments. This explains why only two out of the first nine families are from the original wave of pioneers. But others have taken their places and among the fourteen now on site there are two smallholders, two furniture makers, a car mechanic; and among those with paid jobs outside are a bricklayer and some with computer skills. Indeed, the very complex financial records of the company rely on the enthusiasm of a home computer operator in the group.

As at Lewisham, the gestation period proved the most difficult time in holding the participants together; as at Lewisham, to establish sufficient morale for the group to continue, a start on the physical work had to be risked before the negotiated deal was complete—an act of faith that it would all work out in the end. Even more than at Lewisham, amazing feats of construction work have been done, particularly in installing the sewerage system and draining the land as a preliminary to house building.

Although the land was cheap, it was completely unserviced and much work had to be undertaken to put in roads and a drainage system (they were beyond and below the main sewers) whose outflow had to be to potable standards—more than just a big domestic septic tank, more a mini-sewage works.

After the Gerwyn Lewis house, Robin Heath went on to design an individual house for his neighbour Martin and then a group of five houses for the Lightmoor New Community Housing Association. This came about because late in 1985 Gerwyn described a low-cost housing scheme already at the inception/design stage with brick shells, and suggested that Catterall Morris Jaboor produce a timber-frame version. With the assistance of the Interbuild Fund,

Lightmoor: the loose-fit shell timber-framed houses designed by Robin Heath

they developed a type house design for which they obtained Building Regulations approval.

In contrast to the Walter Segal method, where flexibility of house design is achieved by laying out the rooms on a 600 mm—50 mm (2′0″—2″) tartan grid (see Chapters 22 and 23) and a structural frame is designed to fit the plan, the shell house achieves adaptability by enclosing the maximum amount of space cheaply. Flexibility through 'loose fit'.

The structural system is based on long narrow plans with a standard span across the width of 4.8m (16′0″) and of incremental lengths in multiples of 1.2m (4′0″), this being the convenient size of the plywood structural panels used. At Lightmoor the houses are 9.6 or 12 m long (32′0″ or 40′0″) and each 1.2 m (4′0″) 'slice' costs about £500. When the houses are arranged as a T, cost is slightly increased, by about ten per cent.

The 16-foot width had been observed by Robin Heath as typical of the traditional peasant cottage, just as I had noted it in the Herefordshire cottages around Romilly, and coincides with the width of our own long thin house (see Chapter 3). From the planning aspect, the 16-foot dimension can conveniently be the short side of a large room or the long side of a smaller one, while at the gable end it divides nicely for two bedrooms.

Robin Heath has aligned the houses east-west, giving them a preponderance of south-facing windows. He added outshots at the back for utility rooms and conservatories at the front to make the most of solar gains. The foundations are formed of piers at 8-foot (2.4m) spacings. Robin Heath comments that when a brick 'skirt' is required to make the appearance acceptable to the mortgage company, the lintels then required to span between the piers below ground are an expensive complication.

Architect/self-builder relations

The way the self-builders and the architect related to each other is interesting. Tony Gibson emphasizes the importance of getting this right: ' . . .instead of the [traditional consultation relationship] professionals consulting the residents and then doing what they think would be good for everybody, it is the residents who consult the professionals and then make their own mind up in the light of the information and advice they receive.'[4]

We learned from the self-builders that Robin Heath not only gave them close attention on site but was always willing to reassess the situation and change the design accordingly. He never seemed to be put out by the self-builders changing their minds about features of their individual houses as they went up.

We were intrigued to find that Robin Heath actually makes the structural plywood panels himself and supplies them ready for the self-builders to erect. This does not seem abnormal to Robin, who says he has been fabricating and putting up timber houses for twenty years, on and off, during which time he has collaborated with 'carpitect' colleague Trevor Stevens (he is both architect and carpenter). As for the service so commended by the self-builders, he says that it seems normal to him to deal with clients who want changes as the work proceeds, as most people cannot really appreciate from drawings what a house will be like. He is in agreement with us that an advantage of build-it-yourself is that you get to thinking about changes as you go along and comments that an advantage of timber frame is that it allows this to be done fairly easily.

He does, however, point out the extra burden placed on the architect if each house is to be individually built (the Lewisham mode) rather than the alternative practice of team-building the whole project. It would make for quicker building to utilize the best skills available in the group for the specialized trades and he would have liked to see the group better organized as a 'firm' for the duration of building. A stricter code of conduct at Lightmoor could, he thinks, have speeded the process and kept the interest charges on the interim mortgage to nearer the target. He would have been relieved of his task of virtual project manager had there been a site foreman with an overall view, through whom information could be channelled.

At Lightmoor I saw people living on site busy about their daily lives as well as with building. They were in a variety of caravans and sheds, and in various degrees of house occupation. All this is pleasantly disorderly and vital. Gerwyn was evidently a bit sensitive to what he perceived as the hippie-like image the colony had gained locally, and emphasized how much they in fact had to conform to the harsh realities of our highly regulated and financially controlled society. I hope I will not play into Gerwyn's critics' hands by acknowledging that I was reminded at Lightmoor in 1988 of the Hazelwood Estate in Kent just opposite to where I lived in 1938. Readers of Colin Ward's and Dennis Hardy's *Arcadia for All*,[5] which tells the history of the plotlands movement of those inter-war years, will know that ordinary people, escapees from unemployment and horrible conditions in great cities, particularly London, took advantage of the opportunity to purchase plots of derelict agricultural land and colonized them. At first, accommodation was improvised, conditions were primitive and the structures were thought by tidy-minded people to be a blot on the landscape. The lesson of the book, however, is that from these shanty-like beginnings have developed over the years entirely charming and desirable property, some of it now worth a great deal of money, and what is more important, still loved and cared for by the occupants.

The Hazelwood Estate, looked down on in my childhood days, is now accepted as mainstream village housing, has its own community hall and has been integrated into the landscape with mature gardens and trees everywhere, in a way that the more formal, structured, speculatively built suburban developments in nearby Orpington have not been able to. Margaret Wilkinson assures me that the 'hippie' image is already dispelled at Lightmoor.

Conclusion

Everything at Lightmoor is being done properly. Within the bounds of the liberal planning approval, no other short cuts to simplify bureaucratic negotiations have been found and no hard physical work shirked. There is the all-important factor of a co-operative spirit and a commitment to mutual aid, and first-class professional advice is available.

Lightmoor is a tough place, but it is becoming a happy one as the self-builders overcome one obstacle after another and realize more and more their dream of a self-made community.

The Project revisited

Coming back to Lightmoor in 1995 gave me special pleasure—it all looked so good. I confess that I had been a trifle apprehensive, when I wrote my account five years before, that the future reality might fall short of the visionary predictions of the TCPA. A tough period had to be gone through before, in Tony Gibson's words, 'the new neighbourhood would thrive because opportunities to make a living would be built-in within easy reach of home. Alongside houses and gardens and play areas there would be workshops, livestock sheds, vegetable patches.'

We have had the recession intervening between dream and fulfilment—not a good time to set up small businesses. It seemed that official support was slackening. A major component of the initiating machinery, The Telford Development Corporation has been wound up and its powers diffused between the Commission for the New Town, the County Council and Wrekin District Council, none of them strongly connected to the TCPA.

But I observed houses, workshops, sheds and large productive gardens thriving, all as forecast. It has happened because Tony Gibson was also right when he said a decade ago that it would not be 'like a city materializing on the drawing board in the twinkling of a planner's eye; but one neighbourhood after another each taking its own shape, with the residents who created it generating their own staying-power and their own determination to safeguard what they achieve together.'

They have stuck to it. They have built their houses, cultivated their plots, some have conducted their businesses from Lightmoor and the whole community has settled down happily together. The place has an air of permanence.

As for 'one neighbourhood after another', I recalled that Margaret Wilkinson had not been optimistic that a second phase would actually follow. This time I tentatively enquired of Lorraine Murray, a phase-one pioneer, if anything was likely to transpire. 'Oh yes' she replied, 'it has been built!'

She directed me along the woodland path that connects the first phase, grouped round its village green, to the new group of six houses—more scattered on their site and not easily distinguished from their pre-existing neighbours on the hillside. They are all individual, and with a prospect over a watery dell they make a charming picture, although I thought most of the houses were less interesting architecturally than Robin Heath's designs for phase one. They have been built with obvious love and care by their occupants, one of whom (whose parents he told me, lived in phase one) expressed to me his delight at the opportunity they had been given and with the result they had achieved.

Lorraine told me that phase two is an integral part of Lightmoor New Community Ltd, but has emerged as a distinctly different group within it. Also a constitutional difference has been forced on them. In phase one, when anyone sells in order to move on, the betterment (the increase in the value of the developed land over its initial low cost) accrues to Lightmoor New Community Ltd. In phase two, which also benefited from cheap land, the betterment has to go instead to the Commissioners for the New Towns (the national successor to the disbanded TDC). This is a sad surrender of principle. The TCPA wanted to adhere to Ebenezer Howard's scheme for self-governing communities which rested on the local community retaining and re-investing the betterment of the land it developed. As Colin Ward points out in *New Town, Home Town*,[6] it was the loss of this essential provision that marred the constitution of the post-war New Towns. It is hard to imagine that much of the phase two betterment will ever filter back to be re-invested in Lightmoor.

Lorraine explained to me that Lightmoor New Community Ltd still has regular meetings (now only quarterly) conducting the business of both phases. Not only is Telford Development Corporation not present anymore, but the County and District Councils do not generally attend—it is more a domestic matter to keep things running sweetly rather than to plan further expansion.

That work is now more the concern of the Neighbourhood Initiatives Foundation,[7] an organization closely linked to Lightmoor both through Tony Gibson and Margaret Wilkinson personally, and geographically, as its headquarters, the Poplars, is on the Lightmoor site. The relationship is further cemented by Lorraine Murray, who as well as being a resident self-builder works at NIF.

Where Netherspring at Sheffield and MW2 at Maasport (described in subsequent chapters) have respectively lost and may lose their community facilities, Lightmoor has gained one at the Poplars. This house, which stood derelict at the entrance to the site on my first visit is now completely rehabilitated and fully operational in its dual role.

There I came across Nathan Cox again, chairman of Telford Self-build Group, who I had first met on a self-build course at the Centre for Alternative Technology. I asked him about the origins of his new enterprise. He reminded me that there is a councillor on the District of Wrekin, Graham Bould, who was a keen observer of the Lightmoor development and a long-time advocate of further self-build projects. I recalled going to Telford many years ago and with my colleagues from the Walter Segal Self Build Trust giving a presentation on the Segal-method to councillors and officers. It has taken a long time to prepare the ground, but Nathan has been persistent and now has a group and a site to build on, just up the road at Dawley. Their designs have been drawn up by Wrekin Council's architects with Pat Borer engaged as a consultant for the Segal-method construction. A piece of low-cost land has been carved out of a former greenbelt area, and room made for eight 'affordable' houses (that is, they must be built through a housing association who will ensure that they are permanently available at low rents). With the help of WSSBT, Nathan has succeeded in getting a financial allocation for a scheme and a June '95 start arranged.

I asked him about keeping the resolve of a group of young people over a protracted period, and their willingness to perservere with the task of building houses that they will never own outright. He is confident that the group will maintain its enthusiasm through the building process. They enjoyed the CAT course and have since gone on to prove their ability by taking on the building of a community centre at Wolverhampton as trainees of Bilston College.

Their incentive, he told me, is not so much ownership as the opportunity to get access to good housing that they can afford. They will end up paying little more than half the general level of Telford housing rents, for homes they have designed for themselves.

Nathan has planned a building programme of eighteen months duration, but to counter the risk of disappointment following unforeseen delay he has wisely negotiated with the housing association a two year period for completion.

Nathan has a long waiting list of aspiring self-builders and sees Lancaster Avenue, Dawley, as a pilot project for further schemes.

So the ripples of Lightmoor expand!

Chapter 10

Netherspring, Sheffield

The twin aims of working co-operatively and using the sun to keep warm are achieved

A group of enthusiasts translated social and technical theories into built form and achieved, through co-operative organization and university research, the benefits of low-cost houses with solar heating.

Looking at Netherspring was a particularly enjoyable experience for me. [Brian visited Netherspring in July 1988.] Two things put me in a good frame of mind: I travelled to Sheffield in beautiful weather by motorcycle and then had the pleasure of sleeping in one of the houses before spending a morning being shown around. And the sleep was preceded by a session in the Shakespear pub with a group of the self-builders.

The project

This scheme had come about because of the personal enthusiasm of a group of friends for alternative technology, co-operative effort and an interest in improving thermal efficiency. One was a community worker, one worked in the local authority housing department, one was an architect, another a builder and teacher.

The core group developed their ideas of forming a building co-op that would work as a team for the length of the building contract and then sell the houses to the individual members, who would be responsible for finding their own finance at that stage. The co-operative was registered with the ICOM (Industrial Co-Ownership Movement) model rules, as a building construction company, and each member paid a deposit of £250. Having formulated their ideas, they put one advertisement in the local paper and that generated enough response to justify a public meeting from which the rest of the group was drawn, after interviews.

Land was bought from Sheffield City Council on the steep fringes of some quite large council housing complexes at Gleadless for a fairly modest amount, producing costs of £5,000 per bungalow plot and £3,000 per house (in 1983). Preliminary costings of typical house types produced a figure of £10,000 for materials, £2,000 for roads and services and £2,000 for contingencies. The final costs were close to the original predictions, allowing for the fact that most of the houses increased in size as the scheme progressed. This resulted in the project taking longer than anticipated—three years instead of two. On average, a house was finished and occupied every eleven weeks, which was not bad considering that most people were working in their spare time and all the road, drainage and site works had to be done in addition to the houses themselves.

The group of fourteen households was almost the same size as the Lewisham phase two scheme at Honor Oak Park. The site was rather similar too—very steep and clayey. The Sheffield site had one extra disadvantage: it was entered at the bottom and all the materials had to be brought uphill.

A lot of the structure was timber-framed for rapid construction and to allow freedom of interior layout, but was not in this case built off the ground on stilts. Instead, the north-east and north-west walls were in

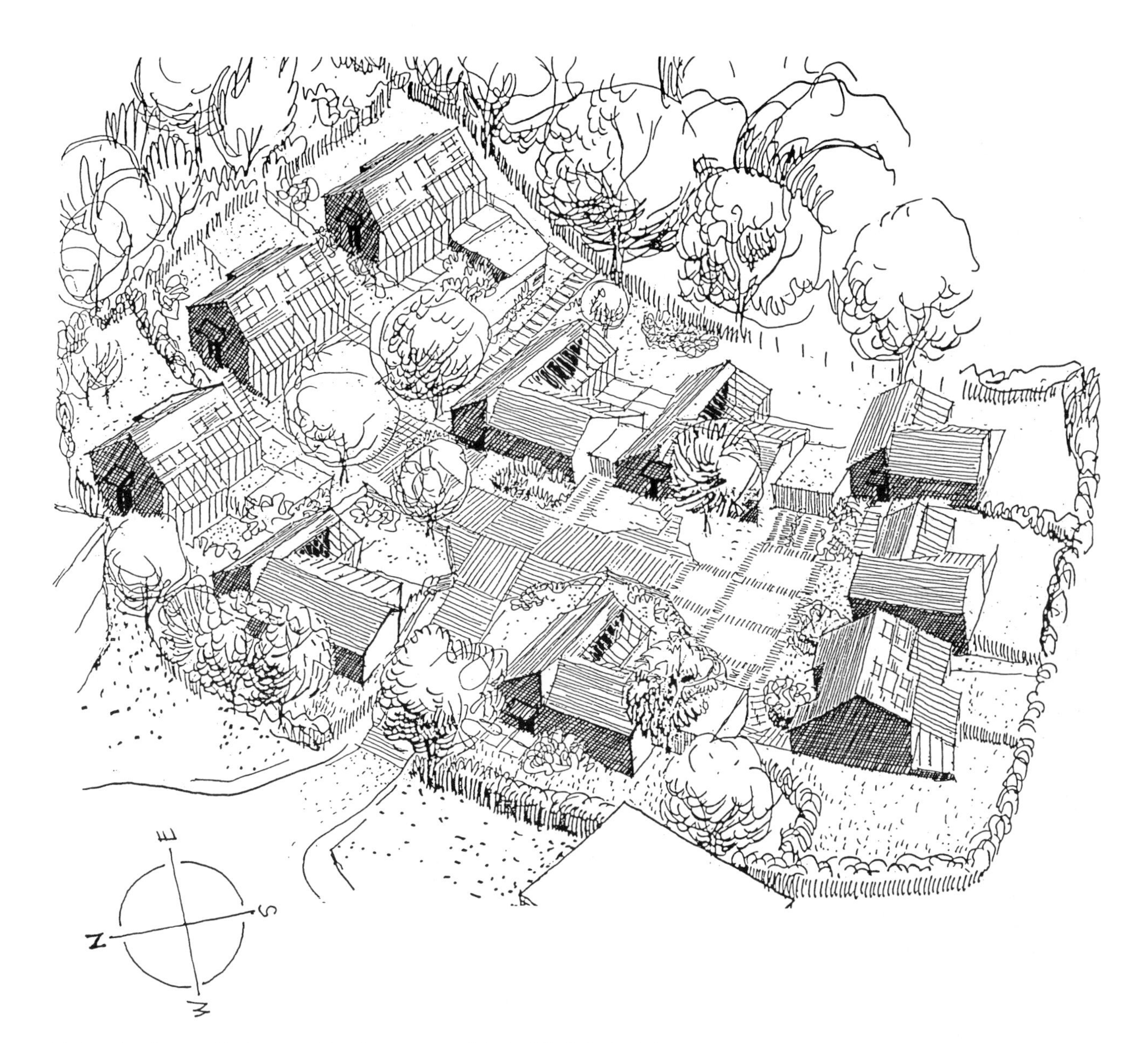

Netherspring: layout of the houses catching the sun on a south facing hillside in Sheffield

insulated cavity brickwork built off concrete strip foundations and with suspended precast concrete floor slabs ('suspended' meaning that the floors were not laid on the ground but spanned from wall to wall and were hollow underneath).

One of the main objects of the Netherspring project was to exploit the solar energy theories of the architect of the scheme, Cedric Green, who had been doing research at Sheffield University. His experiments on a test structure nicknamed the SHED (that was transferred from the university to the Netherspring site as a workshop), proved the viability of solar heating even in northerly Sheffield. Cedric incorporated a number of different devices for using solar heat. The main one was a straightforward cheap glass conservatory on the south side of the building. As the rest

Netherspring: the two-storey houses at the top of the hill

of the house was of heavy, heat-retaining construction, with few and small windows away from the sun and highly insulated against heat loss, it acted as a heat store.

Our initial impression of the scheme on its hilly site with its largely 'wet and heavy' construction was that the self-builders had set themselves a difficult task. Indeed, Cedric Green's initial proposal to the group was that they should build entirely in the timber-frame method he had used for his experimental SHED building. But the group contained two skilled bricklayers who put forward the argument that as the group proposed to build co-operatively, rather than each family building its own house, wall building in brick and block did not pose a problem. This was decided upon, and gave Cedric the opportunity to exploit the heat-retaining properties of the heavy materials in walls and floors.

Even the heavy concrete ground-floor construction proved not to be unduly heavy for the self-builders. Two people could place all the precast beams for a floor in three hours, laying the blocks between as they went. Cedric had then suggested that the internal timber framing be built first, with temporary posts in lieu of the back wall, so that work could go on under the cover of the underfelted roof. Even this arrangement proved unnecessary as the bricklayers stepped up the pace.

All this seemed to run counter to our theory that Segal-method construction makes life so much easier for self-builders, particularly on sloping sites. We willingly concede that there are a lot of ways of setting about building and that each circumstance brings about a particular solution. Formula thinking will not do. Cedric put the proposition that given normal foundation conditions, a co-operative building group with at least one competent bricklayer per eight people and a need for low-energy houses, the case for brickwork external walls and heavy suspended precast concrete floors is incontestable. We hope that you will find encouragement in the fact that different approaches to what seem fairly similar problems

Netherspring: a conservatory

at Honor Oak Park in Lewisham and Netherspring in Sheffield have both brought about excellent results.

Another interesting comparison is with the Machynlleth and Glasgow houses designed by Architype (the design co-operative of which Jon Broome is a member) in collaboration with the Centre for Alternative Technology architect Pat Borer. Although they are of lightweight construction, they too are designed for low-energy consumption and also have the south side enveloped with glazing.

At Netherspring, the heat entering the conservatory is cleverly captured and dispersed round the house and into the domestic hot-water system. Glass-fronted black panels, on the top part of the wall above the fully opening glass-screen wall separating the conservatory from the living room, pick up the sun. This warm air is then ducted down, drawn by a fan, to the space under the concrete floor, where it gives up its heat to the structure, to be released slowly into the living rooms. The cooled air escapes through grilles back into the conservatory to start its journey again.

One of the glass panels fronts a black 'hot box' in which a water tank stands, the contents of which pre-heat the flow into the gas boiler. The result of all this is very low gas bills and very warm houses.

A little extra care with manipulating insulated shutters in the conservatory improves matters still further. A most ingenious device is a set of shutters that swing up horizontally in the summer, making a sort of ceiling in the conservatory so that there is shade underneath. The upper surface reflects the sun, which is thrown on to the absorbing panels. In winter the shutters are set vertical and then fold across the glazed doors separating the living rooms from the conservatory. Cedric's purpose here was to expand in summer into the semi-outdoor room and contract in winter into the more cave-like and highly protected inner room. We rather suspect that in our equable climate this doesn't happen very systematically in most of the houses. The fact is that the conservatory

1 WINTER DAY

- **da** air duct to carry warm air from collector to hypocaust, connected to vertical duct.
- **wd** glazed doors (shutter drawn aside).
- **wn** small windows in north wall of minimum size to reduce heat loss.
- **bi** insulation and earth berm against north wall.
- **cr** single glazed conservatory roof pitched at 32°.
- **ac** returned air to conservatory.
- **sc** solar collector: single glazing, airgap, matt black collector panel, insulation.
- **hf** solid thermal mass in floor slab with hypocaust cavity forming channel for warm air drawn by fan in vertical duct.

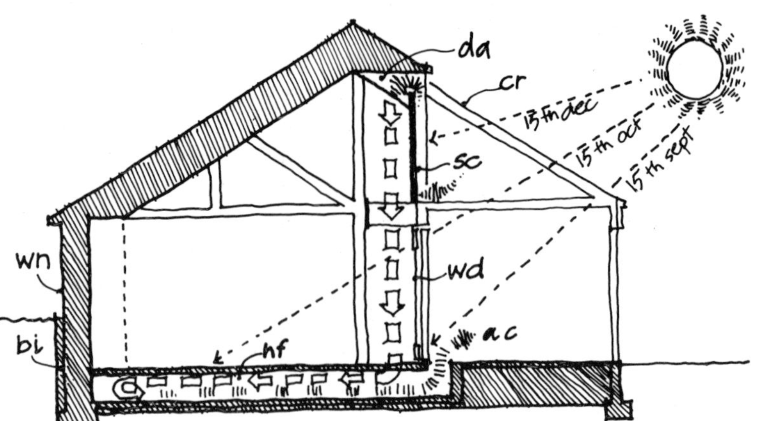

2 WINTER NIGHT

- **da** air duct to carry warm air from collector to hypocaust.
- **ri** heavily insulated roof structure.
- **ns** insulated sliding night shutters over glazed doors.
- **ag** grills in conservatory floor for returned air from hypocaust.
- **hf** hypocaust
- **ci** possible shallow storage cupboards along north wall to increase insulation

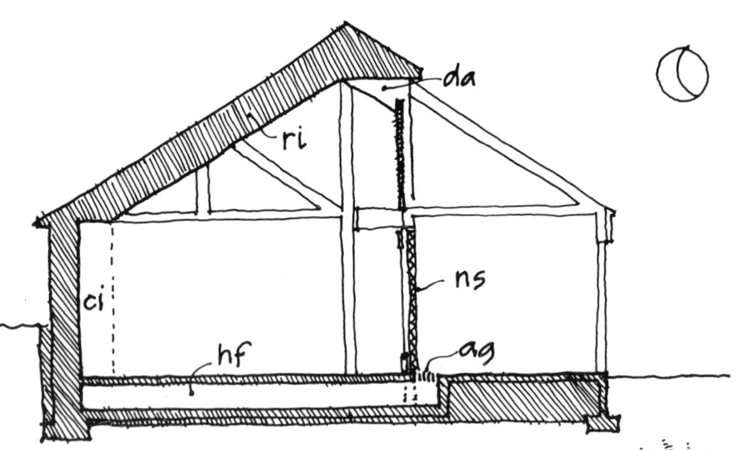

3 SUMMER DAY

- **pi** insulated sliding night shutters hinged up to form conservatory ceiling, shading room and reflecting summer radiation away, or onto solar collector for water heating.
- **cg** vertical single glazing to conservatory with glazed, hinged double doors out to court.

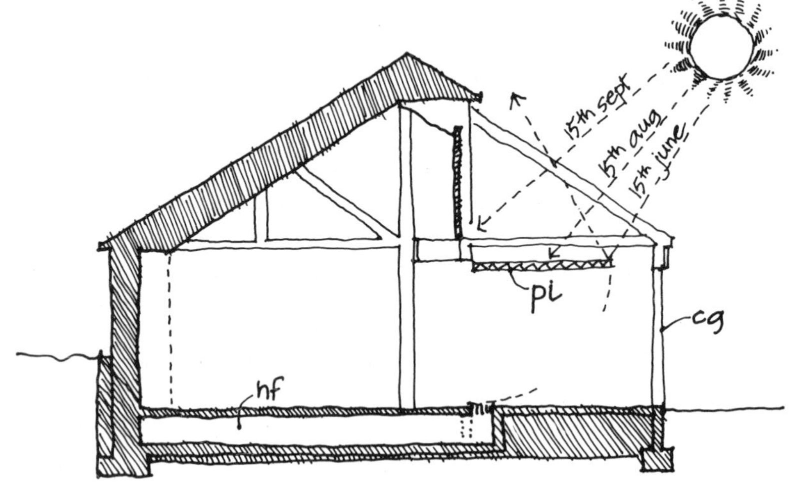

Netherspring: how the solar gains are manipulated

is such a light, cheerful place and its generous space such a boon that people gravitate to it most of the year round.

Indeed, in Gordon's case, he has drastically rethought the whole house plan and has swapped rooms around, taken partitions and glazed screens out, put loft bedrooms in and made his house unrecognizable internally as a derivative of a standard plan. He has kept the summer 'ceiling' as a permanent feature. He has removed the division between conservatory and internal rooms. He has replaced the external vertical glazing of the conservatory with double-glazed sliding aluminium-framed doors. Above the ceiling the solar collector works as before, but below it the apparent room sizes seem to be doubled.

Other people use their conservatories in different ways and it is a fascinating example of how, given the opportunity, people jump at the chance to tailor their built environment to their own particular needs. One idea breeds another and inventiveness abounds.

Externally, the houses are more uniform, there being two basic types: detached L-shaped bungalows and semi-detached houses, most with adjoining garages and all visually linked by the superbly built brick walls, concrete-tiled pitched roofs and conservatory glazing on all south sides.

This reflects the fact that the building was not done by each family individually, as at Lewisham, but was a team operation more like Zenzele, but house by house rather than trade by trade. In this case, there was no training element as such, though skills were picked up. Much of the training burden was taken on by Gordon, the professional bricklayer and qualified clerk of works (aided by an able and at the time unemployed short-course-trained bricklayer), who took on the role of site co-ordinator. There was also a carpenter, and of course having the architect within the group was an advantage, as at Lewisham phase one.

One member took responsibility for quantifying the material that went into each building (the accounting was ably done by several of the women members). This was important with the site being developed house by house. Each one, as it was completed, was occupied and the mortgage obtained, thus keeping to a minimum the bridging finance for the building period.

It might be thought that the enthusiasm of the first-housed group members would wane and that they would become reluctant to keep up the pressure. In the event, although it was a long building period (three years altogether) this did not happen. Provision had been made in the initial agreement for the inclusion of a performance bond of £2,000 held by the co-op that was paid out only on completion of the whole scheme, but enforcement was never an issue. Group spirit was strong and stood up to the disappointment of a prolonged contract period. Two factors contributed to the error in the time forecast. One was the lack of skilled tradespeople, the other the very difficult nature of the site. Not only was it steep and of sticky clay, but it was peppered with springs which were revealed in the excavations, and even two seams of coal! Getting adequate foundations and retaining walls built was something of a nightmare, and even when those problems were overcome the materials handling from the bottom of the steep hill was a continuous problem: the second-hand dumper truck was forever having to be dug out of the mud. Work on the lofty pitched roofs of the two-storey houses at the top of the site, a necessary consequence of the desire to capture the maximum sunshine falling on the site, was a hazardous and arduous operation.

Another delaying factor had been in getting some of the rather last-minute redesigns through the various building controls. An engineer at Sheffield University had to be brought in to supply structural calculations for the timber frames, which were outside the building inspectorate's normal experience. Indeed, some of the heavier heat-storage elements at high level had to be omitted. The thermal capacity of the concrete ground floors had to be relied on alone to even out the temperature difference between the hot air captured in the conservatory roof during the day and the desired evening warmth in the living rooms.

After some skirmishes with the building inspector, who rather gleefully found some of the 'amateur' bricklayers had accidentally omitted some wire ties, Gordon got him to agree that standards were generally much higher than on commercial contracts and the problems diminished.

Now the scheme is complete it seems everybody is well pleased with it. It certainly looks good, there is a good community spirit, the houses were economical to build and are proving very economical to run. Personal commitment to long-term occupation of the houses is less than at Lewisham (or Lightmoor, but that is at an earlier stage still), but more than at Zenzele. These were not starter houses as such, because the members forming the group had to be mature, mortgage-worthy people able to maintain outgoings at the commercial rate (although not of uniform 'middle-class' background). But the houses are not difficult to sell, and in the Sheffield area competitive house prices are not yet so alarmingly high as in the south, so there is not so much pressure on staying put and some are moving on.

A curious feature of the scheme which does not seem so surprising in hindsight, though it caused some disappointment, is the abandonment of two 'communal' features of the original design. There were to be allotment gardens to one side of the site. Interest in these turned out to be slight, and instead the owners of the houses adjoining the allotment area took the land into their own gardens and paid the extra land cost to the co-op. The other communal feature was the site workshop brought over from Cedric's experimental project at the university. The original aim was to expand and convert this into a communal facility, providing a permanent workshop for the residents, with a meeting room and play space. Enthusiasm for this proposal was not universal, and instead two of the self-builders offered to buy the site from the co-op and themselves build an extra solar house on it to be sold on completion. This was done, and of course earned them some money, as well as benefiting the group by cutting their share of site costs.

As the communal building was going to house the computerized equipment that would monitor in detail the thermal performance of the dwellings, this aspect has not been very fully developed. However, some measurements have been taken and bear out the validity of the principles used, though it is not known how much benefit each particular feature has yielded. Instead, the occupants are just getting on with enjoying the well-earned fruits of their labour, neatly summed up by the newspaper headline, 'Sun houses built by co-operative effort'.[1]

More Self-Build in Sheffield

When the Netherspring scheme was completed, a couple of the self-builders were keen to see others have the same opportunity. They organized a meeting in 1988 and established the Woodways Self-Build Group. They enlisted the support of one of the city councillors but met entrenched opposition from the council's officers to the timber frame Segal construction which the group wanted to build. The group persevered, however, and eventually obtained an option on a council owned site. Unfortunately, this site was subject to legal problems and had to be dropped after some years' delay. Their cause was not helped because the only other self-build project in the city, part built by contractors for completion by an all-women group, was not a success. Meanwhile, the one-off Segal-method house illustrated on the cover of this book had been completed in the city. Another site was offered to Woodways, who are starting on site in mid 1995 to build 4 three-bedroom houses for shared ownership, funded by Northern Counties Housing Association. Seven years later only one of the original members of Woodways is still in the group. Interestingly, two of the houses will be built by brothers who were brought up on the Netherspring self-build scheme.

Chapter 11

Brighton Diggers

Low-energy, environment-friendly, Segal-method houses

This project is one of about a dozen developments recently completed in the South of England. The Diggers scheme is singular in its inception, membership and design and therefore unique. Each of these developments is the product of a particular set of circumstances and indeed the very diversity they demonstrate is one of the characteristics of Segal-method self-build. The Diggers project does, however, demonstrate many of the basic principles which it shares with other self-build schemes of this type. Vic Sievey, one of the Diggers, is the narrator.

This is the story of how nine Brighton-based adults, their partners and children, set out to build themselves decent, affordable homes, following co-operative and environmentally-friendly principles.

Building your own home with your own hands, muscles, and brain, is a very satisfying experience. When asked whether I would do it again, my answer is a qualified 'No'—not because the exercise was unsatisfying, or too hard, but simply because it takes a chunk of your life, and to have to repeat it means other opportunities are potentially foregone. The objective, to me, is to build a home, and then to be able to enjoy it!

We were lucky: the majority of the group had known each other for a long time and were committed to achieving successful, self-built housing by co-operative means. I was warned by Jon Broome, when I first met him, before being elected into the co-op, that the building would be the easy bit: finding money and land, obtaining planning permission, and getting to the point where building could commence, would take patience, tenacity, and fortitude! He was not far off!

Funding was eventually obtained from the Housing Corporation by means of a Housing Association Grant (HAG) through the good offices of the SLFHA (South London Family Housing Association) and CHISEL (Co-operative Housing in south-east London). CHISEL is a secondary housing co-operative which provides services to primary co-operatives such as The Diggers. We became contractors to the developing association, SLFHA. The properties, once complete, would be owned by CHISEL, not by us (the Diggers co-op) as had been originally intended, because the 'rules' regarding co-operative registration had changed during the years it took to secure land and finance.

Our group was one of the first to follow the CHISEL 'self-build for rent' model. We would build the houses and then live in them, paying rent to CHISEL. We would receive a 'premium' payment in recognition of our contribution if ever we moved out of the scheme. It is our long term objective for Diggers to own the houses and to manage them communally. This has yet to occur, due to the difficulty of registering a new housing co-op with the Housing Corporation. We hope that this will change, and meanwhile CHISEL itself is run or more or less co-operative lines—we are individual members, and thus have a say in the running of the scheme in the interim.

Brighton Diggers comprises four families with their children (three of them single-parent families), four bachelors and a couple.

We gained three members while building was under way, thanks to the processes of human life(!) and now have seven children ranging in ages from under one to sixteen years. There are two teachers, two gardeners, an artist who designs and sells postcards, a computer consultant, a woodcarver, an accounts administrator and a mature student. Only two of our

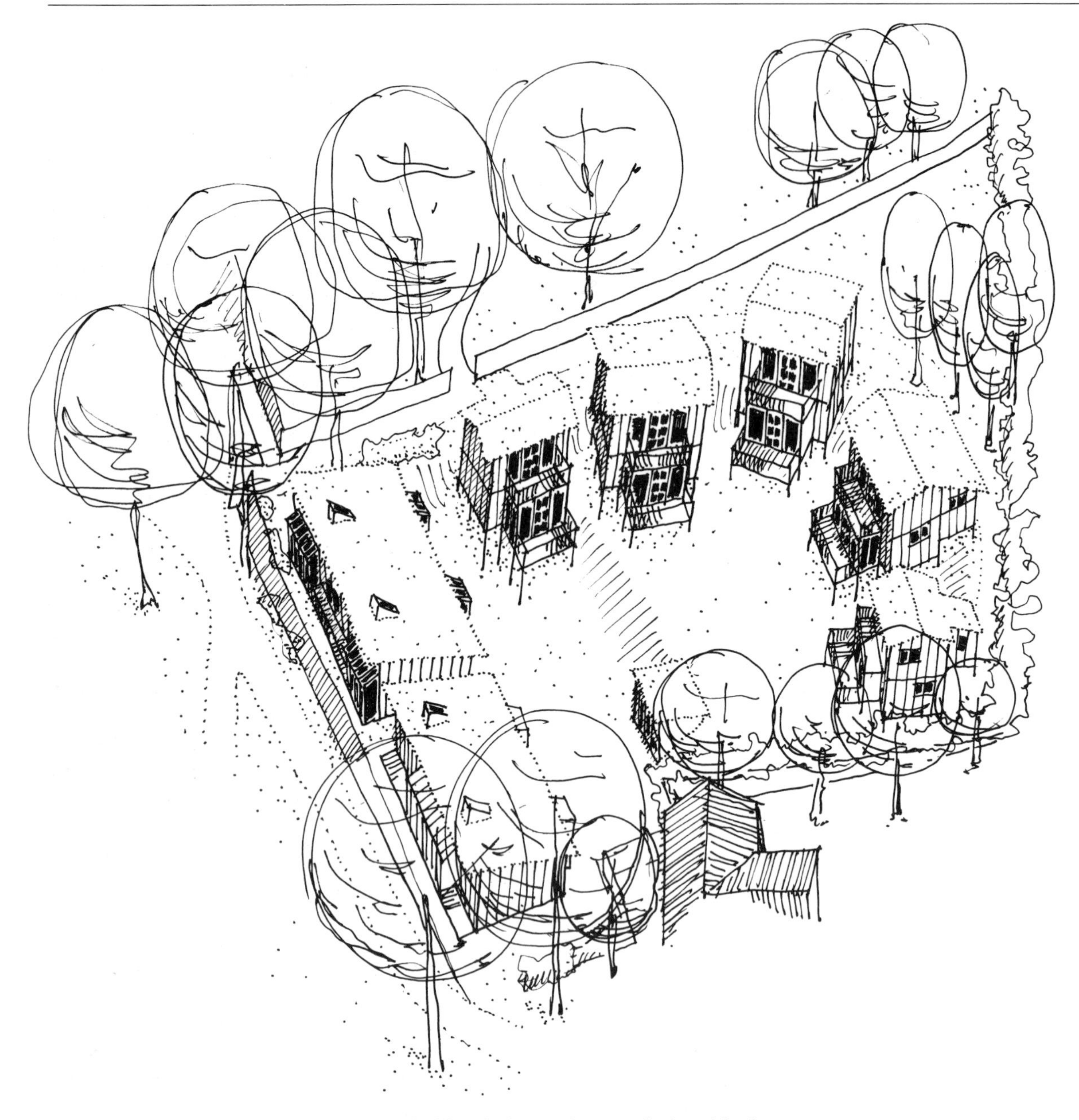

Five split-level houses overlook the low grass roofs of four single-storey houses at the foot of the slope

members had previous building-site experience, the remainder, varying levels of DIY skills. All lived in rental accommodation in Brighton and Hove of varying degrees of unsuitability; be it shared, cramped, decrepit, or simply too costly. Thus one of the main motivations was to provide ourselves with adequate accommodation to suit individual circumstances, and at an affordable rent.

We had chosen the Segal-method of construction because of its straightforward design and construction methodology, and the reduced ground preparation required. Segal-method houses may be built on the steepest of slopes without disturbing existing natural features, flora and fauna.

The site which Brighton Council offered to sell us is on the edge of town, adjacent to a golf course and allotments. Brighton Council had supported a number of self-build groups during the 1960s and more recently the elected members and officers had become familiar with the Segal-method. Our scheme is near to the Sea-Saw project, a group of single, homeless people building twenty-four Segal-method houses for themselves.

Our site consisted of several broad terraces with large mature trees on the perimeter.

We developed a design of houses grouped round a shared open space. Five split-level, detached, two and three bedroom houses are sited on the steep bank surrounding the site. They look over the roofs of four semi-detached, single storey one-bedroom homes towards the west. Each house has a conservatory, a verandah, and a view!

The bedrock of chalk lay just below the surface and so we hired contractors to do the ground works: drainage, paths, steps, and importantly the concrete pads to support the structural columns. They forecast five weeks, and took eight weeks of hard work. We estimated it would have taken us eight months!

We organized ourselves to work twenty four hours per week each. The contract period was set at eighteen months, start to finish. This appeared ambitious at inception, and as we were unable to start building until five months in, it soon became unrealistic. In the final outcome, the house construction actually took twenty four months, which we think was pretty good for part-time amateurs! At times, the programme became quite onerous, as we had not scheduled in any breaks, so holiday periods became catch-up times, especially when winter had limited the available working hours. Weather was not often a problem, only the length of daylight.

Having got everything ready to start, we hired a contract manager, with a broad brief to negotiate material supply, supervise day-to-day construction activity, and to provide on-going training as the project progressed.

We fenced the site in February and March '92; then there was a hiatus for us while the ground works were completed up to August, when the first frames were laid out and completed.

The first five, two-storey houses were to be completed structurally, and externally, prior to beginning the four, single-storey dwellings. As the site has no direct road access, delivery lorries, which had cranes, lifted materials over the fence at the bottom. We then moved them, by hand, up the terraces. The lower part of the site was used to lay-out the columns and beams to construct the structural frames. When squared-up, holes were drilled and bolts fitted to secure the frames at the junctions. These were then moved, needing from four to fifteen people, depending on the size and weight, to the house site, where they were subsequently erected in a 'reverse domino-effect' style. Designed by means of trial and error, and input from a scaffolding expert, an 'A' frame, twenty feet high, with a block and tackle pulley was erected and used to raise each frame to head height. Then 'pusher' and 'puller' timbers were attached to the columns, and the frame manoeuvred to the vertical, where the same timbers were used to secure the frame to the ground, and the adjacent frame.

Once all frames were up and vertical, 3″ × 2″ timbers, pre-cut to the modular spacing, were fixed between, so the whole structure could be squared-up and trued prior to joists for roof and floor being installed.

Having erected the frames for two houses, in October '92, during the Conservative Party Conference in Brighton, we invited Sir George Young, then Housing Minister, to preside at our official launching ceremony and to erect a frame for publicity purposes (having previously lowered it, to allow it to be re-erected!) A host of balloons were released, photos taken, and an excellent buffet demolished.

After frame erection and joists, the sequence was:

- ▷ Roof boards: The first roof, using tongued and grooved 'sterling' board, was installed in November '92 in wet and damp conditions, and proved quite stubborn because the 'tongues' had swollen with the moisture.
- ▷ Wall studs: Vertical and horizontal wall studs to provide a wall thickness of 5″ (3″ × 2″ vertically, and 2″ × 2″ horizontally) were nailed in place to create a frame to which bitumenised softboard was affixed with galvanised nails. Cut outs were made around the window and door openings.
- ▷ Windows: Manufactured, softwood, double-glazed windows were fixed in place having had softwood linings screwed to top and sides. Some fitted easily, some had to be firmly hammered in, others needed

The site plan

to be wedged in as the gaps in some cases were somewhat out of tolerance!

▷ Door frames: The door frames were similarly lined and fitted, the doors being hung considerably later (the openings being covered by plastic sheeting in the interim).

▷ To the outside of the softboard, 4″ × 1″ softwood battens were fixed at intervals to support the exterior 'Glasal' fibre-cement cladding, which was held in place by aluminium profile strips at the base and tops, and pre-painted wooden battens covering the junctions. Mesh at top and bottom protected the air gap behind the 'Glasal', which allowed the building to 'breathe', and drain any moisture away.

▷ Inside 1″ × 1″ battens were nailed to the lower edges of the roof and floor joists. Moisture-resistant plasterboard was cut to fit between the floor-joists, and supported by the battens. To form the ceilings plasterboard was subsequently tacked to the roof-joist battens, after insulating. The floor-joists were drilled to take the central-heating pipes, for which we used Hep$_2$O plastic plumbing.

Concurrent with the above, exterior woodwork was painted; roofing contractors fitted the 'Kaliko' single ply PVC roofing membrane and the grass roofs, which we deemed would be usefully subcontracted to save time and money. In the case of the membrane there would have been the need to train in a specific skill. All the while we were conscious of time and money constraints, and our relatively few numbers.

In August '93, the two-storey houses were nearly complete externally, and we began to construct the frames of the last four single-storey houses, which we brought up to the standard of the other five by the year's end.

The floors in the houses were scheduled to be constructed from sterling-board, like the roofs, but I had hankered after 'proper' wooden floors, so that we could polish them, and have rugs, rather than fitted carpets, for cost, health, and aesthetic reasons. We were advised this would cost three times as much, but Mike, our contract manager, did some negotiating and we ended up with 4″ wide tongued and grooved floor-boards for very little more than the original budget. These were subsequently waxed and/or stained according to individual taste.

The external colour schemes were chosen individually and ranged from white to cream 'Glasal' with contrasting battens and woodwork painted black, brown, beige and dark green, with detail and balustrades in blue, red, grey, and white. The result is a splendid range of contrasting, yet complementary, colour schemes, no two houses looking alike. Mine, for example, has white Glasal, black battens, structural wood, porch and verandah balustrading, with blue doors and windows off-set by red detailing around the frames. All the paint used was either completely 'organic', or environmentally friendly.

At this stage it became very apparent that we were going to take considerably longer than the original eighteen month target, and it was decided to use what appeared to be surplus contingency funds, to hire in labour to do the remainder of the plumbing and to do the plasterboarding. Here we deviated somewhat from the Segal-method by fixing plasterboard in a conventional manner with filler to the joints in some of the houses.

Prior to wiring, plastering, and insulating, the internal walls and floors had to be fitted. At this stage it became very satisfying to see rooms take shape in the space of a few hours. Once the walls and floors were in we could embark on the wiring, which two of us decided we would do ourselves. We started on the lighting circuits in two of the homes, but, because we were being neat and tidy in the cabling, we were slow. We were subsequently out-voted and an electrical contractor was hired. I think that, if we had followed his pattern of work, we could have done three times as much work as we had done, but we had followed training instructions!

The recycled newspaper insulation was supposed to be blown in dry on the floors, but wet on the walls, so it would remain in position while the plasterboard was installed. However, we learned from the self-builders in Greenwich that there was a distinct, and tricky knack to this. On the advice of Excell, the manufacturers, it was decided to blow the insulation in dry, behind a fabric scrim stapled to the wall studs. They delivered the equipment, for shredding and blowing the insulation, which was supplied in compressed bales—825 of them, although in the end we only used two thirds. We were shown how to do the job and off we went. We fitted the scrim, cut slots in each bay, and pumped in the insulation. It was a fairly straightforward job, but needed a bit of a knack to minimize waste. There was the inevitable 'grey snow-storm' in the house during the process, and the operators would emerge looking like grey 'zombies'. We had to finish the insulation in a bit of a hurry so as not to keep the plasterer waiting.

The finance we handled ourselves. At each stage Mike, our Contracts Manager, would quantify the materials used and Robin, our architect, would issue a valuation certificate which was sent to SLFHA, who issued a cheque. I, as treasurer, banked the cheque and was responsible for accounting for the expenditure and paying our suppliers, as well as my share of the construction activity.

We had a budget for kitchen and bathroom fittings which we controlled, and each member was able to specify his own choice of style and colour within the overall figure. Some had hand-built kitchens out of recycled pine. This had been the choice of eight out of nine, but, because of supply difficulties, only two members finally opted for these.

The end result is that no two dwellings are alike internally, or externally, despite the standard method of construction. We were able to alter the dimensions of some of the rooms, the colour schemes, and detail—such as specifying double glazing in the conservatories, so they could be better used, as well as for their primary, passive solar-heating function.

The only hardwood used in the construction was oak for the door sills; plastics were limited to the waterproofing membrane (guaranteed for twenty years), some acrylic baths, and waste water and central heating plumbing. All other materials were environmentally as benign as possible in terms of production and energy costs: European softwood, naturally based paints and waxes, plasterboard, cement-based Glasal, soil and turf roof covering, cork floor coverings in most bathrooms and kitchens, and cellulose insulation.

Heating is provided by high-efficiency gas-fired boilers with computerized, optimizing controllers, to minimize energy consumption. In the single-storey houses instantaneous combination boilers are used, with a target of £50 per annum space heating costs. In the other houses, high efficiency conventional boilers and thermal stores are used.

The buildings are covered under the structural insurance scheme operated by the Housing Association Property Mutual, HAPM.

As I write, in Autumn 1994, having been 'in' for a month or so, there are still things to do to each house, site fencing to be completed, and the site hut to be dismantled and removed, there being no room now to wheel or lift it off-site easily. Overall we have managed to complete the scheme slightly over the original cost estimate, and spent six months longer building than expected. However, we have managed to stay together, avoid serious accidents, apart from a few cuts and bruises, and are still speaking to each other!

The Mayor of Brighton performed an official opening ceremony in October '94, and we were presented with copies of awards for the nature of the construction by RIBA, and by British Gas for the thermal efficiency and environmental awareness of the construction.

It certainly feels like home, and as the weeks go by a 'normal' routine to life is returning.

If in doubt about building your own home, do not hesitate to give it a go. The Segal-method means anybody, and I mean any able-bodied person (and some not so), can achieve what we have done. It takes time, perseverance, and some organization, but it is a long series of simple steps—the first being the 'I want to' step . . .

Chapter 12

MW2, Maaspoort, Holland

Dutch community builds environment-friendly houses

On the Continent, the study of 'holistic health' in relation to the effects buildings have on their occupants is called building biology; it is well in advance of most British practice. Many of the lessons learned by MW2 and embodied in their self-build scheme could be of interest to people in this country who are concerned about the environment. (Brian and Maureen visited Maaspoort in 1988, and again in 1994.)

On our way to a papermakers' and artists' conference in Germany, we had the pleasure of calling on our Dutch friends Rob and Plona in 'S Hertogenbosch. They knew and liked our grass roof in England, so told us excitedly that here in Maaspoort on the edge of the new housing development where they lived were some turf roofs.

We walked round on a sunny morning and I quickly grabbed Plona's camera and exposed her reel of film —the houses were lovely. And something looked very familiar about not only the soft green fringe on top but also the various levels of activity and degrees of completion around. So we asked a woman who was wheeling a barrow down the street what was going on, and she admitted to being not only a self-builder but the carpenter who had organized most of the work. She introduced herself as Germaine and invited us to come back the next day to meet her co-builder Heleen and be shown round. Their house, like all the others in the scheme, was of 'log-cabin' construction—the structure is built up from solid planks halved together at the corners.

The structure was complete and interior finishing was now well advanced, with Germaine turning her carpentry skills towards fitted furniture. We sat down for coffee and Plona helped translate. When they unfolded their story it had a familiar ring to it because although the construction method was different and the Dutch building codes are not the same as ours, there were many similarities with Lewisham, Netherspring and Lightmoor.

Like Lightmoor, the group had come together for social/ecological reasons. They wanted to find a way of living that was kind to people and to the environment. They wanted houses of wood—now quite unusual in Holland—so they went for a Finnish system of precut solid pine planks made to suit the modular designs made by their architect, Renz Pijnenborgh, himself a member of the group, whom I was to meet later.

They chose to build themselves to ensure that they got exactly the sort of houses and workshops they wanted, to keep the price low as the solid timber construction is intrinsically expensive in material cost and to express their commitment to working together as a community.

Like Netherspring, the group built together. The arrangement was that they would work collectively for the first eighteen months, by which time foundations, roads (by the local authority in Holland), drains and house shells would be complete. Then there would be a 'rest' period, while each family finished its own dwelling, and finally a collective effort would be made to bring the communal building to completion. Unlike Netherspring, the commitment to this social hub of the scheme seemed to be enthusiastically maintained, and was expected to be useful to the Maaspoort residents outside MW2*

* MW2 is short for MMWW, initial letters of untranslatable Dutch words meaning 'human- and environment-friendly living and working'.

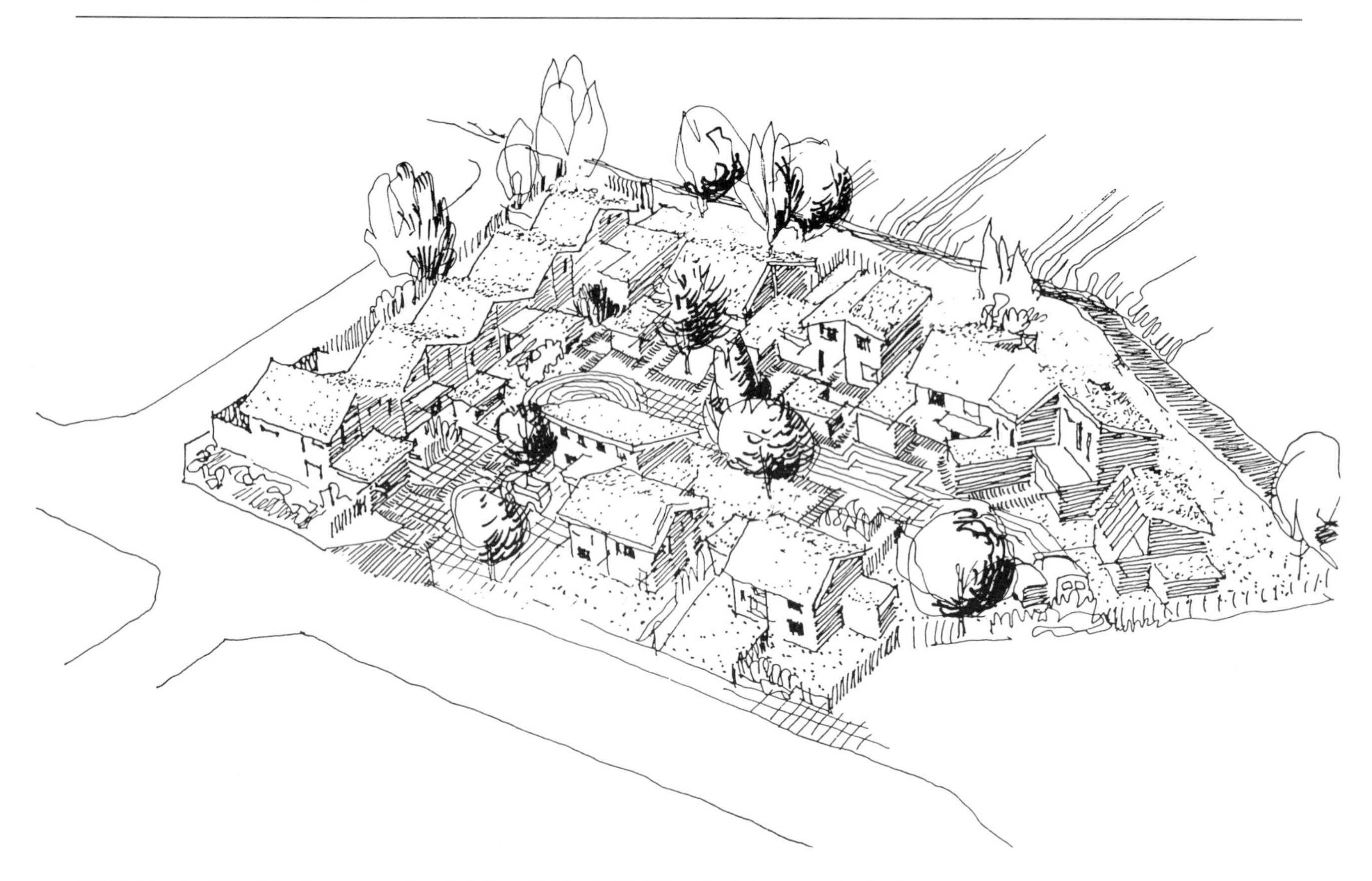

MW2: *Dutch self-builders incorporate the principles of 'biological building' in their community for living-and-working at Maaspoort*

as well.

Like Lightmoor, a fundamental aim of the group is to integrate living and working. The Dutch planning authorities have recently come to realize the social ills done, particularly to women and children in isolated residential suburbs, as a result of rigidly applied planning use zones and are now enthusiastic about the MW2 project, which incorporates various small businesses and workshops with the houses.

All the houses are two-storey, with the main roofs gabled at twenty degrees and with wide boarded fascias trimming the edges of the rubber roof membranes where they emerge under the grass. Some houses are natural-wood colour, some stain-painted white. Various single-storey extension pods and outbuildings have flat roofs, some with mown grass to step out on to.

English people visiting new housing projects in Holland must always be impressed by the standard, and quantity, of the landscape gardening. Every area is cared for, and particular attention is lavished on the horizontal surfaces, which are paved with intricate patterns of brick, tiles and stone setts. This scheme was no exception, though more informal than its neighbours. One imaginative device was the weaving of fence screens with live wands of willow, which take root and become covered with lacy fronds.

What I heard from Germaine and Heleen and Renz, and saw as I took my sketchbook round, was fascinating, and added a dimension to the possibility of self-build that had not been much explored in the British examples we had visited.

Not only was the scheme a way for people on only moderate incomes to get housing, not only was it a social experiment in integrating house and workplace, not only were ecological implications considered, like low-energy use and the avoidance of materials whose production damaged the environment; it went beyond these to pay particular attention to the biological aspects of the house—its effect on the whole health of its occupants.

I came to this scheme with only a scant awareness of

Maaspoort: timber houses with turf roofs

the movement, quite strong on the Continent, which studies the effect on humans of radioactivity, electromagnetic fields, gases given off by toxic chemicals, noise and so on, and devises methods of building and choice of materials to limit the potentially harmful effects of these things.

I was struck by the common sense of much of the theory, while remaining somewhat sceptical of the unfamiliar concepts, such as the concern with cosmic and terrestial forces which, it is claimed, can flow into buildings and reduce the wellbeing of the occupants but which, they say, can be identified and controlled. Some dowsers are sensitive to the geology of a site and can detect energy currents which are likely to have undue influence. Apparently, underground watercourses can have bad effects on people

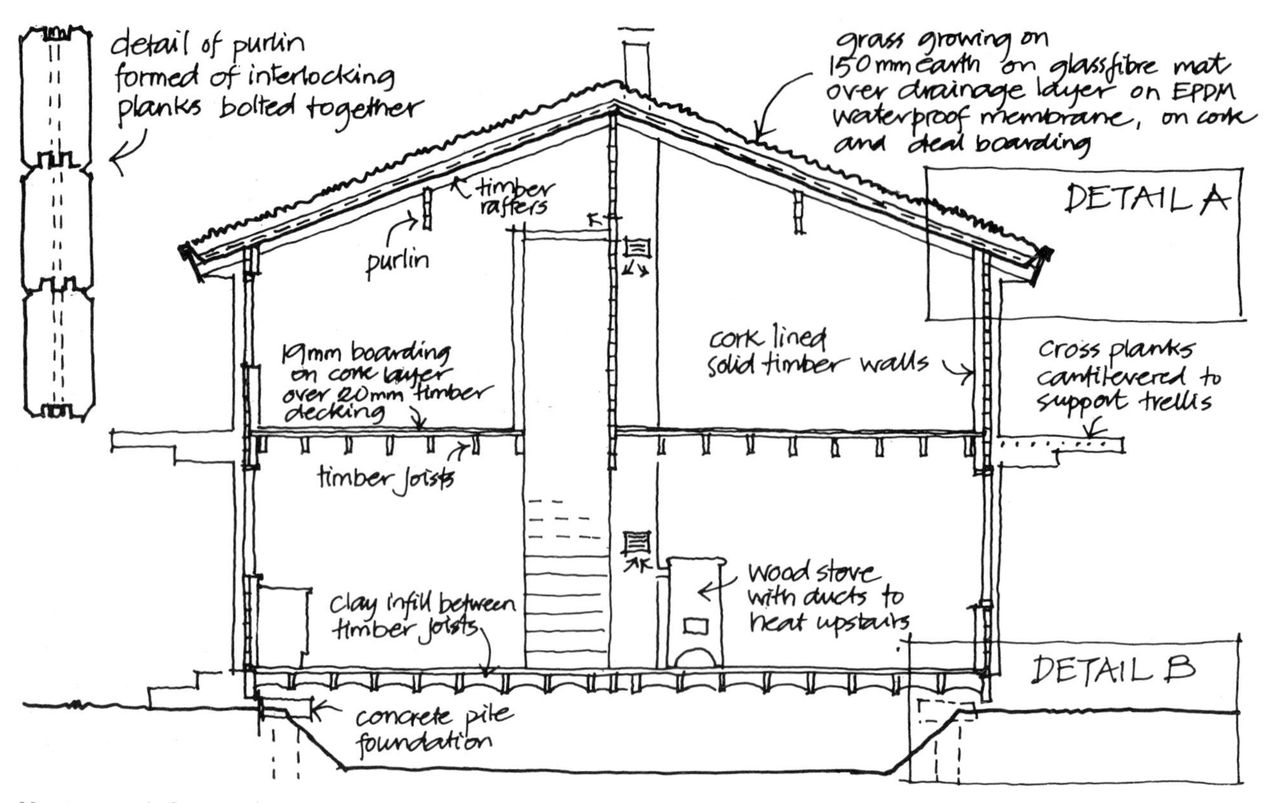

Maaspoort: typical cross section

sleeping over them, and electrical fields, if too strong, are harmful. (It has since become accepted, in the USA and later in Britain, that dwellings should be sited well clear of high-voltage electrical equipment). Renz Pijnenborgh is sensitive to the dangers to health that manifest themselves in orthodox building and has

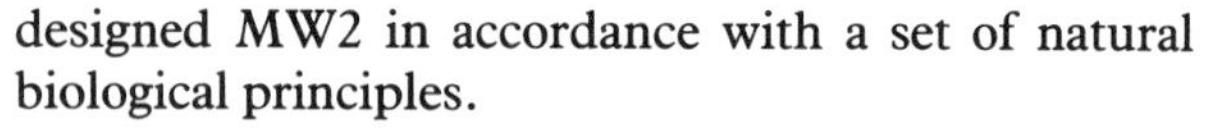

designed MW2 in accordance with a set of natural biological principles.

For instance he regards the envelope of the building as being like human skin—protective but not isolating you from the external environment. Roof, floor and walls should not be completely impervious

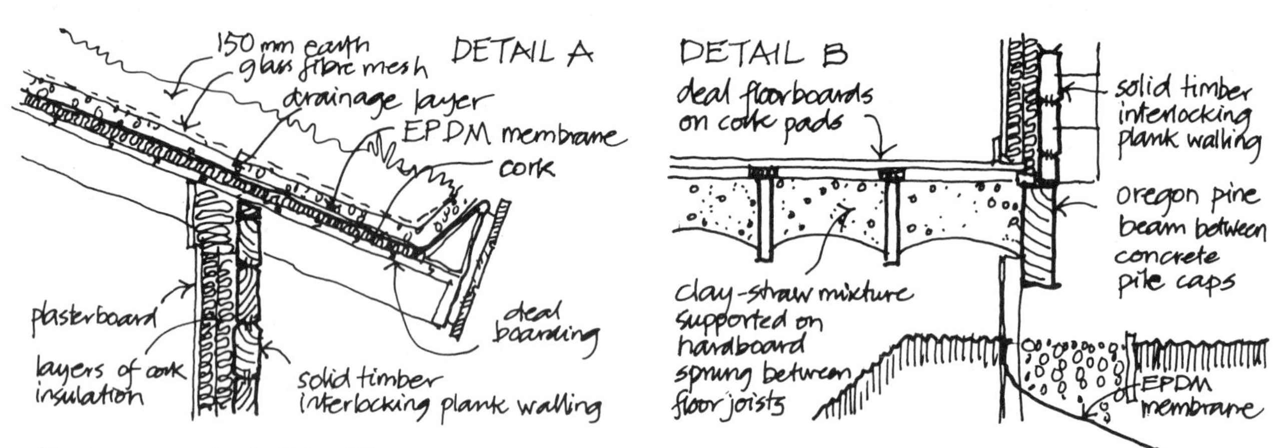

Maaspoort: construction details A and B

but be able to breathe, while resisting water penetration. Renz does not like the modern practice of incorporating vapour barriers to control condensation. Rather, humidity should be controlled by using building materials with the right degree of hygroscopy, or absorbency.

We go into ecological matters a little more thoroughly in Chapter 19, but mention here the manifestations of biological building principles evidenced in a quick appraisal of this lovely Dutch scheme.

First Renz showed me the site survey on which were plotted the findings of the dowser, and then photos of the measures they had taken to modify the lines of force which he thought to be unduly strong; this aspect is worthy of special study and I can mention only that it was done, without providing an explanation.

Then the site, being on typically Dutch made-ground, with soft sand foundations, was piled by an outside subcontractor—six piles per house, 7m (23feet) deep, and extra piles driven for future extensions. The caps of the piles were not linked with reinforced concrete beams in the usual way because Renz wanted to avoid a ring of steel round the base of the house, so timber pads of greenheart were put on top of the piles and big Oregon pine baulks formed the ring beams. Below the Oregon pine a skirt of corrugated sheets retained the outside soil and controlled airflow under the floor.

The ground floor was of timber, well insulated with a filling of a special clay mixture between the joists. Under the floor was some storage space, and Germaine and Heleen had boldly excavated the sand in the middle of their house to make a huge cellar.

Given a standard bay multiple of 4.2 m × 4.2 m (14 feet square), each family designed its own house layout and Renz prepared detailed drawings and schedules. The construction is solid timber Scandinavian plank (or log-cabin) construction: a large number of firms in Finland manufacture the planks, machine the joints, bore the bolt holes and cut them to exact lengths, so that the structure arrives in kit form and is easily put together without further carpentry work.

The only protection for structural timbers at this stage is immersion in borax solution, done on site. Only non-toxic timber finishes, clear varnishes and a translucent white stain are used subsequently. There is a firm in Holland specializing in these products.

Renz's advice here is very similar to David Lea's (an architect who has used Segal's type of construction in a particularly sensitive way), that if timber construction is properly detailed, there is normally no need for additional protection. The occasional wetting of wood does not harm it; only prolonged dampness and airlessness are harmful. So any rain that enters the junctions between wooden members must be able to drain out again, and the whole structure must be well ventilated. I asked about insect attack, and Renz said that wood with a moisture content of below twenty per cent is not prone to it.

The clear organic coatings they use need regular—perhaps yearly—renewal, which is a lot of work, but as Heleen said, it ensures that every part of the house gets frequent close scrutiny and any deterioration would be spotted in time to do something about it. The white stain needs to be applied every five to eight years.

Something that has to be allowed for is the shrinkage across the width of the planks that takes place before they are fully seasoned—something up to 3 millimetres per board, which adds up to something like 150 mm (6″) in the height of a two-storey house. Germaine showed me the careful detailing this requires at door openings and so on to allow the movement to take place without damage.

The timber joists and boards of the first floor are covered in their house with a soft, anhydrous plaster screed to add mass, sound insulation and fire protection, and finished with cork tiles.

The roof is pitched at twenty degrees, decked with timberboards, insulated with 50 mm (2″) cork slabs below the waterproof membrane (which is a rubber sheet that Renz assured me was not synthetic butyl, as on my own roof, which it resembled, but some modified form of natural rubber) and covered with a specially prepared mixture of soil and various seeds, and topped with thin grass turves. He is so confident and enthusiastic about his grass roofs that he has set up, as a sideline, a firm to specialize in them commercially.

Another instance of the use of natural material is at the party wall division between adjoining houses. The few inches that the solid plank walls are apart after they have been faced with cork is filled with a special clay mixture, rammed in. This again gives mass and sound insulation.

External walls are not quite well enough insulated by being of solid timber 75 mm (3″) thick; not on the side exposed to the weather anyway, so here the lovely appearance of the wood internally has to be lost and cork insulation is applied, covered with a special plasterboard made from natural mined gypsum. The gypsum made as a by-product at power stations on the Continent is liable to contain too much radioactivity. (British plasterboard, being made from mined gypsum, is apparently safe.)

Heat losses are low enough for each house to be heated by a single wood-burning stove of the Continental type, which is a tiled heat store which extracts every bit of heat from the flue and carries warm air through various channels to the upstairs.

Something that was entirely new to me was the extreme care taken to reduce intrusive electromagnetic fields generated by the mains wiring. All the cables and fittings were protected by what Renz called a Faraday box. This involved twisting the insulated live and earth wires together and wrapping them with metal foil before sheathing them again in some form of plastic. Renz assured me that in Holland there are electrical firms who understand these principles and can carry out such installations.

A final touch that pleased me very much was a big brick paved circle at the heart of the layout that, when I looked closely at the pattern, I recognized as a replica of the labyrinth in the floor of Chartres cathedral. The hours of patient, loving work that must have gone into this embellishment expressed the care and dedication of this admirable group of self-builders.

Holland revisited

On finding ourselves in Maaspoort again in 1994, Maureen and I called on Germaine and Heleen to see how they and MW2 were getting on. The buildings looked even lovelier, with gardens and trees growing to maturity and the timber house-walls freshly stained and varnished. Indoors, Germaine and Heleen had built in some more fine furniture for their enlarged family—they now have a baby boy—and everything was wearing well.

We walked round the precinct with them, noting that the communal building was closed that evening even though there were some youngsters hanging about with not much to do. It seems that, as at Netherspring, it will after all go to make another house. The group is just too small to be able to run it for themselves and the hoped for involvement of the wider community has not yet happened. They were hanging on for another year hoping it can be made to pay.

Further on we came to one of the few houses that had changed hands and as we walked by the new couple who had joined the community arrived on their motorbikes from a holiday in Norway and immediately asked us in. We started swapping biking experiences and soon found a shared interest in building biology too—which is what of course had attracted them to MW2. What was more, Ann actually had a job in Den Bosch at VIBA-Centrum,[1] a Dutch organization devoted to the subject. She invited us to meet her there the following day.

We found her in a large building in the industrial quarter of the city. We were received in an impressive office and library and could see that the Vereniging (Association) for Integral Biological Architecture took their business seriously. We were nevertheless quite amazed on entering the exhibition to see how extensive it was. There were dozens of displays, some of them by architects showing drawings, photographs and data relating to housing projects all over Holland, by engineers and environmental consultants, by contractors specializing in biological building and by many suppliers of materials and equipment.

They had stands demonstrating piled foundations, floor systems, walls of sand lime and earth blocks, mud renderings on reed lathing, bamboo gutters, timber constructions of all kinds, roof membranes with turf and wildflower covering and a variety of natural insulating materials. We were interested to see that as well as expanded clay granules there were examples of cotton fibre and sheeps-wool quilts (treated as the British Warmcell shredded paper loose-fill is, with borax). There were ceramic products for drainage, floor and roof tiles and tiled stoves.

We passed stand after stand of water treatment plants, composting toilets, high efficiency boilers, photo voltaic electricity generators and solar water heaters—all laid out in abundance in this permanent, freely accessible exhibition hall backed up with an information counter which supplied us with literature, contact addresses and even prices. We really could do with something like this in Britain!

Part Three

An action guide

We now leave encouragement, exhortation and example and come to practical matters. We do not offer you a how-to-do-it manual such as you can buy to make a sailing dinghy from a kit because we do not recommend standard solutions. Your house should be unique. Rather, this is a how-to-think-it guide that helps you find your own way through all the necessary stages before building can commence. We leave the technicalities of orthodox building to the many excellent construction textbooks and DIY manuals already available.

Chapter 13

The strands of the self-build process

A self-build project requires that a number of things come together in the right order, at the right time, in the same place and in harmony with each other:

- ▷ One or more people with a housing need, the desire to build, the time, the dedication and the right skills;
- ▷ Land;
- ▷ Money to pay for land, materials and fees;
- ▷ Input from a number of specialists such as an architect and a solicitor;
- ▷ A design with its documentation;
- ▷ Permission to build;
- ▷ Materials and tools.

The planning stages can be unsettling because you are dealing with a range of intangible matters. You will need to know a certain amount about a range of different things. You will be dealing with a variety of different people, each with their own particular points of view. Enjoy the process of putting the whole jigsaw together and try not to allow the inevitable frustrations that will surely occur along the way to overwhelm you. Self-build extends people to their limits, but that can be exhilarating, as one obstacle after another is overcome.

The strands of the development process

The process of building is a complex one, but one can think of it as having a number of strands, all of which have to be developed through a number of planning stages before the actual building can start. These strands involve different people at different stages. The process is more involved if you are building as a group and we include this aspect in this part of the book. It will probably be the case that a certain amount of what follows can be simplified if you are building on your own. We have identified SEVEN strands in the process, each of which is detailed in a separate chapter.

1. The people building. Who are they? What skills do they have? How are you to organize yourselves? What are your needs? Are you going to build for ownership or rent? This aspect will assume greater importance if there is a group of people building but even if a single family is going to build there are important questions, such as is the whole family going to take part? See Chapter 14, 'ORGANIZING YOURSELVES'.
2. Input from specialist consultants. What professional help do you need? Where do you get it? How much does it cost? This may involve dealing with an architect or a solicitor, for example. See Chapter 15, 'PROFESSIONAL ADVICE'.
3. A place to build. Where is it going to be? How big does it need to be? How do you find a site? What restrictions are there imposed by planning policy? This may involve negotiations with estate agents, local authorities and solicitors, amongst others. See Chapter 16, 'LAND'.
4. Money. How much will it cost? Is a loan necessary? If so, where will it come from and how will it be repaid? You may be dealing with bank managers or building society surveyors. See Chapter 17, 'FINANCE'.
5. Plans to build from. What is the concept of the house? How big, what rooms, where is the sun, which way is there a view? What type of construction? How do you record and communicate your design? See Chapter 18, 'DESIGN AND DOCUMENTATION'.
6. Materials. What materials to use? What is their effect on the environment? How are you to conserve energy? What type of heat source are you going to use? See Chapter 19, 'MATERIALS AND ENERGY CONSERVATION'.

7. Permission to build. What permissions are required? How do you obtain them? What are the conditions for the supply of water, electricity and gas that you must meet? See Chapter 20, 'PERMISSIONS'.

The stages of the planning process

All these strands need to come together before you can start building. They will not progress neatly, one following the other; you will need to work on some or all of them at the same time. They will also become intertwined to some extent. Don't despair. These planning stages are the most difficult part of self-build; by comparison, the actual building is straightforward.

At the beginning of the process you will be thinking mostly about general principles for each of the strands. Only once you have made these general decisions can you move on to the more detailed considerations. As you progress, the decisions you will have to make will be within a narrower framework. We have identified SEVEN stages from your decision to self-build through to completion. Each of the strands identified above will have to progress through the stages of the development process, which do run in sequence. The stages are as follows:

1. INCEPTION. Draw up a brief and assess feasibility. How much can you afford, what kind of house, what locality?
2. OUTLINE PROPOSALS. Test out alternative ideas and make the basic decisions concerning organization, site, sources of funding, costs and professional advice.
3. WORKED-UP SCHEME. Develop these ideas and make final decisions. It involves securing firm offers of land and finance and obtaining planning permission, among other things.
4. DETAILED INFORMATION. Detail the proposals; recording and communicating the final scheme.
5. PROJECT PLANNING. Bring all the completed strands together and prepare a plan of action for the building works.
6. WORK ON SITE. Build the project, involving obtaining materials and money to pay for them and regular inspections of the work.
7. COMPLETION. Finalize long-term finance and wind up a housing association if appropriate.

These last three stages are examined in Chapter 21.

The diagram on the opposite page summarizes the seven strands of activity and the seven stages of progress charted in the action guide.

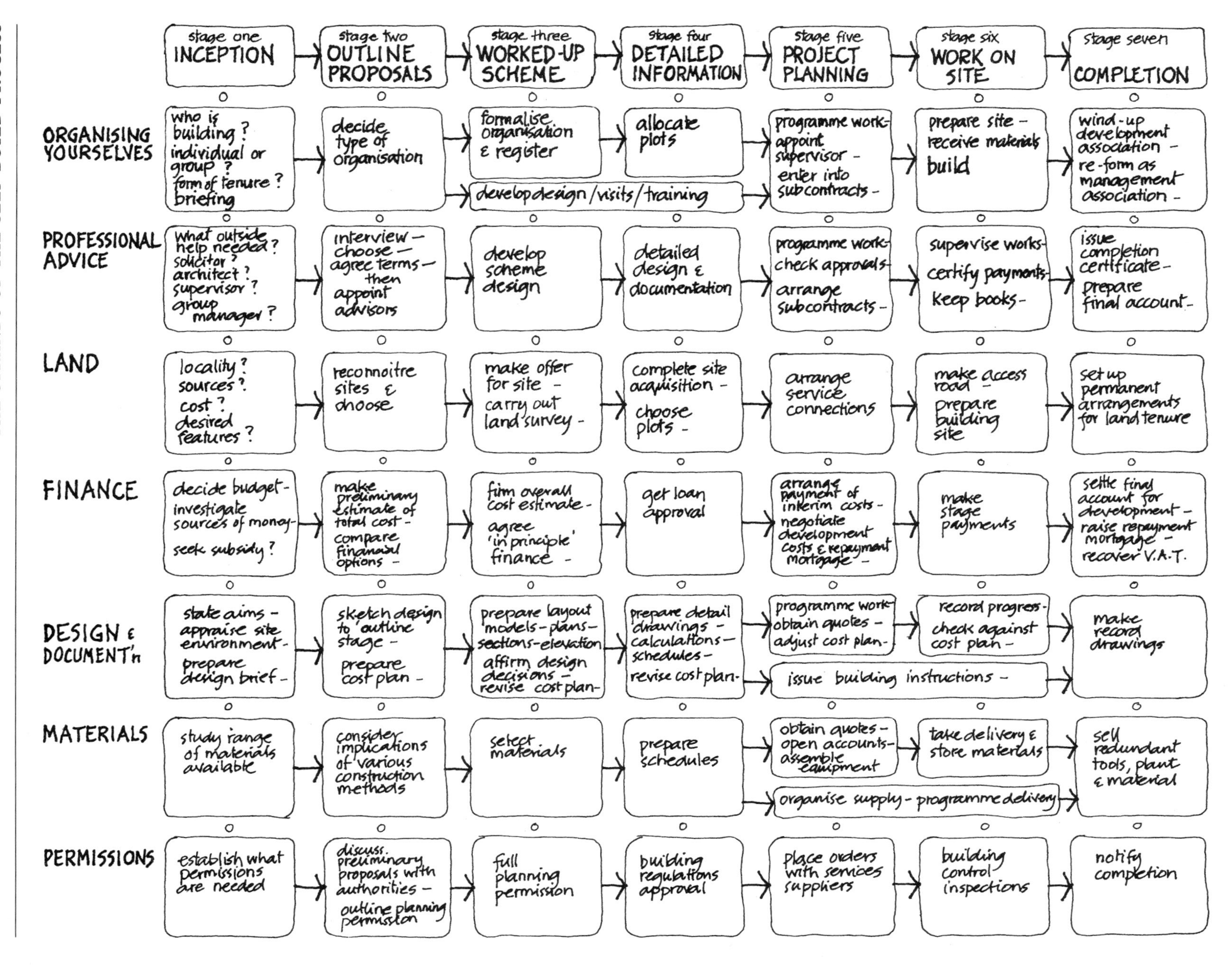
stage one INCEPTION
stage two OUTLINE PROPOSALS
stage three WORKED-UP SCHEME
stage four DETAILED INFORMATION
stage five PROJECT PLANNING
stage six WORK ON SITE
stage seven COMPLETION
ORGANISING YOURSELVES
who is building? individual or group? form of tenure? briefing
decide type of organisation
formalise organisation & register
allocate plots
develop design / visits / training
programme work - appoint supervisor - enter into subcontracts -
prepare site - receive materials build
wind-up development association - re-form as management association -
PROFESSIONAL ADVICE
what outside help needed? solicitor? architect? supervisor? group manager?
interview - choose - agree terms - then appoint advisors
develop scheme design
detailed design & documentation
programme work - check approvals - arrange subcontracts -
supervise works - certify payments - keep books -
issue completion certificate - prepare final account -
LAND
locality? sources? cost? desired features?
reconnoitre sites & choose
make offer for site - carry out land survey -
complete site acquisition - choose plots -
arrange service connections
make access road - prepare building site
set up permanent arrangements for land tenure
FINANCE
decide budget - investigate sources of money - seek subsidy?
make preliminary estimate of total cost - compare financial options -
firm overall cost estimate - agree 'in principle' finance -
get loan approval
arrange payment of interim costs - negotiate development costs & repayment mortgage -
make stage payments
settle final account for development - raise repayment mortgage - recover V.A.T.
DESIGN & DOCUMENT'n
state aims - appraise site environment - prepare design brief -
sketch design to outline stage - prepare cost plan -
prepare layout models - plans - sections - elevation - affirm design decisions - revise cost plan -
prepare detail drawings - calculations - schedules - revise cost plan -
programme work - obtain quotes - adjust cost plan -
record progress - check against cost plan -
issue building instructions -
make record drawings
MATERIALS
study range of materials available
consider implications of various construction methods
select materials
prepare schedules
obtain quotes - open accounts - assemble equipment
take delivery & store materials
organise supply - programme delivery
sell redundant tools, plant & material
PERMISSIONS
establish what permissions are needed
discuss preliminary proposals with authorities - outline planning permission
full planning permission
building regulations approval
place orders with services suppliers
building control inspections
notify completion

Chapter 14

Organizing yourselves

If you decide to build your own home, you will have to take a series of organizational decisions along the way. Should you build alone or with other people? If you decide to build with others, what structure and legal basis should the group have? How will the group take decisions? How will you organize working together? This chapter looks at the background to these decisions and outlines some of the options.

Building individually

The fundamental question is whether you want to build a single house on your own or with other people. It is estimated that in 1987 about 12,000 houses were self-built in Britain. Of those the great majority were built singly by an individual on a small plot of land. They were built generally in rural areas by people who could afford to buy a building plot outright or who had access to land in some other way—maybe through farming or having a piece of land owned in the family. They either felt confident in tackling much of the work themselves or had enough money to employ local builders and tradespeople to carry out all or part of the work under their direction. The advantages of this approach are:

▷ It offers complete freedom to do exactly as you want. You can design a house exactly as you wish.
▷ You can build it at a pace that suits your particular way of life and fits in with your other commitments.
▷ You can tailor the scope of the project and the timetable for construction to suit your particular financial circumstances.
▷ You can pick the locality and perhaps the particular site that you want.
▷ You will avoid the work involved in setting up and managing a group and the potential conflicts that may arise.

There are a number of limitations, however, which include the following:

▷ The cost of a building site is almost certainly going to be significantly higher for a single plot than for part of a larger site.
▷ The cost of access roads and drainage will be reduced if it is shared between a number of houses.
▷ The cost of overheads such as insurances is likely to be reduced if it is shared out between a number of houses.
▷ There is the opportunity to buy or hire relatively expensive tools and plant, such as a radial arm saw, if the cost can be shared between a group.
▷ An individual is negotiating with local authorities and other institutions from a weaker position than a properly constituted group.
▷ A group is likely to possess a wider range of skills and experience and therefore is unlikely to need to buy in so many skills.
▷ Single house plots are very few and far between in many areas, such as the cities.

Other considerations are to do with how much you enjoy working with other people or prefer to be in complete control and not have to rely on others, the extent to which you consider it a good or a bad thing to know your neighbours very closely; whether you would feel more confident having other people around to share the responsibilities or would find relying on others a burden. There are no clear answers to these questions, because largely they depend on how well the particular group of people gets on together. This in turn depends to some extent on the degree of freedom that each individual has within the group, which depends on how it is organized.

We are open-minded on the issue of building alone or together; we have done it both ways. In reality the distinction blurs. The lone self-builder is not likely to be entirely alone and the group member still has scope, particularly with certain types of group structure, to do things how he or she likes. We want this book to be useful to both. If we devote more space to group building it is because the arrangements have to be that much more complex.

Particularly in this part of the development process, the individual self-builder will have a clearer path. You can without delay take stock of your situation, plan and build within your means. But there is the rub. If your means are slight, you will have to be very lucky to be able to go ahead on your own. So you may need to gain the financial, social and political strength that a group can provide.

The structure for a self-build group

Should you decide to build as one of a group you will have to consider which type of structure is appropriate. In doing this you will need to bear in mind that not all the models will be available to everyone. For example, many people on low incomes will not be able to afford to build a house for outright ownership. Conversely, people who already own their own home or who could afford to do so will not be able to choose a model which involves public subsidy. What you can reasonably afford will be very important in helping you to narrow down the choices.

Alternative ownership models

The way you organize yourselves as a group will vary according to what type of tenure, or ownership, you are going to have once the properties are finished. There are a number of alternatives: outright ownership through a self-build housing association; shared ownership through a self-build housing association; or collective ownership for rent through a self-build housing co-operative.

The basic mechanism of a self-build housing association, co-operative or co-ownership society is very similar for the different models; this is dealt with later in the chapter.

Outright ownership

A project of this type is set up as a self-build housing association which lasts for the duration of the scheme and is wound up once the properties are completed. The Zenzele project in Bristol (described in Chapter 8) is an example of this type of scheme. This is also the type most favoured by the majority of the commercial self-build consultants. Until the recent development of the alternative models devised to make self-build affordable for people on low incomes (discussed below), this was the almost universal form of scheme.

The association obtains 'development finance'—that is, a loan from a commercial lender to cover all the building costs, and in most cases the land costs, during the length of the development. On completion, the homes are owned individually and each self-builder obtains an individual mortgage. This will repay the development loan taken out by the association. Mortgage Interest Tax Relief (MIRAS) is available both on the development loan and for the individual self-builders' mortgages, although the government intends to phase out this form of financial assistance. MIRAS is available at 15% on the first £30,000 of borrowing from April 1995 but this is likely to be phased out completely over the following three years.

No public subsidy in the form of a Housing Association Grant (HAG) is available for this type of scheme. You may be able to obtain a commercial development loan arranged through the Housing Corporation for up to 40% of the cost of the group scheme provided that you can demonstrate that the members are in some housing need.

In some cases it is possible for all or part of the cost of the land to be deferred. In this case the self-builders pay for the land when they sell the property and move or by increasing their mortgage at some time after they have moved in.

Shared ownership

In this arrangement when the individual self-builders take out their separate mortgages on completion they do so for only part of the 'equity' or value of the house. This 'equity share' will be composed of the assessed value of their labour—their 'sweat equity'—plus a proportion of the remaining equity which the

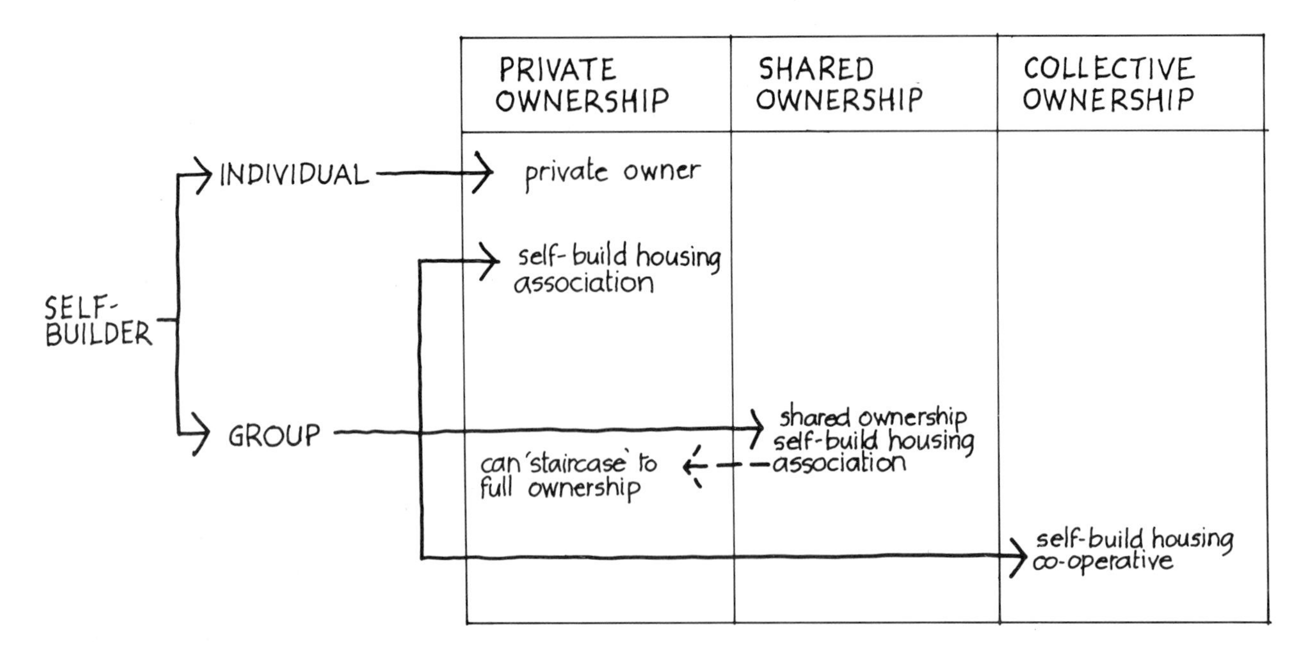

self-builder can afford to purchase. The self-builder has the option to purchase the remaining equity at a later date. The self-builder pays rent on that portion of the equity they do not own, which is held by a housing association.

A scheme of this type can attract Housing Association Grant (HAG), which pays for a proportion of the rented part of the equity. The balance is held on a group mortgage, paid for by the rents charged to the self-builders. This mortgage will be discharged when the self-builders have all purchased 100 per cent of the equity or when the loan has been fully repaid.

To obtain HAG the self-build housing association will have to enter into an 'agency' agreement with an established, registered association, which would receive HAG on their behalf and own the balance of the equity not owned by the self-builders (this is explained more fully in Chapter 17). There are limits on the maximum value of dwelling that is eligible for shared ownership HAG.

Due to the relative shortage of finance to provide homes for rent or shared ownership, there is very heavy demand for HAG and this is particularly acute in some parts of the country. The members of your group will have to demonstrate that they are in housing need and enlist the support of the local authority. Even then it is quite possible that you will not be successful in getting finance agreed for your scheme in the first year that you apply.

A development loan would be made available through the Housing Corporation and on completion each self-builder would take out a mortgage for the proportion of the equity that they are going to purchase. These individual mortgages would be eligible for MIRAS in the normal way but the group mortgage held by the association is not. Unemployed people should normally be able to get interest paid on the mortgage; housing benefit is payable on the rent.

A shared ownership arrangement is flexible and allows self-builders to buy what they can afford at the outset and increase it later if their income rises. This was how the Lewisham schemes described in Chapter 7 were financed, except that a local authority rather than the Housing Corporation provided the money. Most local authorities probably would not enter into this type of arrangement under current housing finance rules. Shared ownership has financial advantages only if HAG can be obtained.

Collective ownership for rent

There are a handful of projects being built for rent with Housing Corporation subsidy. The idea is that a conventional housing co-operative structure is applied to a self-build scheme. It is funded using standard Housing Corporation procedures comprising a mix of HAG and private loan, known as 'mixed funding'. The difference is that the private loan component is largely replaced with the self-builder's free labour. It does not rely on obtaining land at a discount from a local authority. The self-builders are recompensed for their free labour if and when they move out of the scheme with the payment of a 'premium'.

The self-build group would register under 'fully mutual' co-operative model rules published by the NFHA. (Fully mutual means that *all* the self-builders comprise the *entire* membership of the co-operative.) The self-builders build the scheme for a 'notional contract sum' agreed with their housing development agents. This notional contract sum is based on what it would cost to build the development using a contractor in the normal way. The housing association pays for the materials and overheads for building the scheme. The difference between this cost and the notional contract sum is the premium which is divided by the number of co-op members. The value of the premium is index linked. On completion the self-builders are granted a 'premium assured tenancy' by the housing association who own the development to whom they pay a rent which covers management and maintenance costs, the cost of repaying a small element of private loan and a contribution to a reserve fund. The rent would generally be in the order of 20% below the affordable rent level set by the NFHA. The housing co-operative can enter into an agreement with the housing association and assume full management and maintenance responsibility, or they can get the housing association to carry out this work.

This arrangement offers the self-builders low rents, the local authority full value for its land and the housing association affordable housing of high standard and quality at something like 80% of the cost of normal housing association development.

Another approach to a co-operative self-build scheme for rent has been initiated by Birmingham City Council. They have obtained the consent of the DoE under the Local Governments Act to allocate part of their general fund to a self-build co-operative as an interest-free development loan. There is no grant aid in the form of HAG, so rents are higher than in the HAG funded co-op model described above and there is no provision for the self-builders to receive any reward for their free self-help labour. The self-build co-operative obtains a group mortgage on completion, which pays off the development loan.

Collective ownership for rent provides homes for rent for those that prefer this type of tenure and makes the houses available for rent to people on low incomes for all time, because it is not possible for individual self-builders to purchase shares in the equity of their property, as they can in shared ownership schemes. This arrangement would be attractive to a local authority concerned that houses self-built on their land would otherwise pass into private ownership and the private market.

The legal structure of a self-build group

It is important to have an agreed constitution for a self-build group. The main benefits of a formal organization are:

- ▷ to have some form of legal entity for the group so that banks and other institutions will be able to enter into legal agreements with you;
- ▷ to avoid certain tax liabilities;
- ▷ to limit the liability of the members;
- ▷ depending on the type of organization, to obtain certain public subsidies.

The most common form of legal structure is a self-build housing association. The association is formed for the duration of the scheme only and is generally wound up when the houses are completed. Groups that are organized in this way are usually building so that they will own the houses individually on completion. (There are exceptions to this. For instance, the two self-build housing associations in Lewisham were for shared ownership.)

Setting up a self-build housing association is a relatively straightforward business. The NFHA publishes a set of model rules. Advice should be sought from them on the details of the registration procedure. A minimum of seven members is required (this can include both partners in a couple). The registration fee is £912 at the time of writing, which includes affiliation to the NFHA for a period of three years.

The model rules have been developed to provide a framework for a conventionally organized group. They are detailed and provide for a committee with a chairperson, secretary and treasurer. It is possible to simplify these arrangements, as was done by the members of the Lewisham groups, who decided that they wanted a simpler and more democratic arrangement where there was no separate committee. All decisions were made by a majority of the membership at a general meeting. One possible pitfall in the model rules is the provision that membership is granted to one representative of each household building in the group. This is more often than not the male partner in a couple and consequently women tend to be under-represented. It also means that many people who are involved in the enterprise have no formal say in the way that decisions are made. The Lewisham groups had all adults in each household as members and thus avoided the resentments that have plagued some other groups.

Under the model rules most self-build housing associations adopt two other documents. They are:

1. A LICENCE TO OCCUPY AGREEMENT. This sets out the terms under which a self-builder may occupy his or her dwelling for the period between its completion and the completion of the last house on the scheme, at which time the self-builders simultaneously buy the houses from the housing association and commence to pay individual mortgages.
2. A set of WORKING REGULATIONS. These govern the hours that each self-builder is required to work. These regulations follow a fairly standard pattern and are to be found in the *Self-Build Manual*, published by the NFHA. They have been devised for a conventionally organized scheme. Such schemes are generally organized around a succession of traditional trade skills: bricklayer, plasterer, plumber, electrician and so on. The group is formed with at least half and often three-quarters of the members possessing a building skill and this work team then builds each house in turn. In order for this arrangement to work, it has been found necessary to devise a very strict pattern of rules. Each member contracts to do a fixed number of hours' work each week, commonly twenty to twenty-five hours, and there is a system of fines, commonly £10 per hour of non-attendance, to ensure that members pull their weight. There are regulations concerning sickness and holidays. The regulations also state that the houses shall as far as possible be of the same size and value.

In fact, of course, it is very difficult to legislate to make people do something that they do not want to. Successful working arrangements rely far more on people's goodwill and enthusiasm. Fines tend to lead to ill will and are often unenforceable. This single aspect of self-build has probably caused more trouble than any other—in an extreme case, we have seen a group of people in a site hut at one end of the site not talking to the other half of the group in their own hut at the other end of the site.

The Lewisham groups were able to avoid this restrictive regime on site because they employed a method of building that enabled them to organize the site work so that each household was responsible for building its own house from start to finish. Each household was thus able to set its own pattern of work to suit its own way of life and other commitments. There was voluntary co-operation on the common parts, such as drainage and roadway construction, and a high level of voluntary mutual co-operation on building the houses as well, because, as we all know, it is often better to work with someone else, especially if they have a particular skill to offer. All the time the builders were in control of their own particular building programme and could decide what standard of finish was appropriate to them and how long to spend achieving it. This arrangement also had the advantage that the self-builders in Lewisham did not feel compelled to build identical houses for reasons of fairness and thus, in our view, miss one of the most important opportunities that self-build has to offer, that of having your house exactly as you want it.

If you decide to adopt a more flexible set of principles than the standard self-build housing association regulations, you may find that you will have some hard persuading to do with the officials of the funding agencies: the building societies and the Housing Corporation. The standard regulations have been devised to serve the objectives of ensuring that the scheme is completed as quickly as possible and with a minimum degree of risk; they are not there to make the process an enjoyable one leading to individually designed and crafted houses.

We believe that there are advantages to a midway path, making each household responsible for the design and construction of its own house but organizing the work on site into teams of people building together at least to the point at which the houses are weathertight, with structure complete, roof on and external walls and windows in. This means that everyone can benefit from the lift of working with others and the weaker members can learn from the others at the outset while leaving the time-consuming finishing stages, which are most critical to the finished result inside the house, completely under individual control.

Most associations apply to the DoE for a Section 341A approval at the time of registration. This has the following effects:

- ▷ The association is not liable for Corporation Tax on any gains accruing to it on the disposal of land to the members of the association.
- ▷ Any rent charged to the members who occupy houses under licence is disregarded for tax purposes.
- ▷ MIRAS is available to the association on the interest charged on the development loan during construction.

Other legal structures

Other than a self-build housing association, the constitutional arrangements that are used have been mentioned above. The procedures for setting up a shared ownership housing association or housing co-operative are similar to those for a self-build housing association, and the NFHA publishes the appropriate model rules for these types of organization.

Organizing a group

What follows are comments on the conduct of a self-build group and they apply to whichever particular type of structure is appropriate for your circumstances. Self-build groups are formed in three main ways:

1. By self-build consultants setting up a scheme by obtaining an option on a site, preparing a design and obtaining planning permission, at which point they advertise for members, interview them and select a group that has the right level of income, mix of skills and age range. This is how most commercial self-build schemes are initiated.
2. By an institution such as a local authority or housing association sponsoring a scheme by identifying a housing need, earmarking a site and carrying out a feasibility study. At this point they may advertise or circulate people registered on their housing waiting-list, hold a public meeting and invite people to form a group. There may be selection criteria imposed in relation to housing need. This is how a group in Greenwich, south-east London, was formed.
3. By a small group of individuals who wish to build their own houses coming together. This core group will investigate the feasibility of a scheme with respect to site and finance. They may decide to manage the whole enterprise themselves or they may decide to employ either a housing association or a self-build consultant to act on their behalf. A group formed in this way may have to justify their membership policy in terms of housing need to obtain local authority support and they may have to demonstrate that they have the necessary skills to obtain finance. A self-build group in Islington, north London, came together in this way from a group of people living in short-life accommodation.

The time factor

It is sometimes the case that developing a self-build scheme can take two years or more before starting on site because of the way the housing system is set up at present. Your group will have to work hard to try and ensure that the procedures do not take too long. Although it is important, in our opinion, to make sure that all self-builders have a say in the decisions that will affect them, there may be an advantage in not involving a lot of people before some of the ground-work has been done. Because it can take a long time to get a self-build project underway, the feasibility can be tested by a core group who then bring in the rest of the membership when the scheme can be seen to be going ahead. The majority of self-build consultants do everything, including designing the scheme, before advertising for members. This has the advantage that the self-builders do not have to go through the time-consuming planning stages. The big disadvantage is that they often have very little say in the end result.

Meetings

Once a group is formed, whether by advertising, holding public meetings, word of mouth or local authority nominations, the work of organizing begins.

A group will hold regular meetings and one of the first things to get clear is how decisions are to be made. It is important that the members have full control of the process and that there is a clear method by which decisions are reached; it is important to agree on who has a vote and how they are to use it—to take majority decisions or to decide by consensus. Are you going to have a committee structure or will decisions be made at general meetings of the whole membership? You will need some method of organizing the meetings; there could be an elected chairperson or a different member could chair each meeting in turn.

You will need some method of ensuring that members attend meetings. Some groups have an attendance requirement, which means that a member is expelled if they do not attend more than a certain number of meetings without prior notification of absence. You will then need some form of waiting-list to fill vacancies in the group. This may be organized on a first-come-first-served basis, or be subject to interview, or you may be taking nominations from the local authority.

Officers

You will need a secretary, whose duties will include keeping accurate minutes of meetings and noting decisions reached. He or she will be responsible for registering the group as a housing association and ensuring that the rules of the group are complied with.

A treasurer will be required to deal with opening a bank account and keeping accounts. You will have to appoint an auditor and the treasurer will be responsible for ensuring that audited accounts are submitted every year. The auditor will advise on the preparation of accounts.

You could, at this stage, also be thinking about appointing particular group members to be responsible for the ordering and delivery of materials, a safety and security officer and a first aid officer, who may need to engage in some training.

Funds

A group will require some working capital at the outset for:

- ▷ housing association registration fees, planning and building regulations fees;
- ▷ stationery and postage;
- ▷ publicity and advertising for a public meeting;
- ▷ feasibility work on one or more possible sites, together with cost estimates;
- ▷ paying for a place to hold meetings;
- ▷ site survey showing dimensions, levels and trees.

This may amount to a few thousand pounds. This start-up money can be raised in a number of ways:

- ▷ By requiring each member to put in a sum of money as a loan, for which they are issued loan stock to be redeemed when funding is secured.
- ▷ By each member paying a regular subscription.
- ▷ If you are a co-operative, the Housing Corporation has a fund which you can draw on for start-up funding.
- ▷ If you are a black or ethnic minority group, the Housing Corporation has funds to promote such groups. A Housing Association with a development

programme would have to be appointed to act as development agent on behalf of the group and it is they who would apply for this funding.

- ▷ If you are aiming to provide houses for people on low incomes in housing need, you may be able to obtain a grant from a charitable trust. The housing association movement has one, the Housing Associations Charitable Trust, (HACT), or otherwise refer to the annual handbook of charities in your local public library. This lists those with a particular interest in housing or with a brief to support organizations in a particular area.
- ▷ You may be asking any professional advisers you may appoint—architect, land surveyor, development agent, solicitor—to work 'at risk', on the understanding that they will be appointed and paid if and when funding is secured.
- ▷ The Royal Institute of British Architects (RIBA) has a fund, administered by the Community Architecture Group, from which it may be possible to obtain money to pay for fifty per cent of the cost of having an architect prepare a feasibility study. This money is repayable if and when the scheme goes ahead.

Publicity

Potential funders and local authorities will want to know who they are dealing with. It is worth spending some money on producing a brochure or pamphlet explaining:

- ▷ who your members are;
- ▷ why you want to build your homes;
- ▷ what your aims and objectives are;
- ▷ why you should be supported by the local authority and others.

Find out something about the housing situation in your area and the council's policies.

As far as possible, match your objectives to the council's. If the membership of your group is not full, you may want to offer places in the group to people from the council's housing waiting-list. If you plan to house people whom the council plan to help, such as homeless people, ethnic minorities or first-time buyers make a feature of this in your publicity material.

Education

The group will need to make a great many decisions about the scheme over the next few months: how to set up a self-build group, how the scheme will be financed, the design of the site layout and so on. People will need to acquire the background knowledge necessary to make informed decisions about the different aspects of the project. This is best done at the appropriate stage as each decision comes to be made—the factors governing layout have to be considered before the relative merits of timber or aluminium windows, for instance.

There are a number of ways of carrying out this learning process. One of the best is to arrange visits to a number of completed projects. The members of the group disperse and talk to the people who are working on or who have completed the various projects and get a good first-hand view of their experience. These points of view can then be compared in the minibus on the way home.

Other groups have organized a training programme of evening classes which cover the various aspects of the self-build process—talks on the factors that affect design, demonstrations on how to read technical drawings, practical sessions in a workshop on how to use hand tools, power tools, plumbing, electrical wiring and simple surveying techniques. The local adult education institute, if there is one, or technical college will often be delighted to offer their facilities for a truly practical purpose like self-build.

Plot allocation

You will need to decide how building plots are to be allocated between the members of the group. Our experience has been that if people put their preferences in order it is surprising how many people actually get their first choice, some wanting a plot near the road so they can see everything that is going on, others opting for the plot tucked away at the back, out of the way.

Consultations

The group will have to develop the design in its many aspects: the site layout, the design of the individual houses, the choice of construction and materials. This can be done by a combination of general discussions with the architect and private meetings with each household in turn. You will need to build up a close relationship with the architect and get him or her to produce alternatives for discussion.

Maintaining enthusiasm

Often the hardest part of the whole process is keeping the morale of a group up during the planning stages of a project. It can seem as if nothing is happening, just endless discussions with little to show for it. The early stages of finding a site, getting agreement that a self-build scheme should go ahead on it and obtaining an agreement in principle to provide finance for a scheme can take months and years. A training programme can be a useful way of keeping enthusiasm up. It may also be necessary for the group to organize as an effective campaigning body to persuade local authorities and government departments to take decisions.

The business of building and the formalities after completion come as a great relief after the months of planning and waiting. But before we get on to that, we have to consider the matters of land and finance in more detail, together with the expert advice that you may require in dealing with them.

Chapter 15

Professional advice

You will almost certainly need some advice on different aspects of a self-build enterprise. This chapter discusses the question, 'Do you need professional advice, and if so, how much and who from?'

Management and technical skills

There are two main areas where self-builders seek help:

1. **The project management of a group scheme.** It is a complex process, steering a project through the many decisions of a wide range of aspects at different stages of the development of a scheme.
2. **The more technical aspects of design, financial and legal matters.** Legal contracts will have to be drawn up, accounts audited and plans and calculations approved by the planning and building control authorities.

Choosing professional advisers

The really crucial question here is how much control you want to have over the outcome of your self-build enterprise. There is a danger that the more you hand over responsibility to others, the more they may assume control. This is particularly true of the project management function. Many, if not most, professionals are still trained to think and act as if they know all the answers. They will tend to mystify their knowledge, wishing you to believe that they know best. What you need is one of a growing number who are concerned with explaining what the situation is, helping you to decide what you want and helping you to get it. You need people with experience of self-build, experience of working with groups and preferably people who are local. This is particularly relevant when choosing an architect.

Costs are also something that you need to consider, especially whether the professional service you are being offered is really going to be value for money. You should have a clear indication of the services to be provided, the cost, and when fees will be due so that you can budget for these payments. If a group appoints all the advisers discussed below, their fees could be as much as twelve per cent of the total cost of the scheme. Fortunately, fees paid to professional advisers can be included as part of the total cost of the scheme for mortgage purposes. But the more advisers you use the more you will have to borrow and the higher the eventual cost of living in the houses will be. On the other hand, if mistakes are made because advice has not been sought, the costs could be higher still. One simple rule to control costs is to get your advisers to agree (in writing) to work 'at risk' until the land is bought and funding has been secured. This means that you agree to pay for their work at the feasibility stages only if and when the scheme is definitely going ahead.

When choosing any professional adviser, be guided by recommendation. This could be personal or could come from one of the bodies working in the self-build field. Another alternative is the relevant professional institute: the Law Society, the Institute of Chartered Accountants or the Royal Institute of British Architects (although many qualified architects do not join the RIBA). Another possibility in this area is the Association of Community Technical Aid Centres (ACTAC), whose membership includes people working in the self-build field. The Walter Segal Self-Build Trust offers free advice about setting up groups and finding funding. It is also in the process of establishing a network of people around the country educated in, and with experience of, self-build, and the Segal method of building in particular.

The Community Self-Build Agency offers a similar service and has representatives in the North of England and in Scotland. The Association of Self-Builders is a network of self-builders set up by self-builders for self-builders. It aims to provide a forum for self-builders to meet and discuss common issues through regional groups. They publish a newsletter. Finally, the Individual Housebuilders Association is a trade association of companies supplying materials and services to the individual self-build market who can provide lists of house kit suppliers etc.

When appointing project managers or architects always interview more than one. Before the interview think through all the tasks you want them to do and determine to what extent you want to be involved in, for example, making decisions. You can then go through these points at the interview and you will be able to gauge their reaction to your requirements. It will also provide a clear basis for the consultant to give you a quote as to how much they would charge. Ask them for references from former clients and contact them for an opinion. Visit some of their completed schemes and talk to the residents. It is worth taking a good deal of time at this stage making sure you employ people you are happy with, to avoid trouble later on. Finally, with each and every one, make sure you have a clear agreement on the services required and fees payable before asking them to do any work. This last advice sounds obvious but it is surprising how often it is ignored.

Project management

You can employ someone to co-ordinate the self-build process for you or you can choose to do this yourself. Your decision will be affected by whether you are going to build alone or as part of a group, and if as a group, the type of scheme you are planning.

Self-build management consultants

If you are building as a group, you could consider appointing a self-build management consultant, or joining one of their advertised schemes. They offer an all-in service that includes:

- ▷ purchasing the land;
- ▷ arranging for main services to be connected to the site;
- ▷ organizing a self-build group and taking responsibility for its legal affairs;
- ▷ advertising for and selecting group members;
- ▷ commissioning an architect to prepare a design and seek planning permission;
- ▷ providing a detailed budget and work programme for the group;
- ▷ negotiating finance for the project and helping individuals to raise mortgages;
- ▷ issuing monies to the group for the building as and when required;
- ▷ keeping the group's accounts and handling VAT returns on behalf of the group;
- ▷ finalizing accounts by arranging mortgage transfers, legal fees and interest charges;
- ▷ advising on plant and subcontracts;
- ▷ arranging an independent surveyor to certify payments;
- ▷ providing a contracts manager to be responsible for progress and work on site;
- ▷ informing the group of any cost increases;
- ▷ attending group meetings.

There are clear advantages to putting yourself in their hands. A great deal of the time-consuming and difficult planning work has already been carried out before you get involved, and experienced people who know where the problems lie and who can give good support and advice are dealing with the project. *There are serious drawbacks, however.*

First, their services are not cheap; commonly eight per cent of the value of the house. Second, they are all for keeping things simple from their point of view. For this reason the variations in design are kept to a minimum as catering for individual requirements causes complications and takes time and effort. The range of people that can join a group set up by consultants is usually quite limited, because they are very selective. They aim to assemble the most effective work team possible and to this end they are looking for the right mix of skills —three bricklayers, two plasterers, one electrician and so on—the right mix of ages—not too young, not too old, mostly young fit men in their thirties, with maybe one older member to lend a steadying

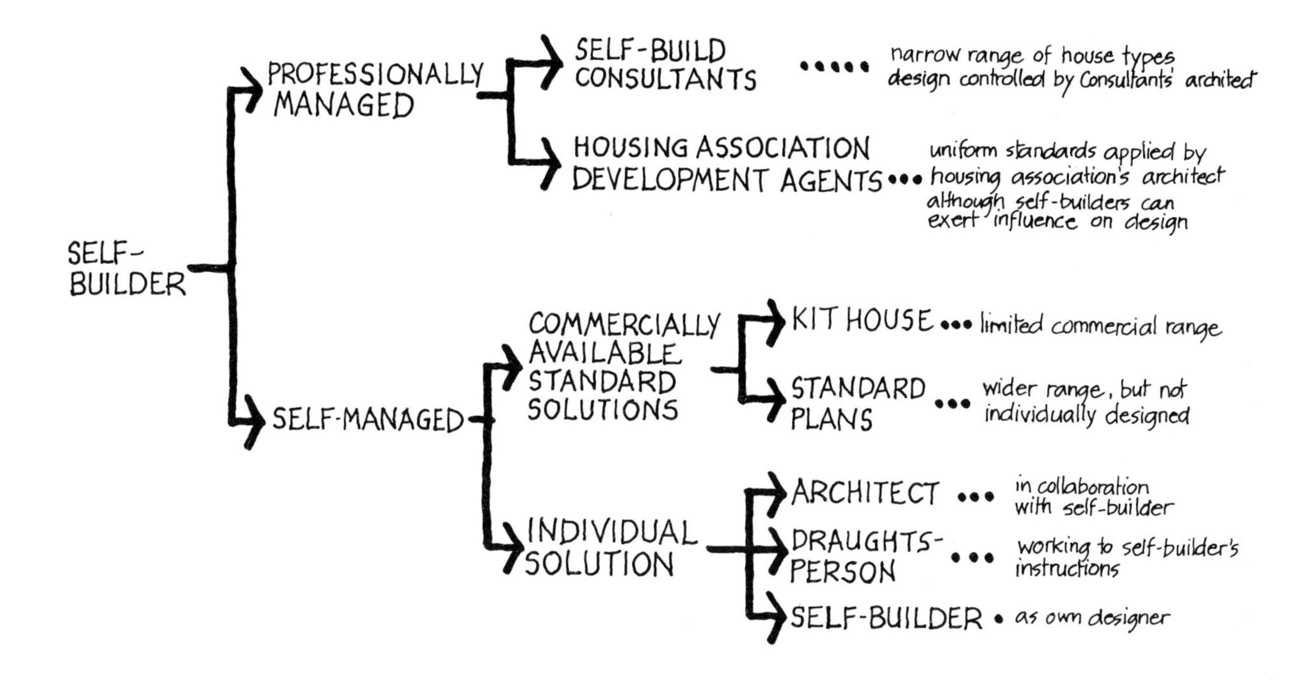

influence but preferably with a couple of strong teenage sons.

The self-build consultant's approach has been devised so that schemes proceed as quickly as possible to a satisfactory conclusion. The financiers require risk to be kept to a minimum and indeed will often insist that a consultant be employed by a group that has set itself up before they will lend money. These days this type of scheme is often not so much a way for people on relatively low incomes to get a house for themselves, as was the case in the 1950s and 1960s, but rather a way for people who are already owner-occupiers to get a bigger and better large four-bedroom, detached house which is to all intents and purposes indistinguishable from the developer-built article. This derives in part from the fact that many people who build through schemes of this kind do so with a view to selling relatively quickly, thus realizing the value of the labour that they have put in to build the house. They can then afford to buy an even bigger and better house, having used their self-build project as a springboard into the housing market. This approach requires that the self-build house be as saleable as possible, as conventional as possible.

This framework satisfies a particular market but has little to offer many people—women, older people, young people without resources of their own, unemployed people, people without particular building skills, people without access to a mortgage or a bank loan, people who do not already own a house. It offers professional back-up, but at a price and at the cost of not being in control of the process; the consultants are the ones who run the project.

Development agency

Should you decide that you will employ someone to manage your scheme because you do not have the time or feel that you do not wish to take the responsibility, then there is an alternative to employing a self-build consultant and that is to employ a housing association or secondary housing co-operative (that is a housing co-operative that provides services to other primary housing co-operatives but does not own houses itself). About twenty-five or thirty associations have been encouraged by the Housing Corporation to provide project management services to self-build housing associations in different parts of the country. These organizations should be acceptable to funding agencies. Your regional office of the Housing Corporation will be able to advise you of suitable development agencies in your area.

It remains to be seen how effective housing associations will be in this role, because self-build is a specialized field involving a level of individual consultation that is not part of most associations' way of operating. Self-build may also be low on their list of priorities. Secondary housing co-operatives, on the other hand, specialize in providing a high level of consultation and may be good partners for self-build groups. There may be a secondary in your area but many have recently been taken over by Housing Associations, although sometimes they still operate as a separate unit within the larger organization.

If you have decided to self-build for rent or shared ownership with finance from the Housing Corporation, you will almost certainly have to use the services of a local housing association or secondary housing co-operative. Compared to commercial self-build management consultants, you will have considerably more flexibility and more ability to be involved in making decisions and appointing the architect. In this case the association or co-operative will obtain a fee for their work from the Housing Corporation.

Self-management

You may decide not to involve project managers and to appoint your own professional advisers as and when required; an architect and a solicitor may be all that are needed. You can thus ensure that you get what you want, although in return you will of course be taking on a considerable amount of work which may involve sophisticated training for members. You will have full responsibility for organizing every aspect of the scheme from start to finish, including buying land and obtaining finance. You will also almost certainly save yourselves money not paying a management consultant. You will be able to develop the design from the beginning just as you want it and you will be able to have a variety of house layouts if you want. Not a great many groups at present decide to manage their own project, largely because the overwhelming advice from the NFHA, the Housing Corporation and the building societies is to employ a self-build consultant. If you decide that you don't want to employ consultants to project manage your scheme, you will have to persuade these bodies that you are competent to run your own affairs.

Technical advisers

If you do decide to manage the project yourself, you come to the question: do you need to engage professional consultants for the technical (rather than managerial) functions? The main area you have to think about is who will design the house(s) and do the work necessary to obtain planning permission and building regulations approval? There are a number of alternatives open to you.

First, you could do it all yourself. This would be in the spirit of self-build and perhaps you can. Some people have enough experience to be their own architect; others may be in a group where those skills are well enough represented to be shared.

Second, appoint an architect.

Third, you could abandon the idea of an individually designed house and opt for buying a house kit or a set of standard plans. We believe that this is to forsake one of the great opportunities of self-build. These will only ever produce a standard house, not one tailored to your particular site.

Fourth, you could consider employing an architectural draughtsperson (somebody with skill in preparing technical drawings but who is not qualified as an architect) to draw up some plans for you under your direction.

With all of these options, the local authority will have to be dealt with for the planning and building regulations approvals. The drawn and written information that has to accompany the applications for the various building approvals is quite complex. Although common sense will take you a long way, at least a certain knowledge of planning law and the building regulations is necessary. If a lay person is presenting this information, it is unlikely that they would be able to get it complete and correct first time, but provided that an intelligent effort has been made, the officials will certainly help with advice on how to bring the documents up to standard. Otherwise, an unqualified person can get some help from family or friends who have the knowledge to share but who are not in a position to be the appointed professional agent of the applicant. Undoubtedly, the officials prefer to deal with a qualified person who knows the ropes and can talk their language. Most people would be well advised to opt for the alternative of appointing an architect. Do not be downcast. You can still be the designer of your own house if you remain in control of the project, but you will have professional help in getting it right.

The architect's role

Many of the most crucial judgements to be made about your project fall into the architect's domain. You will probably need advice very early on regarding the suitability of a particular site. How many houses can it accommodate, what are the soil conditions, where are the services, how much will it cost to develop, are there planning restrictions? You will need to decide on your requirements, explain what you want the house to be like, decide on the specification of the materials to be used and set a realistic budget. The job of the architect at this stage is to use his or her experience to bring these elements together in proper balance in a single design. The process is largely one of trial and error, with alternatives being discussed along the way. Then the drawings, specifications and calculations necessary to make applications for planning and building regulations approvals will be produced.

There are limits to the abilities of many architects, particularly when it comes to the engineering aspects. Many architects, having forgotten their college training in the mathematical analysis of structures, rely on rules of thumb or published tables of safe loading in arriving at a 'design'. This is satisfactory for most building construction, and if there is a problem, a specialist engineer can be brought in to advise. If you decide to build using the Segal method of construction, described in detail in Part Four, you will need to think particularly carefully about your choice of architect, who must be able to prepare the documentation and undertake the necessary calculations.

You will need advice on programming the work on site, obtaining competitive quotations for doing parts of the work, preparing any subcontracts that may be required and resolving the problems that will arise on site. You may rely on a professional issuing regular certificates that the work has been properly carried out to obtain money to pay for materials from the funding authorities and an architect's completion certificate may be required to obtain a mortgage on completion.

In addition, some architects who have experience with self-build schemes will be able to provide advice on setting up an appropriate group organization and on the financial arrangements that are possible. The builders of the Lewisham schemes and other Segal-method developments also obtained training from their architects, who attend evening classes to go through the drawings and schedules and give a step-by-step description of the building process.

You should also check whether your architect will be available to be on site at times when you are working. These will tend to be outside normal working hours—at the weekend, for example. It is also useful if they are contactable at home in the evenings, as it may not be possible for you to phone during your working day.

The cost of an architect

The majority of self-builders probably do not employ architects, preferring either to use the services of an unqualified draughtsperson, to buy a kit house or to buy a set of standard plans from a company to save money. The level of service is obviously in line with the price they pay: a mass-produced set of plans can be obtained for a few hundred pounds; a draughtsperson might charge something between £1,000 and £2,000 for a set of one-off plans. If you employ a qualified architect, you will pay a bit more; but not only can you obtain the individual design you want, but you will also have safeguards against any loss due to professional negligence. An architect will charge according to the service provided but what is described as the full service outlined above might cost between £2,000 and £3,000 per dwelling and will be money well spent.

Other advisers

You will need to employ a solicitor to handle land purchase and all the other legal documentation. If possible find one with experience of self-build and low-cost land purchase if appropriate. They will be able to advise on:

- ▷ drawing up the rules of the group;
- ▷ registering it with the appropriate organizations;
- ▷ preparing the appropriate wording for working regulations;
- ▷ preparing contracts of employment for professional advisers;
- ▷ agreeing the terms for the development loan and mortgage finance;
- ▷ preparing deeds of sale for the houses.

It is important that your solicitor is appointed separately from your other consultants. If they recommend a solicitor, ignore their recommendation and find your own. If you should need to take legal action against a consultant, you will want to be sure that your solicitor is backing you and not the consultant.

You will need to appoint an auditor to audit your group accounts every year. The auditor must be a qualified accountant and must be appointed within three months of registration of the group. The auditor will advise on tax matters and how to present the accounts. The auditors' fees will depend on the amount of work they have to do. It will be cheaper to employ a part-time book-keeper than to use the auditor to sort out messy accounts. You will be able to get a list of suitable auditors from the NFHA or the regional office of the Housing Corporation.

Depending on how comprehensive a service your architect can provide, you may also need an engineer to calculate the structure and a quantity surveyor to provide cost estimates and schedules of materials. You may need to commission a land surveyor to get the dimensions and levels of your site.

Many groups employ a foreman with building experience to direct operations on site. Your architect will be making visits to the site to inspect the work at intervals of between one week and one month. A foreman would provide advice at the other times you are working on site. For this reason he has to be prepared to work unsocial hours at the weekend. He could be a retired contractor's site agent or clerk of works, or you might decide to contact an employment agency that specializes in the building trade. You may decide to get your foreman to provide training on site.

Chapter 16
Land

In this chapter we hit potential hazards. Depending on your luck, the ebb and flow of market forces and the political climate of the day, you may sail through easily or find this strand the most troublesome.

▷ How do you go about finding a site?
▷ Who has land for sale?
▷ How much will it cost—or put another way, how much can you afford?
▷ What help can you get from a local authority and how do you go about getting it?
▷ What are you looking for in a site for self-build?
▷ Are there planning or legal restrictions on its use?
▷ How do you go about the actual purchase?

Finding a site

This is probably the most difficult aspect of the whole process, obtaining a suitable site at a cost that you will be able to afford. Building land is in short supply and its use is heavily regulated. In order to start thinking about how to find a site we list the possible sources of land:

▷ the open market;
▷ local authorities;
▷ utilities, such as British Gas or British Rail;
▷ churches;
▷ housing associations;
▷ developers.

The open market

Buying land on the open market may be a difficult and risky business and you may find prices unaffordable. Before buying land, you will probably need professional help to establish whether it is suitable (we cover this aspect later in the chapter). Your first step should be to decide on an area within which to concentrate the search and get to know it well. Then:

▷ Contact the local estate agents.
▷ Place an advertisement in the local papers saying that you are looking for land.
▷ Scan the magazines that cover the property market:[1] *Estates Gazette* and the *Chartered Surveyor Weekly*. These will tell you of any possible sites being sold through the commercial sector of the market. They will also inform you of any possibilities coming up for auction or tender. Auctions in particular are difficult and risky for people who are not in the property business (we cover this at the end of the chapter), but they may offer just the opportunity you are looking for.
▷ Get on your bike with a large-scale map and go round the area looking for bits of land with no obvious use—bits that are more or less derelict for one reason or another. You will then have to try and find out who owns them. It is some handicap, and an extraordinary fact, that there is no comprehensive register of land ownership publicly accessible. There are proposals to make the Land Registry open to the public, but until that time you will have to rely on discreet enquiries of the neighbours. Try and get in touch with the owners in this way and see if they are willing to sell and on what terms.
▷ Ask the planning officers, who often know their areas very well and may be able and willing to help you identify possible sites.

The cost of land on the open market

The cost of land bears a relationship to the value of the development that can be built on it. This derives from the selling price of the completed development, less the cost of building it and the commercial developer's profit. The balance is the price that a developer (sometimes referred to as a speculative builder) can offer for the land. For a residential development, the price of a plot is typically between thirty and fifty per cent of the value of the finished house.

Thus if a house of the size and type that you are about to build in a particular area would have a market value of £100,000, you would expect to pay, say, £40,000 for the land. Add to this your building cost, which we will assume is £40,000 (the cost of materials, fees and overheads but not including your free self-help labour) and your total cost would be around £80,000. The value exceeds the cost by around twenty-five per cent.

While we do not advocate that you self-build in order to maximize the value of your house, you will have to be aware of the values of comparable properties in the area because you will probably have to borrow money to build. Lenders will want to be satisfied that the house has sufficient market value to protect their investment as represented by the ratio of cost to value.

As with house values, the actual land cost will vary widely according to the area. For example, in October 1989 the average value of a detached house in Greater London was about £204,000, while it was only £57,000 in Northern Ireland and £100,000 in the Midlands. It will also vary depending on the desirability of the particular locality within those regions and the immediate surroundings of the particular site you may be considering. Values also vary with the state of the market. This can lead to complications because, although land costs are lower if the value of the finished house is lower, the cost of building is relatively fixed. For example, if the completed market value was £80,000 then the land cost might be £32,000. Assuming the same building cost, your overall cost would be lower at £72,000 instead of £80,000 as in our first example. In this case value exceeds cost by only eleven per cent. Paradoxically, you may have difficulty in securing the loan although you are asking for less.

One possibility is for you to reduce the cost of the building, but of course this will affect the value of the finished house. You may have to juggle the type of house, its value, the cost to build it and the cost of land until they are in balance. This is a simplification of the situation but we hope it serves to illustrate the link between land cost, building cost and value.

Building land is in short supply and there is considerable competition in the market, so you may find the cost prohibitive. You do have one or two advantages over a commercial developer: your costs will be lower and you will not be seeking a profit. Against this you do not have their financial resources and experience. The only possibility of reducing the cost of land in the open market is to build on a site that is not commercially viable and is therefore cheaper, but you need to be aware that this will carry greater risks and development costs will probably be higher to overcome such problems as poor ground conditions or lack of services. We will now pass on to the most important source of cheaper land for building: the local authority.

Local authorities and self-build

Most councils own vacant land that would be suitable for residential development. Some local authorities support self-build by making sites available to groups or by selling single plots, and information is available from 'Build-it' and Individual Homes'. The government is putting pressure on councils to dispose of their vacant land and they are now required to register all sites over a certain size with the DoE which keeps a public register of vacant land owned by local authorities and other bodies such as Railtrack and British Gas.

Many local authorities are anxious to use some of their land to alleviate local housing problems and are willing to make special disposals of land to organizations that can help them. So if your group plans to house local people on low incomes and is willing to allow the council to nominate people for membership, you stand a better chance of persuading the council to sell you land. The Community Self-Build Agency, set up by the Housing Corporation, is also prepared to approach councils on behalf of self-build groups.

Talking to the council

Before you can deal effectively with the council, you need to know who to talk to. You will want to enlist the support of the councillors who represent the area in which you want to live or in which you have identified a suitable site. You will need to convince the councillors who sit on the decision-making housing and planning committees and those who chair those committees in particular. The staff at the town hall will tell you who these councillors are and how to contact them. Remember that they are often busy people, so send them a letter of introduction, together with the publicity for your group, explaining your proposals. Couple this with a request to meet them at a time convenient to them. If you do not hear from them, turn up at their regular 'surgery', when they meet people in their area, bringing a copy of the correspondence. The town hall staff dealing with councillors' affairs will be able to tell you where their surgeries are.

The other approach you should take is to talk to the paid officers of the council. In a large housing department there will often be an officer responsible for dealing with housing associations and this is a good place to start. If not, there may be someone with this responsibility somewhere in their job description. Go as high as you can and talk to someone with overall responsibility rather than someone who deals with day-to-day matters. The officers make recommendations to the committees and so their support is very important.

You will have to convince councillors and officers alike that self-build is a worthwhile activity to support by using arguments such as these:

- ▷ Self-build offers a good way for people to get good houses that they can afford.
- ▷ Self-build should be part of a comprehensive housing policy for the area.
- ▷ Self-build will create small neighbourhoods of satisfied, self-confident people living in high-quality housing that they will continue to look after and value.
- ▷ Half or all of the places on the scheme could be offered to people on the council's housing waiting-list.
- ▷ The council will obtain an income from the sale of the land while still putting the land to a use where the emphasis is not so much on creating profit as on fulfilling a social need. This is unlikely to be the case if the land is sold to a developer to build housing for sale at the highest price.
- ▷ Self-build is a good way of giving expression to the energy and talents of people who are out of work and who might otherwise be doing nothing useful.
- ▷ Self-build offers an excellent opportunity for people to acquire skills that will be of lasting use.
- ▷ Self-builders can make use of sites that would otherwise be uneconomic to build on.

You will, of course, have to confront sceptics on the political front—people who, on the one hand, view council sponsorship of self-build as an unacceptable use of public resources for private gain or who, on the other hand, oppose the removal of the opportunity for profits to be created by estate agent, developer and contractor. There will be practical objections too—that self-builders do not do a good job, that self-build represents an inefficient way to build and so on. (In fact people building for themselves generally do a much better job than any building contractor employing people who have little interest in the finished result and self-builders wanting a house to live in find that it is the normal system of development that takes too long.)

What can the council do to help?

As a housing authority the council can make land it owns available cheaply by applying to the Secretary of State for the Environment for consent to sell at less than the full market value. They will be arguing that the benefit of creating low cost homes for people off their waiting-list is of equivalent value to the community. However, this assistance is counted as council expenditure and counts against the spending limits imposed by central government. This means that the council will be able to spend less on other priorities such as improving council properties. Furthermore, recent legislation allows the government to reduce spending limits by the amount of any cash receipt from the sale of land. However, following pressure from many housing organizations, the government has said that this penalty will not apply to the sale of land to social landlords (such as housing associations or co-operatives) which are providing subsidized rented housing or housing for shared or outright ownership for people in housing need. We cannot predict what the overall effect of this is likely to be.

An attractive possibility is a 'deferred purchase' agreement. The idea is that a local authority makes land available for a self-build scheme at no initial cost. The land is paid for only at some time in the future when and if individual self-builders want to buy their particular plot or if they want to sell their house and move elsewhere. A formula is agreed on at the outset which fixes what the council receives when that happens. It may take the form of the local authority getting the same proportion of the total value represented by the land at the time of sale as at the outset. There are many ways of formulating an arrangement of this kind to suit particular circumstances.

You should also be able to get land on a 'build now, pay later' arrangement whereby you build on land made available by the authority on a licence so that you pay for the land only when the houses are complete. This saves the interest charges on money borrowed to buy the land over the building period, which can add up to a surprisingly large sum.

Councils have the power to make grants, loans and loan guarantees to independent organizations that are providing subsidized housing for sale or rent. This too counts against the councils' spending limits like the sale of land at less than its full market value and it may require persistent argument to persuade a council to make assistance of this kind available within the current local authority spending rules. One council has, however, granted an interest-free development loan from their general fund for a self-build co-operative as described above.

The council may be promoting a large housing association development on their land and may persuade the housing association or consortium of associations to include self-build in their plans.

As a planning authority the council has certain powers to make land available for low-cost housing. Recent DoE guidelines allow a rural authority to specify the housing needs, such as low-cost housing, that particular sites must be used for. The council, however, cannot specify the type of organization, tenure or building method, such as self-build, that is acceptable. However, if the council has previously specified that a particular site must be used for low-cost housing, this may reduce its market value and make it easier for a housing association or self-build group to buy the site. The allocation of privately owned land for low-cost housing may in some cases, such as where the land was previously zoned for agricultural use, enhance its value and make its owner more likely to put it on the market. In addition, rural authorities, previously restricted by 'green belt' controls, can now grant planning permission for housing development on land that has now been earmarked for housing use in the local development plan. This power allows rural authorities a similar level of planning flexibility to urban councils. In both rural and urban areas, councils can use the planning system to persuade private developers to make land available for community-based housing projects, such as self-build. Section 52 of the Town and Country Planning Act allows the council to enter into an agreement with a developer who is applying for planning permission to develop a site. Such a 'planning gain' agreement could impose restrictions on the use of the land, such as using part of it for low-cost housing such as self-build. Some developers will resist any such imposition on their plans but others may be willing to enter an arrangement that offers a community benefit in return for being granted planning permission without problems.

As well as the housing and planning departments, you may be involved with the valuation department, which deals with the council's land, and the legal department, which will draw up contracts of sale and any deferred payment agreement. Do not assume that the different departments communicate properly with one another and do not assume that enthusiasm shown by the officers in one department will be matched by those in another. If bureaucracy causes delay you may need to enlist the support of the councillors to apply pressure on your behalf. Besides the council, there are other places to try.

Utilities

Organizations such as Railtrack and British Gas own large amounts of land and sell off redundant land from time to time. Their landholdings of large sites will be in the DoE register mentioned above. They must obtain the full market value for their land but it is worth finding out what they own in your area and whether they plan to sell any of it.

Churches

Churches, especially the Church of England, often own redundant land or have had land left to them by local parishioners. Many vicars, priests and ministers are involved with their local community and are aware of the difficulty young people have in finding a house in their local village and the problems faced by young people finding a place to live in the cities. Some churches are already involved in initiatives to provide low-cost housing with housing associations so it may be possible to persuade them to provide land for self-build. It will be difficult to obtain a site at less than market value from the Church of England, whose Church Commissioners control their property.

Housing Associations

Housing Associations own land for residential developments and although they will usually have plans of their own, they may be willing to make a site available for self-build or they may be developing a large scheme as part of a consortium, in which case they may be willing to include a self-build association in the consortium.

Developers

There may be developers operating in your area who would be interested in including a self-build component in a larger scheme as a 'sweetener' to win public support for their commercial development.

Is a site suitable?

Once you have identified a possible site or some alternatives have been offered for consideration, assess their suitability:

- ▷ SIZE. How much land will you need, given the density of development that is permitted? If you are looking for land for a group of dwellings, you should have in your mind that about twelve detached houses with gardens can be reasonably accommodated on one acre of land. This will give a typical suburban type of layout. You could build up to, say, sixteen houses on one acre, which would be quite a tight layout. At the other end of the scale, a house standing on enough ground to allow a productive garden that could make a significant contribution to a family's food needs would occupy, say, half an acre.
- ▷ ORIENTATION. In which direction does the land face? A southerly aspect is desirable for sunlight.
- ▷ OUTLOOK. Is there a view and is the site overlooked by adjoining properties?
- ▷ SHELTER. Is the site exposed to chilling winds and driving rain?
- ▷ LOCALITY. Is transport available close by? Are there shops and schools in a convenient place? Are there noisy or smelly factories, chicken houses or motorways nearby?
- ▷ GROUND CONDITIONS. Has the land been used for industry and therefore is there a risk of ground contamination? Have there been houses on the site before and might there be old foundations or basements in the ground? Is there a need for demolition work? You may be able to obtain City Grant or Derelict Land Grant to cover a part of the site preparation work. Most types of soil do not present a problem to build on, but there are obvious difficulties with marshland and not so obvious ones with sites on clay soils, which shrink when they dry out in summer. The local building inspector will advise you of any particular problems.
- ▷ SERVICES. Are mains services—electricity, water, gas, telephone and drainage—available? You will have to contact the relevant authorities for this information. If services are not available, how much will it cost to bring them to the site? Or alternatively, by how much will this reduce the cost of the land?

▷ ACCESS. Is there a limitation on the access for vehicles?

Planning control

You will need to know whether the use you propose for the land is acceptable to the planning authority. You will need to ask such questions as how many dwellings will be permitted, will there be restrictions on the form of the buildings and the materials to be used, and are there limitations on road access? Although considerations of this kind will be dealt with at a detailed stage later on you will need to establish the feasibility of doing what you intend to do at an early stage by talking to the planning officer responsible for the area.

If you are buying a site on the open market, it will often be sold with planning permission. This may be an outline permission which establishes the principles of the use and number of dwellings. The detail of the layout and materials of the buildings will have to be the subject of a later application for detailed permission. Alternatively, you may be buying with detailed permission. This is permission to build a specific building, approved in all its detail. If this is the building you want, all well and good. If it is not, you need to establish whether you will be able to vary the existing permission or whether you will have to submit an alternative application and whether that is likely to be approved.

If you are considering a site that does not have planning permission, you will need to investigate its previous planning history. It may be that there is some fundamental planning reason that would prevent you from obtaining permission. You should note that it is possible to make a planning application on land you do not own, but you do have to notify the owner.

It may also be necessary to commission an architect to prepare some initial drawings if you are dealing with anything other than outline matters, and even then if you are proposing anything even slightly out of the ordinary. Planning permission takes at least six weeks to obtain, and can take as long as six months, and so an offer for a site is usually made subject to obtaining planning permission (see Chapter 20 for more detail on this aspect).

Financial matters

Estimate the probable costs at this stage to assess how much you can offer for a site and how much finance you will need. You may decide to enlist professional help for this. Insist on detailed estimates to prevent professionals playing safe and saving themselves work by giving broad estimates with an over-large contingency sum built in. On the other hand, beware of over optimistic estimates from the inexperienced professional anxious to please and so secure a job (see Chapter 15 for advice on appointing professionals).

Legal matters

Are there covenants on the land which place legal restrictions on what you can and can't do on the site? Is the site subject to a compulsory purchase order? Are the title deeds in order and so on? You will need to employ a solicitor to establish this as well as to prepare the contract and register the transaction with the Land Registry. Good legal advice is particularly important if you are entering into a complicated deferred purchase arrangement with the council or something of that kind.

Acquiring your site

Having found a site and investigated its feasibility, you are in a position to buy. There are three main ways in which land is offered for sale. The first is a conventional sale: you make an offer and it is accepted or rejected—no particular difficulties there.

The second is sale by tender. Either this can be a formal tender whereby the person selling invites tenders on a set of conditions and is duty bound to accept the highest tender which fulfils the conditions; or it can be an informal tender whereby the successful tenderer is invited to negotiate a sale. This places less stress on the buyer, who does not need to have all the financial resources mobilized before submitting a tender.

The third is at auction. This is particularly difficult for someone not in the property business. You do not get very long from the notice of auction to make the necessary checks on feasibility, costs, planning and legal restrictions, so there are risks attached. If you are successful at auction you:

- ▷ are legally bound once the hammer goes down;
- ▷ must pay a deposit, usually ten per cent on the day;
- ▷ must pay the balance within a specified period, usually four weeks;
- ▷ will lose your deposit if you are unable to complete the purchase.

In other words you need to have access to good advice and finance before considering bidding at an auction.

Chapter 17
Finance

This chapter outlines the key factors you will have to consider when raising finance. You must be able to afford the long-term cost of your self-built house. You will probably expect to gain some financial reward for your efforts. You must be able to raise enough finance to meet your costs. You may need to obtain some form of subsidy to make the house affordable. We suggest how to approach such questions as: how much can you afford? Will your entitlement to benefit be affected? How will you be rewarded financially? How much will it cost to build? How can you reduce costs? What method of arranging finance will be appropriate? How do you avoid paying VAT?

How much can you afford?

There have been some tragic cases of people building and then being unable to afford to move in when they have finished. In thinking about whether to build on your own or as a member of a group, and if in a group whether to build for full ownership, shared ownership or rent, you will have to consider the question of what you can afford.

There is a certain amount of discussion in housing circles about what constitutes affordable housing costs. A common assumption is that one should be prepared to spend up to thirty per cent of one's take-home pay on housing. Many people spend more. But a survey of new housing association tenants suggested that most of them could afford only twenty per cent of their income.

It will be up to you or your group to decide what housing costs you can afford to pay. You will have to discuss the likely level of mortgage payments or rent for each person when the house is occupied. Each individual self-builder will have to make sure that they can meet these costs. Have preliminary interviews with your building society to find out how much they will be willing to lend you.

Will your entitlement to benefit be affected?

If you or other members of a group are claiming income support or housing benefits, you will have to ensure that the project does not threaten your entitlement.

Your entitlement to income support benefit if you are unemployed depends on your remaining 'available for work' during the construction period. The Department of Social Security (DSS) have proved flexible on this point if approached in the right way. The members of the Zenzele scheme in Bristol undertook to take a job if one were offered during the building of the scheme. They would then build in the evenings and at the weekends. This principle has now been accepted generally. If you are unemployed the DSS can authorize the payment of mortgage interest. Where you are planning to pay a mortgage you must make sure that these payments are eligible. Zenzele again pioneered this approach by persuading the building society to grant 'interest only' mortgages for a number of years.

If you are employed but your income is low, you may be eligible for housing benefit. This will pay your rent (but not mortgage interest payments). You will need to ensure that your rent falls within the housing benefit 'ceiling' set by the rent assessment officer in your local council.

Theoretical example of typical cost of one house in a group of twenty

	Cost per dwelling £
Group overheads (as described in Chapter 13) — £1000	50
Housing association registration fee — £900	45
Land	40,000
Site survey — £400	20
Legal charges, including fees, stamp duty and Land Registration	1,000
Planning application, standard charge	140
Building regulations application, sliding scale	175
Architect's fees, including some input from a structural engineer and a quantity surveyor	3,200
Accountant's fees, over two years	300
Subtotal, before work on site	**44,930**
Insurance	400
Tools and plant hire	800
Temporary services, electricity and water	150
Supervisor, if required (assume two and a half days per week for one year)	800
Materials, including foundations and external works	25,500
Service connections: water & sewerage charges	1,300
electricity	180
sewer connection	240
Subcontract labour, including brickwork and plasterwork	6,000
Total	**80,300**

How will you be rewarded financially?

Most schemes compensate self-builders financially for the labour that they put in. It is unlikely that many people would be sufficiently motivated to undertake to build their own house without this compensation. The different funding arrangements seek to do this in different ways, as follows:

▷ Building for private ownership recompenses the self-builder when he or she comes to sell by allowing him or her to keep the full tax-free value of the completed house.

▷ A deferred payment arrangement allows the self-builder to keep the full value of the completed house less the value of the land (which he or she has not paid for) when and if the house is sold.

▷ In a shared ownership arrangement self-builders are given a share in the equity of the property at least equivalent to the value of their labour; this would be realized when they sell the house.

▷ In the co-operative model for rent self-builders are granted a 'premium tenancy'. The premium is equivalent to the value of their labour and this is paid if and when they move away. This premium is index-linked to preserve its value.

How much will it cost to build?

It is difficult to be specific about costs as there are so many variables. Even for constant items, estimates soon become out of date. Quoting actual costs is difficult and may be misleading. Our advice is therefore general, but we do give you a worked-out example for a house to illustrate how a budget is built up.

The cost of your house will depend on a number of factors:

▷ where you are going to build;

▷ the size, type and standard of house that you are planning;

▷ whether you are building on your own or as a member of a group.

As well as the more obvious outlays on land and materials, there are others to be considered: there will be the costs of setting up a group, of professional fees, of planning and building regulations approvals, of subcontract labour, of overheads and of services.

We will give as an example the 'capital' costs of building a typical middle-of-the-road house, so that you can form an impression of what expenses you are likely to incur and their relative order of cost. This cost will have to be fed into the particular funding arrangements that you are using to estimate the likely level of outgoings once you are living in the house. A balance will have to be struck between capital cost and outgoings.

Our example is for a medium-sized, three-bedroom, detached house of sound quality and conventional brick construction to building regulations standards. It is on a medium-sized plot in a medium-cost area—on the fringe of a Midlands town, for example. It is one of a group of about twenty houses, giving some benefit in shared costs, such as road access. The figures quoted are for a base date of October 1989. A house of this type might be worth up to £100,000 on completion and we have derived an estimate of the cost of land of £40,000 from this, as described in Chapter 16. Costs are separated into those that you will incur before you start on site and those of the building operation itself.

To the total you will have to add the cost of borrowing over the development period. This may amount to several thousand pounds.

You will not have costed your self-build labour at this point. If the value was £100,000, with costs of £80,000 in our example, this would give a return of £20,000 for your labour. The balance of the difference between the self-build costs and the value represents what would be developer's labour costs, profit and overheads if the house were built by a developer.

The cost of materials for an equivalent Segal-method house would be similar but much less subcontract labour would be needed. A roofer is all that would normally be required.

You should also consider the long-term expenditure involved in using the house. Running costs for gas and electricity would be around £285 per annum. Charges for water and sewerage vary but might amount to around £100 per annum. Maintenance, which would include redecoration every three years, together with a provision for major repairs after fifteen years—this might include replacing the boiler, for instance—would be covered by a prudent allowance of £550 per annum. This amounts to nearly £1,000 per annum or £30,000 over the design life of the building of thirty years. To this you have to add the cost of financing the building, which amounts to a substantial amount every year. The interest charges over twenty-five years might be two and a half times the amount of the actual loan itself at current rates of interest, or more than £200,000. There is a clear advantage in keeping borrowing as low as possible.

Reducing costs

One way of reducing borrowing is obviously to reduce costs, but there may not be a lot you can do about reducing many of the costs involved. We described a number of potential ways of reducing land costs in Chapter 16 and these could be very important given the high cost of land. Other measures are:

- ▷ Do as much as you can yourself: take on as much of the management of the project as you feel competent to tackle and for which you have the time; employ a minimum number of subcontractors and do as much of the building as you feel capable of.
- ▷ Build as quickly as you are able. The savings in financing costs can be considerable.
- ▷ Consider the type of construction that you are proposing to use. If you choose the Segal method, you can reduce the cost of materials and subcontract labour as we have mentioned, but you can also save time and therefore money. You can possibly reduce land costs too, by using steeply sloping land that is uneconomic to build on using conventional methods.
- ▷ Build small and add on later when funds are available. The term 'starter home' is popular with politicians but usually means starting life in a tiny box that you will soon have to move out of. With self-build, and with the Segal method in particular as it is very easy to extend, you can start with the essential core of a house and build on later. Again, the adaptability of the Segal method lends itself to this approach.

Types of funding required

The funding of a self-build scheme falls into three distinct, consecutive stages.

Start-up finance

This is to pay for the initial costs of starting a group, for preliminary feasibility studies and for other costs prior to securing the main development finance. Methods of meeting these costs are outlined in Chapter 14.

Development loan

This is the short-term loan required to pay for land, fees, insurance, tools, any subcontract labour and materials up until the time that the houses are completed. At this stage long-term mortgage loans will be obtained to pay off the development loan. The sources of development funding will vary depending on the type of scheme (as discussed later under Funding Models). The amount of the development loan will be reduced if you are able to secure a Housing Association Grant (HAG). Interest will be charged on the development loan. This is normally 'capitalized' and added to the amount of the loan to be repaid by the long-term mortgage when the houses are completed.

Long-term finance

All the common financial arrangements available require an element of individual or group long-term mortgage finance on completion. This is used to pay off the development loan. If you are building for rent, the development loan will be paid off with a 'group' mortgage for the whole cost, which will be paid for entirely by the rents charged.

Another long-term cost you will face is for ongoing repairs and maintenance. This outlay will be the responsibility of individual owners in schemes for outright or shared ownership and will be paid for by an element of the rent charged in co-operative self-build schemes.

Finally, if your self-build group continues to exist after the houses are completed, as in schemes other than those for outright ownership at the start, you will incur management costs for continuing financial and legal administration. These might include paying for a part-time book-keeper, stationery and telephone charges. They would be covered by an element of the rent.

Funding models

We offer a description of the basic alternative models that are available.

The arrangements for development loans, mortgages and subsidy will depend on which of the three basic types of ownership you are planning: (1) outright ownership (2) shared ownership or (3) collective ownership for rent.

Outright ownership

If you are building for outright ownership, the group's development loan will be paid off by a series of individual mortgages. The individual mortgages will reflect the cost of each house—£80,000 in our earlier example. The development loan will be obtained from a bank or building society. If you are registered as a self-build housing association, you may be able to obtain a development loan to cover 40% of the cost arranged through the Housing Corporation. You will also qualify for income tax relief on the interest payable on the development loan if you go on to register with the Department of the Environment. Individual members will qualify for income tax relief on their individual mortgages up to a limit currently set at £30,000. Note that this financial relief set at 15% in 1995 is due to be phased out over the following three years.

Shared ownership

If you are unable to afford outright ownership, you may be able to obtain HAG for a shared ownership scheme if you have appointed a registered self-build housing association to act as your development agent. The grant will pay for a proportion of the development costs, reduce the mortgage required and thus reduce the amount of your long-term housing costs. If you are building for shared ownership, members will purchase a proportion of the equity with individual mortgages. The balance of the equity will be purchased by the housing association with a 'group' mortgage, the cost of which will be met by the housing association charging a rent to the individual members. This rent will be set to include an amount for any ongoing management costs. The housing cost to the self-builder is the sum of their individual mortgage payments and their rent. Shared ownership is generally only advantageous if you are able to obtain HAG or some other subsidy. Otherwise you are no better off than you would be if you were to purchase outright.

The way that self-build shared ownership with HAG funding works is as follows:

The individual self-builders have to acquire at least twenty-five per cent of the equity (or value) of the property. They may be granted a certain amount of the equity 'free' as a reward for their labour; this is sometimes referred to as 'sweat equity'. The value of this sweat equity is the difference between the cost and the value of the property and was £20,000 in our example. If the market value is greater than the cost by twenty-five per cent or more, the self-builder will not have to take out an individual mortgage. If this is not the case, as in our example, then the self-builder will have to take out an individual mortgage for the difference:

- ▷ value: £100,000;
- ▷ twenty-five per cent of value required to be purchased by self-builder: £25,000;
- ▷ sweat equity: £20,000;
- ▷ difference between sweat equity and the twenty-five per cent equity to be taken out as an individual mortgage: £5,000.

HAG payable on a proportion of the balance of the equity not acquired by the individual self-builders:

- ▷ value: £100,000;
- ▷ equity acquired by self-builder: £25,000;
- ▷ balance of equity: £75,000;
- ▷ HAG payable at a rate of, say, fifty-eight per cent (this is the average and will vary according to the location): £43,500.
- ▷ The group will then have to take out a group mortgage to pay off the balance of the development loan: group mortgage equals £31,500 (£75,000 less £43,500).

The self-builders' housing costs will then amount to the mortgage repayments on their individual mortgage of £5,000, plus the rent, which will be paying for a mortgage of £31,500, plus any management costs—a total mortgage commitment of £36,500.

The individual self-builders have the option to purchase more of the equity if they wish either at the outset or at some later stage when they can afford it.

The group mortgage will be discharged either when every self-builder has purchased the whole of the equity in their house or when the housing association has completed all the loan repayments.

The self-builders are eligible for income tax relief on the interest on their individual mortgages but there is no relief on the group mortgage.

Collective ownership for rent

This will generally reduce the self-builders' housing costs still further. You will normally register as a housing co-operative and in this way qualify for HAG for a proportion of the full amount of the development costs. The self-build for rent model developed by CHISEL and applied to a number of developments in the South East incorporates the payment of a premium payable to the self-builders if and when they move out in recompense for their free labour. This is index-linked so that it rises in value in line with the Retail Price Index. The self-builders can withdraw their premium when they leave the scheme as a reward for their labour in building the house. The housing association will raise a group mortgage on that part of the development costs not covered by HAG. The co-operative is eligible for mortgage tax relief on this group mortgage. The housing association will charge each self-builder a rent which will cover the repayments on this mortgage together with amounts for management, maintenance, long-term repairs and a fund to be set aside as a reserve. To return to our example:

- ▷ development costs, £80,000, plus premium for labour, £20,000, gives a total development cost of £100,000;
- ▷ HAG payable at fifty-eight per cent as before: £58,000;
- ▷ balance of development cost covered by group mortgage is development cost £80,000, less grant £58,000 equals £22,000 per property.

The self-builders' housing costs will be the rent necessary to cover the cost of this group mortgage, plus management and maintenance allowances.

The actual housing cost to the individual self-builder in a shared ownership as against a rental scheme will vary depending on a number of factors, which include the difference between cost and value, the level of premium, the level of management and maintenance allowances and individual benefit entitlement. You will have to make calculations based on certain assumptions to determine the likely level of housing costs in any particular case.

Sources for development loan funding

Development loans will generally be obtained from a bank or a building society. They will want to satisfy themselves that the value of the properties will exceed the amount of the loan in order to protect their investment if the scheme is not successfully completed. In that event they would have to sell the land and the part-finished houses, so they will want to feel confident that they would be able to get their money back. The savings of your 'free' self-help labour should ensure that value exceeds costs and enable the lender to take certain risks, such as lending to a group that includes people who are out of work and lending one hundred per cent of the development costs.

Even if the value is greater than the size of the loan, they will want to minimize the risks of something going wrong by satisfying themselves that the self-builders have the necessary skills, that the scheme is properly managed, that the group has good professional advice, that costs have been properly assessed and that the loan will be paid back by long-term mortgage finance when the time comes. They will need to know that the scheme can be completed on time, as one of the most common causes of difficulty is when a project takes longer than anticipated and the costs of materials and borrowing escalate beyond the means of the self-builders.

You will have to provide information to potential lenders that will include details of the scheme, costings, a cash-flow prediction, a programme of work and details of the constitution of your group, of membership of the group and of your advisers. If you are

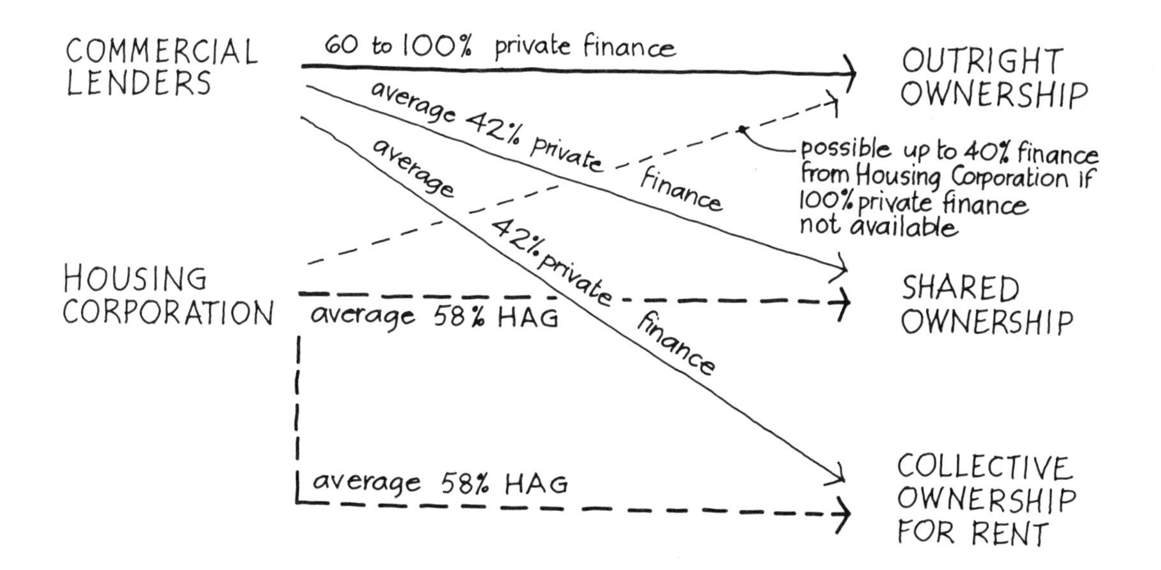

building for outright ownership or shared ownership, both of which involve self-builders taking out individual mortgages, you will have to provide information on people's earnings and career prospects.

The lender will have to have 'first charge' on the property. This legal term means that the lender can sell the property to recoup their investment in advance of any other party, such as the Housing Corporation, being able to recoup their investment.

If you are building for private ownership and if the commercial lenders are not willing to lend the full amount needed, you may be able to obtain a 'top-up' loan from the Housing Corporation for the balance required. At the time of writing, the maximum amount that the Corporation may make available has been raised from twenty to forty per cent of the total, to reflect the current difficulty of obtaining funding for self-build from the building societies. This is subject to the availability of funds.

One difficulty for individual self-builders has been that commercial lenders have not generally been willing to lend on the site alone. A number of possibilities offer themselves:

▷ people with a house already have been able to obtain a second mortgage on it in order to buy a plot of land;
▷ alternatively, they have sold the house and used the proceeds to buy a site on which they have parked a caravan to serve as temporary accommodation during the building period;
▷ otherwise people have had to resort to a bank loan at high rates of interest.

This situation has changed recently because a number of building societies have launched schemes specifically aimed at the need of individual self-builders. They include a development loan and long-term finance for the cost of the land, materials and professional fees. One hundred per cent finance is offered which does not need to be repaid until the house is completed. At this stage the interest on the development loan is capitalized and included in the long-term loan.

In common with others, this type of loan is payable in stages as the work progresses, against a valuation carried out by your architect. Some societies get their own surveyor to visit the site and make a valuation. Some societies make the loan available in, say, six stages, in which case there may be a cash-flow problem to finance the current stage as you get paid only for work that has been completed. This situation is eased if you are able to take out a loan paid in monthly instalments. Otherwise you may need to obtain a bridging loan or, better, an overdraft agreement from a bank. This is more convenient and accrues interest only as and when you make use of the facility.

Grants towards development costs

The principal grant that is available is HAG from the Housing Corporation. It is available only for rent or shared ownership schemes. It is not available for schemes leading to outright ownership. HAG will meet a proportion of the development costs of a scheme and the balance must be sought from a commercial lender. HAG is payable according to a complex scale that depends on the area of the country and other variables; for the financial year 1995/6 it will cover an average of fifty-eight per cent of the cost of land, fees and materials although this proportion is likely to be reduced in following years. In order to obtain HAG, your self-build association must have an agency agreement with an established housing association that will receive the grant on behalf of your group and use it to pay your development costs.

The main drawback is that demand outstrips the HAG available by about six to one. It is allocated on an annual basis and you will almost certainly need the active support of your local authority, who are consulted by the Housing Corporation to find out their priorities for the area, to have any chance of success.

Because of recent severe cuts in government money for housing consideration is being given to avoiding the need for HAG altogether by reducing costs and by obtaining land at nil cost from a local authority. This will usually mean that people's housing costs will be somewhat higher, however.

Local authorities are empowered to give grants to organizations providing low-cost housing in their area, but government financial restrictions make it very unlikely that they would grant-aid a self-build scheme at the present time. However, a number of authorities do operate 'Tenant Incentive' schemes whereby some existing council tenants can obtain a cash payment as an incentive for them to vacate their rented home. It is not available if you are participating in a HAG-funded scheme for shared ownership or rent but it may render self-build for outright ownership affordable.

Long-term mortgage finance

Mortgages will generally be obtained from a bank or a building society. You will already have had to demonstrate that you fulfil their lending conditions when you obtained your development finance; at that stage you will also have obtained 'in-principle' agreement on long-term mortgage finance.

Mortgages come in a variety of forms these days. Individual mortgages can be a normal repayment type, an endowment mortgage linked to life insurance, a fixed-interest mortgage, or interest-only to qualify for benefit. In addition, most building societies offer low-start mortgages for first-time buyers. These come in many guises, but their common feature is that they reduce payments in the early years of the loan and make up for it by either charging higher than normal repayments at a later date (when it is assumed that you will be better off) or lengthening the repayment period. Low-start mortgages do not work out less expensive in the long run. You will have to seek the advice of the branch manager or a mortgage broker on which is the most suitable form of mortgage for your particular circumstances.

Value Added Tax

This is the final financial matter to be considered. VAT is charged on most materials and services, such as architects' fees, but much of it can be reclaimed.

Individual self-builders can reclaim only when the building is complete. You will therefore have to budget to pay the VAT during construction. You need to obtain the necessary forms from your local VAT office. All the materials used in the building must be itemized and original VAT receipts, together with a copy of your planning permission and evidence of completion of the building must be provided. Individual self-builders can reclaim VAT only on the materials used and not on services, such as professional fees. Individual self-builders who are registered can claim quarterly through their business in the normal way only if the house is financed through the business.

A self-build housing association may be able to register for VAT and apply for the VAT to be reclaimed monthly rather than quarterly. In this way you can claim for VAT on all invoices received during the month whether they have been paid or not. It may be that you can obtain your VAT refund before paying the account, which will ease your cash flow. A group may be able to reclaim VAT paid on professional fees as well as on the materials.

For both individual self-builders and groups, VAT cannot be reclaimed on fixtures and fittings, such as cookers, refrigerators and carpets, but can be reclaimed on integral installations of equipment such as central heating and double glazing.

Chapter 18

Design and documentation

This strand concerns developing the design and preparing the documents necessary to obtain the mandatory permissions, order the materials and build on site. We discuss the choice between heavy and lightweight construction methods and the impact of your choice on the environment.

Design

Some of the strands that have been occupying your attention so far may have seemed difficult and at times even tiresome. Here you come to the exciting part, and can give full reign to your imagination. The wealth of ideas about your house that will have been developing in your mind all this time can now be converted into a practical and detailed scheme. But beware of short cuts.

Standard plans and package deals

Many people already have an idea of how they want their ideal home to look—perhaps from having seen an attractive design in a book. They then try and find a site for it. But the choice of sites available is limited, and it is unlikely that a stock design will be right for the particular conditions of the particular site they end up with. In other words, the site comes first and such matters as the direction of the entrance, the fall of the ground, orientation for sun and view and privacy are the factors that determine good layout. A standard plan from a book, devised without a site in mind, cannot relate to all these variables, nor can it do justice to all your particular needs and wishes.

Another apparently attractive alternative to designing your own house is to buy a package-deal timber-frame kit. There are companies who prefabricate timber panels in their factory and supply (and erect at extra cost) the basic structure. This could be quicker than doing it yourself, although you still have to wait for your house to take its place in the queue at the factory. You still have to tackle the time-consuming work of foundations, roof covering, cladding the walls, finishing the interior; plumbing, electrics and fittings; and external works such as services and driveways. More importantly, you will not be able to vary their standard layout very much.

To go for a mass-produced plan or package is to miss the key advantage that self-building offers—the opportunity to achieve the unique solution particular to you.

You the designer

If, after having weighed up the pros and cons of engaging professional help (as discussed in Chapter 15), you have appointed an architect, do not then make the mistake of leaving all matters of design to him or her. Your part in the process is crucial.

What constitutes this design process? We see it as assessing all the contributing factors—practical needs, aesthetic judgements, personal preferences, costs, constraining regulations, ecological effects and available resources—and manipulating them all until they are juggled into a proper balance. It is clear that even with professional advisers involved, the owner-builder plays the predominant part. The architect cannot function until you have, at the least, set out all your requirements and preferences, explained your financial situation and the budget that can relate to it and assessed your other resources.

The timeless way

Design starts with the most general questions about where the building should go and proceeds through many levels of further refinement, taking into account more detailed factors as you go and adjusting as necessary to keep the whole in balance.

In Chapter 3 we told of our pleasure in discovering the twin volumes assembled by Christopher Alexander, *The Timeless Way of Building* and *A Pattern Language*, and took you through a particular house, identifying some of the patterns used. We suggest that this is a good way for self-builders to approach design, controlling the interplay of related factors by going always from the general towards the particular. As each broad issue is resolved, and a general pattern emerges, opportunities are revealed for the next range of decisions, which when taken allow yet further developments to flow out of them. It is an organic, progressive process and does not depend on the ability to conceive an entire, perfect concept at the outset. As in life, though, there will be low and high points, mental blocks will be followed by sudden and unexpected flashes of inspiration and insight.

We believe that when you have read Alexander, whether you are designing by yourself or 'through' an architect, you will feel confident that you are on the way to arriving at a delightful solution.

However, it is still necessary to supplement Alexander's theories with your own observations. Visit completed buildings and estates when considering planning the layout, measure rooms with which you are familiar when planning the house, visit the showroom when considering the type of windows to have and so on. Models can also be helpful in trying out different layout alternatives, particularly in a group situation, visualizing the building on its site and the spaces inside the building.

It is important to keep costs in view at all times when developing a design to avoid the disappointment of having to abandon overambitious and unrealizable dreams. You should take advice on the sort of average cost per square metre for materials for a domestic building. This, when compared to your budget, will suggest roughly how much floor area you should be able to build at a certain level of quality. If you subsequently improve the specification—or add features such as pergolas or balconies, you will have to either increase the budget or reduce floor area accordingly.

Design considerations

The aspects that the design will have to bring into a balance will include the fundamental ones of the accommodation that is required and the overall budget for the project. The appearance of the building will be affected by the choice of materials, which may be conditioned by their need for maintenance, their ecological implications, and planning control. The planning authority will also be concerned about overlooking of neighbours. You will be concerned with privacy, view and orientation to the sun. The arrangement of the site, the slope of the land, the position of access, services and drainage, trees and other existing features will all influence the design. If you are planning a group of houses you will have to consider parking and the design of roads, their width, turning arrangements, visibility at junctions and access for refuse collection and the fire brigade. Inside the house you will be concerned with sunlight and its energy—saving possibilities, daylight and artificial lighting, ventilation, heating and hot and cold water, the provisions for bathing, cooking, eating, sleeping, working and storage. You will need to consider the fittings, fixtures and finishes. The building regulations establish standards for sound and thermal insulation, fireproofing and structure. You will also be interested as a self-builder in the ease of construction. Externally, think about the layout of the garden and its relation to the inside of the house. You may also want to plan for later extension.

Choice of construction

While considering the basic design, you must have in mind how it is to be constructed. You have to:

- ▷ decide what it is to be made of;
- ▷ devise a structure that suits the disposition of spaces that you have arrived at;
- ▷ meet the requirements of the various regulations applied to building;
- ▷ obtain the necessary permissions.

The choice of material and construction for house building is wide, and you will be guided by such considerations as the place you are building in, the way you want the building to perform, your own skills and what you can afford.

Weigh up the pros and cons of the commonly available materials and the two basic ways of using them: in what has come to be known as 'wet-and-heavy' construction—generally load-bearing brick or masonry walls with concrete strip foundations and possibly floors; or the equally strong but less massive 'lightweight' frame and panel construction, probably in timber-frame or possibly steel, with separate pad foundations.

The building techniques available vary widely and while we want this book to be helpful to you in a practical way, we cannot possibly take you through each one in detail. You will see from the Bibliography that there is much published information to which we refer you, particularly applicable to what is termed 'conventional' or sometimes 'traditional' construction. In Part Four we make our contribution to the range of methods available by spelling out in detail how to build in the way Walter Segal devised specifically for the self-builder without trade skills or building experience.

First, we raise some hitherto neglected ecological issues.

Responsible building

It is only now becoming customary to consider the environmental impact of your building, both on the global front and on the health and wellbeing of its occupants. Your self-build project gives you a wonderful opportunity to create a built environment that is ecologically sound.

'House building is both a creative and a destructive process.'[1] The building industry pays little attention to the destructive side. It is taken for granted that site clearance for a building project devastates the landscape, at least temporarily, that the production of materials involves quarrying, polluting industrial processes and that timber is felled regardless of its rate of regeneration. Even on the creative side, the finished buildings continue to apply a heavy load on the environment. Standards of modern house building, particularly in Britain, are low. Materials are not sufficiently well chosen nor used to their best advantage. Many of our houses perform poorly in terms of comfort, efficiency and longevity.

Matters such as orientation for best benefit from sunshine, siting for shelter from exposure to harsh weather, good sound and thermal insulation, long-lasting and easily maintained material and construction methods, energy-efficient equipment and the avoidance of products that have caused undue harm in their manufacture, all have received too little consideration.

We discuss the ecological implications of building materials and measures to conserve energy in Chapter 19, but one decision that has a fundamental effect on your design is whether to build using 'heavy' construction methods, such as masonry, or 'lightweight' ones such as timber-frame.

Light or heavy construction

The ecological advantage of light construction is largely in the economy of structural material required and speed of construction, which shortens the time that the surrounding environment is disfigured by building operations. The subsequent energy requirements of a light building can be larger or smaller than a heavy one, depending on how it is equipped and used. Generally, it will not retain stored heat, but on the other hand, it will be quick to warm. There may be a saving if the building is in intermittent use.

Heavy construction in Britain is almost bound to use more ecologically demanding materials. But many of the traditional heavy building materials are acceptable environmentally and have as an advantage high thermal capacity. This can be used, as so well demonstrated at Netherspring (see Chapter 10), to store heat gained and even out the temperature fluctuations between night and day. For sound insulation there is nothing, apart from isolation, like sheer weight to stop airborne sound.

On appearance, there may be an instinctive feeling that heavy, permanent construction sits more happily in the landscape than the more apparently fragile and transient lightweight. But medieval timber-frame belies this and there can be a particular grace and elegance in poising a building off the ground, rather than embedding it in it, that is equally satisfying.

The site

Whether your building is light or heavy, to enlist the friendship of the environment you have to be sensitive to place.

'To create true harmony, every building and every feature of the landscape, natural and artificial, has to occupy the position appropriate to its nature and human requirements. To be at the right place, facing the right direction, performing the appropriate action at the proper time is to be both ceremonially correct and practically efficient,' says Nigel Pennick in *Earth Harmony*.[2]

Few self-builders will be able to choose the perfect site, but whatever degree of choice you have, use it to place your house where it will get the most benefit from the sun, be sheltered (and give shelter around it) from cold winds. Shelter it as far as possible from outside sources of noise, traffic disturbance, foul air, excessive electromagnetism from high tension cables and visual intrusion. Note the structure of the soil and the way surface water flows over it. Discover, if you can, any underground water courses. When it comes to preparing the site for building, approach the work with due respect.

The typical conventional commercial builder's practice is obnoxious. He strips the topsoil from the site, removes all trees and plants in the way, and leaves the house's eventual occupant to create a garden from scratch from some crudely levelled-out pile of subsoil. The self-builder can form a sympathetic relationship with the site from the outset. A surprising amount of work can be done by hand, gently and bringing with it the advantage of gaining intimate knowledge of the terrain. If the construction period is long, as is likely with heavy forms of construction, and the self-builder perhaps lives on site in a caravan or tent, then gardening can develop alongside building with obvious advantage at occupation time. If you use lightweight construction, like Jon's Number 6 Segal Close, using the Segal method of timber construction described later (see Part Four) the amount of earth-moving necessary is minimized, trees are left in place even if close to the building and the trauma of a machine-torn, ravaged site is avoided.

Much of this book is concerned with making known the advantages of Walter Segal's method of timber-frame and panel building to self-builders, and we may err on the side of stressing the advantage of the light touch.

Substructure

A key factor in choosing between wet and heavy and frame and panel is in the implications for foundations. Solid brick and block must be supported by utterly rigid footings, upon which they weigh heavily. This inevitably means large excavations and correspondingly large masses of concrete. Otherwise disastrous cracks appear as unequal settlement takes place. This particularly affects internal plastered finishes and external renders.

Lightweight framed structures are tolerant of movement and need far less substructure. For this alone they are ecologically preferable.

Brick construction

A lot of people will want to build in brick as many have successfully done before them. The path is well trodden and there is plenty of literature to which the inexpert self-builder can refer. We have discussed the merits and demerits of heavy construction and have described successful self-build schemes at Zenzele, Lightmoor and Netherspring, where it has been used. Some of the attractions of this approach are that:

- ▷ It looks like many people's idea of how a house should look.
- ▷ It is regarded as a sound investment by valuers and the building societies.
- ▷ Good quality brickwork is maintenance free and lasts well (but beware of cheap sandfaced bricks that suffer from frost attack after twenty years).

However, from a self-build point of view there are drawbacks:

- ▷ Bricklaying and then the application of the plaster inside are both building operations that need skill and plenty of practice to carry out properly. Many self-builders subcontract this part of the work out, but this does reduce the cost savings possible through building your own house.
- ▷ It is an inherently heavyweight form of construction, which can make moving materials around a big job.

- ▷ It is a relatively slow building operation, which can lead to higher costs than necessary, human as well as monetary.
- ▷ Extensive groundworks are necessary to create the substantial foundations needed to eliminate even the smallest movement that would cause cracking of the house.

Timber construction

Timber construction can avoid these difficulties and is, in addition, easy to insulate thermally to a high standard. Timber construction has been around for a very long time and much of our heritage of vernacular, domestic architecture (known affectionately as half-timbered, because only the frame, not usually the cladding, is in wood) bears witness to its attractiveness and longevity. More recently, another method of timber construction, stud-framed wall-panel construction, has become widely used in Britain. It was developed in North America and is now used extensively in Scandinavia and northern Europe.

In Britain a veneer of brickwork is often built around the outside of the timber-frame structure. There are technical hazards with this practice, however, because the brickwork is not waterproof and moisture is trapped in the cavity between the brickwork and the timber structure. This is a recipe for trouble in our damp climate. If the vulnerable damp-proof membrane that is required to protect the timber is not complete for any reason—damage during construction, inadequate joints, etc.—then the timber structure will be exposed to the continually damp conditions that lead to rot developing. Old timber buildings have survived because air can get to them easily, ensuring that continuously damp conditions will not persist.

Documentation

Whichever type of construction you decide on, you will need to have drawn and written information to elaborate, record and communicate your design. These documents will be required to:

- ▷ develop the ideas for the design and evaluate alternatives;
- ▷ describe the proposals to the approving authorities for planning permission and building control (as described in Chapter 20);
- ▷ provide the information to order materials;
- ▷ give the information necessary for building on site.

The documentation for a Segal-method building is detailed in Part Four, but whatever the type of construction, in general terms you will require:

- ▷ sketch plans, elevations, site layouts and three-dimensional views to work up the design;
- ▷ worked-up versions for a planning submission;
- ▷ more detailed information for a building regulations submission and for construction on site. These will include:
  - ▷ drawings:
    - ▷ site plan showing services and drainage;
    - ▷ floor plans showing dimensions;
    - ▷ elevations showing vertical dimensions;
    - ▷ sections showing construction;
    - ▷ foundation plan showing setting-out dimensions;
    - ▷ plans of the floor structure;
    - ▷ roof plan showing the structure;
    - ▷ details of the construction showing how the roof joins the walls, etc.
  - ▷ a specification which describes the materials to be used;
  - ▷ possibly calculations of any structural members that cannot be determined by rule of thumb;
  - ▷ possibly a schedule of materials.

Chapter 19

Materials and energy conservation

We point to ways of conserving energy and methods of providing it. We outline the factors to be taken into consideration when selecting materials and consider their effect on the environment.

Energy conservation

Comfort temperature

The first consideration is that of appropriate indoor temperatures. The hotter we keep the house when it is cold outside, the more energy we consume. Better perhaps in merely cool weather to wear a pullover. If the house is well insulated, there will be a much reduced level of cold radiation (or warmth absorption) from cold surfaces. This has the effect of making the indoor climate appear warmer, which in turn allows you to reduce the actual air temperature and still feel perfectly comfortable. So good thermal insulation must be an important consideration.

Orientation for passive solar gain

You will obtain a large measure of free energy from the sun if you arrange for the majority of the windows to face as near south as possible and restrict the size of windows to the north. It may even become necessary to prevent overheating by the provision of a wide eaves overhang to shade the windows in high summer, and you will also need good ventilation. The siting of the building is critical to prevent it being shaded from the sun by other buildings or trees. You will still obtain useful solar gains if the building is positioned so that the windows face from south-east to south-west. The same orientation will suit a conservatory, which will greatly enhance the solar gain. It will also provide a buffer between inside and outside and a source of preheated ventilation for the house. Incidentally, a conservatory is inexpensive additional space that will be comfortable without additional heating for most of the year, as well as being useful for growing plants.

Shelter

It is greatly beneficial to shelter the house from the prevailing winds. Trees planted for shelter will reduce wind speeds by fifty per cent within a distance of five times their height. Climbing plants such as ivy will also shield the fabric of the house.

House planning

It is good to plan a draught lobby at the entrance instead of opening the front door straight into the hall, which funnels cold air into the heart of the house. Place the living areas on the south side.

Thermal insulation

Incorporate as much insulation as you can reasonably afford at the outset as it is difficult to incorporate more later. Insulation standards in Britain are still infamously low compared to North America, Scandinavia and Europe. The minimum standard required by the building regulations was raised, however, in April 1990 from an average U value* of 0.6 w/m^2 degree Centigrade to 0.45 for floors and walls (including the windows) and 0.35 to 0.25 for the roof. (We remind the reader that a good U value is a low one!) In addition, the area of windows and roof lights

** U, the rate of flow of heat through an element of construction, is measured in watts, per square metre, per degree centigrade difference of temperature between inside and out. The calculation of U values has to include the effect of 'cold bridges' from July 1995. The thermal properties of various materials and the method of calculation is set out in 'The New Metric Handbook' published by Butterworth Architecture.*

should not exceed 22.5 per cent of the total floor area. Should you wish to vary these provisions you can adopt one of two alternative calculation methods, either a 'target U value' method or use the 'Standard Assessment Procedure' SAP). A house in our climate would be considered to have a high degree of insulation if it had U values of around 0.25 and 0.15 for walls and roof respectively, and we advocate that you should aim for this.

Where should this insulation be put? In heavy construction the choice is between positioning the insulation on the outside or the inside of the structure. If on the outside, this will give a high thermal capacity to the building. This means that the structure of the building will warm up and store heat, which will tend to reduce fluctuations in the internal temperature. The difficulties associated with this are keeping the insulation in place and protecting it from the weather and mechanical damage. If the insulation is on the inside, the building will respond more quickly when you put the heating on, but you lose the effect of thermal storage. This suits intermittent occupation. In lightweight construction it is more difficult to provide thermal storage but it may be easier to envelop the structure with a continuous insulating layer, thus avoiding the danger of 'cold bridges'.

The danger of condensation within the structure must be guarded against, either by the provision of a vapour barrier or by ventilation in the appropriate places. This is too technical a matter to go into thoroughly here, but it is well covered in *Thermal Insulation: avoiding risks*, published by the Building Research Establishment.[1]

Window glazing

Windows play a key part in energy conservation. They are capable of capturing significant solar gain if the orientation is favourable but are also a major source of heat loss in other conditions. Single glazing can be adequate if supplemented with insulating curtains and/or shutters. Inexpensive lift-off double glazing is a big improvement but may allow annoying condensation between the panes. Sealed double-glazing units need to be factory-produced and can be nearly twice as good as single glazing if the gap between panes is 19mm or more, but they are expensive. Low-emissivity glass, which has a microscopic coating to reduce radiation passing out of the window, and units with an inert gas succh as Argon will both further improve performance.

Airtight construction

A significant proportion of heat can be lost through infiltration—air passing through small gaps in the construction, particularly in windy weather. It is particularly important to guard against in a lightweight, dry construction and is the reason that a layer of polythene was incorporated into the wall construction of the low-energy Segal house built at the Centre for Alternative Technology. Compressible rubber seals can be incorporated into panel joints in timber panel construction. The penetrations in the building envelope are often overlooked—where the soil pipe enters the building from a ventilated underfloor void or where the boiler flue passes through to the outside, for instance. Opening windows and doors should be draughtstripped and sealed where they are built into the walls.

Controlled ventilation

Windows can be thrown open in warm weather to get a good circulation of air. At other times ventilating the house to remove stale, humid air is not best achieved by opening the windows, which can be wasteful of heat. Warm, stale air will rise through the house—the 'stack' effect—and you can arrange for it to escape at high level and admit fresh air to replace it at a low level in a controlled way, such as through the use of 'trickle' ventilators in windows. Or you can rely on mechanical means: it is now a requirement of the Building Regulations to install fans in the bathroom and kitchen, where a large part of the humidity in a house is produced. These can be controlled by a humidistat which activates the fan when the humidity reaches a critical level.

Condensation

Internal condensation should not be a problem in a well-insulated house if it is ventilated properly because it will be easy to keep warm and the temperature of the internal surfaces will be high so condensation will not form. However, another effect of a high

ANNUAL ENERGY COSTS FOR A 3 BEDROOM, SELF-BUILD, SEMI-DETACHED BUNGALOW FLOOR AREA 68m²
WITH DOUBLE GLAZING, GAS BOILER, ROOM THERMOSTAT AND PROGRAMMER

*THE BASE CASE IS INSULATED TO 1995 BUILDING REGULATIONS STANDARD OF U VALUES: WALLS & FLOOR 0.45, ROOF 0.25 W/m²/°C

	NATIONAL HOUSE ENERGY RATING NHER UP TO 10	STANDARD ASSESSMENT PROCEDURE SAP UP TO 100	CARBON DIOXIDE PRODUCED CO_2	ANNUAL COST OF HEATING	ANNUAL COST OF HOT WATER	ANNUAL COST OF COOKING	ANNUAL COST OF LIGHTING & APPLIANCES	TOTAL ANNUAL ENERGY COST INCL. £86 STANDING CHARGE
BASE CASE	7.2	60	4.4	£145	£87	£23	£165	£506
PROVIDE SHELTER ON ALL SIDES — COSTS VERY LITTLE								£502
PASSIVE SOLAR GAIN — LESS THAN AVERAGE OVERSHADING + CONSERVATORY								£466
IMPROVE AIRTIGHTNESS · 2 AIR CHANGES/HR @ 50 PASCALS PRESSURE — COSTS VERY LITTLE								£458
IMPROVE INSULATION TO U VALUES FOR ROOF & FLOOR – 0.11 W/m²/°C AND WALLS 0.21 — GIVES GOOD RETURN								£397
FIT CONDENSING BOILER WITH THERMOSTATIC RAD. VALVES — GIVES GOOD RETURN								£368
IMPROVE HOT WATER CYLINDER INSULATION + INSULATE PRIMARY PIPEWORK — COSTS VERY LITTLE								£362
USE LOW ENERGY LIGHTING								£340
CAREFUL USE OF LOW ENERGY APPLIANCES								£306
IMPROVE DOUBLE GLAZING TO ARGON FILLED UNITS + LOW EMISSIVITY GLASS								£302
ALL MEASURES	NHER 10	SAP 96	CO_2 2.3	£29	£49	£18	£119	£302

MANY SMALL MEASURES TAKEN TOGETHER MAKE A WORTHWHILE SAVING OF 40% BASED ON 1995 BUILDING REGULATIONS STANDARDS

standard of insulation is that the temperature of the exterior of the building will be lower, because less heat will be passing through the inside to warm it, and there is an increased danger of 'interstitial' condensation forming within the thickness of the construction. This can be designed against by installing vapour barriers (making sure that they are well sealed at penetrations and junctions) or by ventilating the construction (making sure that the interior of the building remains draught-free) or by a combination of the two measures. This is a complex technical issue but you should be aware of the importance of preventing condensation as it can cause extensive deterioration if it is not taken care of.

Efficient heating

A well-insulated house requires much less heat input and considerable savings can be made to the extent and cost of the heating system required, which go a long way towards paying for the insulation. Central heating will probably not be necessary as heat will distribute itself within the well-insulated envelope of the house. The efficiency of heating appliances has been improved significantly recently. The development of the condensing boiler, for instance, makes use of much of the heat that normally goes out of the flue. Other measures are more sophisticated controls and hot-water storage tanks with integral insulation.

Low-energy lighting

A typical low-energy light bulb uses 18 watts to produce the same quantity of light as a conventional 75 watt tungsten light bulb.

Sources of energy

Having reduced the consumption of energy with all or some of the measures above, you will still need an energy input into the house for lighting and domestic appliances, hot water and cooking as well as heating. The choice of an appropriate fuel will depend on your particular circumstances. Do be warned that electricity is not a primary source of energy, only a means of transmitting it, and although 'clean' at the point of use, most commercially produced electricity is extravagant and polluting. It is obvious to use it for lighting and mechanical power—jobs which electricity does cheaply and well—but heating is another matter. If you are building in the country and are attracted by the idea of wood burning, a renewable resource, bear in mind that much storage space has to be devoted to it. Wood must be obtained well in advance of requirements, seasoned and kept dry.

Appliances

The energy efficiency of appliances and the way in which you use them are important; only thermal insulation and efficient heating systems offer higher potential savings. Few people wash their clothes by hand these days and more and more use machines to wash the dishes.

Operating costs

The annual energy costs for a detached three-bedroom bungalow of 68 m² (730 square feet) with gas heating are given above, showing the benefits to be gained from the measures just described. Note that adding insulation and installing an efficient heating system offer the highest savings. Note also that some measures such as providing shelter, improving airtightness and the insulation of the hot water tank produce savings for very little cost.

Further measures

You may consider it desirable to incorporate other devices to improve performance and reduce energy costs further, and even move towards the concept of the 'autonomous house'. We refer you to other sources that give details of wind and water power, solar power from photo-voltaic cells and heat recovery from ventilation air and waste water, all interesting and worthwhile topics but not likely to be so cost-effective as the measures already described.[2]

One device that is worthwhile is the active solar collector for the heating of domestic hot water. The costs of commercial manufacture have been reduced to a level whereby the appliances will pay for themselves in reduced fuel consumption over a fairly short period. If you make your own solar panels from black-painted steel panel radiators in a glazed box with insulation on the back, they will very soon pay for themselves. The Centre for Alternative Technology has easy-to-follow directions available for the construction of such a DIY system,[3] which Brian followed at Romilly as described in Chapter 4.

Materials for self-build

There is little distinction to be made between self-build materials and those used by the professionals. You will go to the same sources as they do—the builders' merchants. Think ahead with the aid of a comprehensive catalogue and make your order as precise as you can. Open an account with your local merchant and you will get a discount. It is often worth making a bulk purchase to improve on this, but for many things it is better to use their storage facilities and draw the materials for use on site as you need them. That way it is easier to keep them secure and in good condition.

Depending on where you are and the quantities you need, it is sometimes possible to go straight to the manufacturer, or at least the bulk distributor, and obtain better prices still. Find out if transport is included; if you have a pick-up you can save substantially by collecting material yourself.

Apart from the lowest price, what qualities should you be looking for?

Durability

You want materials with good resistance to the weather, mechanical damage and wear and tear and that will require the minimum of maintenance. Obtain the highest quality you can afford as it will cause a great deal of inconvenience and cost a great deal of time and money to renew defective components later. Avoid where possible paint finishes, which have to be prepared, primed, undercoated and finish-coated as often as every three years. Particularly troublesome are softwood window sills, which rot as soon as the paint film fails or if they are built into damp brickwork.

Locality

Look for local materials that belong to the vernacular tradition of your area. There are quite clearly defined zones across Britain with traditional materials and techniques. They stem from such things as the availability of abundant timber, clay for brick and tile making, slates for roofing, quarries for walling stone of different types and so on. Easy transport and centralized mass production has eroded these local traditions to some extent but you may be required to use certain materials by the planning authority to fit in with the locality, especially if it is of particular character or in a conservation area. Manufactured substitutes are often permitted in place of more expensive traditional, natural materials. Examples are the use of reconstituted stone blocks—really concrete blocks —to imitate stonework or concrete tiles that resemble Cotswold stone slates. They always look like cheap imitations, so use the authentic material if you can.

Ease of construction

This aspect has not been given a great deal of attention in self-build to date, but the careful selection of materials and techniques could make your life easier. We have already mentioned the fundamental choice between heavy and light construction and the advantages of a dry form of construction. Even if you decide to build using heavy, wet techniques, there are still ways of making it easier. You should always think through the exact steps that are required to build a particular part of the building as you are deciding on the materials and the details of construction, identifying possible problem areas. There may be ways round them. For example, you may consider dry-stack blockwork. This is a method of building concrete block walls without the need for skilled blocklaying experience. Specially designed blocks that key together rather like Lego are stacked up without mortar. You can adjust the wall to obtain an accurate line and vertical at your leisure. Steel reinforcing rods are placed in the voids in the blocks both vertically and horizontally. The blocks are designed to locate the rods accurately. Then the wall is filled with concrete up to 2.4m (8′0″) high at one go. There are also various dry-roofing techniques that are quite manageable for unskilled builders to replace conventional tiling and slating.

Environmental implications

We first raised this aspect in Chapter 18 when discussing design. Now that we come to focus on materials and energy conservation we return to these implications at more length.

The building industry has been slow in coming to grips with the need to conduct its operations in a way that is not unduly damaging to the environment. Your self-build project gives you a wonderful opportunity to set higher standards and create a built environment that is in harmony with nature. We all feel an obligation to try and ensure a sustainable future, but it is quite difficult to assemble the necessary information to build responsibly. The building trade cannot or will not divulge what degree of malignancy to the environment is caused by commonly used materials. The purveyors of energy to be used for heating, lighting and motive power do not come clean on the issue of the true cost to the environment of the exploitation of their particular interest, be it gas, oil, coal or nuclear fission.

John Seymour, who first suggested to Brian that we write this book, has himself written, with Herbert Girardet, *Blueprint for a Green Planet*,[4] which addresses the broad question very effectively. It is a householder's guide to living Green and invokes the power of the purse as a means of obtaining improvement. Few citizens will ever have another opportunity to wield so much purchasing power with discrimination as when they self-build. In the several chapters devoted to building matters, Seymour and Girardet impart an enormous amount of general information about constructing and servicing buildings in an ecologically responsible way.

The sheer hazard to the occupant of a house arising from the materials it is made from is the subject of a book edited by S.R.Curwell and C.G.March, called *Hazardous Building Materials*.[5] The research for it was financed by Godfrey Bradman, in order, as he says, 'to help in the selection and evaluation of more safe and suitable builders' materials. I am pleased to share it with all those who are building the next generation of homes . . .'[6] It is aimed at the professional building industry, but much of it is highly relevant to self-builders.

The Green Consumer Guide,[7] by John Elkington and Julia Hailes, has a section on the hardware and DIY store that is of interest to self-builders and they too invoke the power of the purse.

Another source we have used is the issue of *Environment Now*,[8] number 10, November 1988, where Robin Murrell and Avril Fox observe that almost every aspect of house building is environmentally destructive, yet Green thinking has so far had little influence on building industry attitudes and practices. They comment on the difficulty of finding information about the environmental impact of different methods of construction and so provide a brief but useful set of guidance notes about which materials are environmentally benign and which malignant, upon which we have drawn here.

There is no single Green way

We are not able to give definitive recommendations to guide all your choices. We offer what advice we can, derived from the sources mentioned and from our own experience.

All aspects of ecological building are interrelated: one cannot talk about heating without bringing in insulation, or insulation without talking about construction, or construction without considering materials, or materials without thinking about their manufacture. Sometimes desirable aspects conflict with one another. There is no single Green way.

Building biology

You will be spending a lot of time in contact with various materials while building and after you move in. Take care that they are nice to be with. We can learn from the Dutch self-builders of MW2 in Maaspoort (described in Chapter 12), who are part of the Biological Building movement, which is already quite strong on the Continent and beginning to have a presence here. They treat the building like an extension of your clothing, your own skin even. They say the building should be able to breathe. The surfaces around you should be comfortable to touch; warm enough not to cool your body by radiation; absorbent enough to accept the dampness you create in the atmosphere. No potentially harmful chemicals are introduced—the timber, for instance, is not treated with powerful insecticides or fungicides but is preserved by immersion in a borax solution prepared on site. The air is kept free from toxic substances. The building shields you from external noise and vibration but does not isolate you. Natural light and harmonious colours augment well-proportioned spaces. Great care is taken to control electromagnetic 'smog' from power circuits and appliances.

The whole issue of healthy building raised by the Rudolf Steiner inspired 'Biological Building' movement on the continent is now acknowledged to be of mainstream interest in this country, generating research and discussion throughout the architectural and allied professions.

One of the several bodies engaged in promoting it, The Association for Environment-Conscious Building,[9] publishes a products and services directory, Greener Building, and a quarterly magazine 'Building for a Future' which rounds up the news of current research and features examples of good practice.

Ecological implications inherent in the choice of building materials

In our evaluation of the 'eco-characteristics' of the commonly used range of materials, we have remarked on their effects on the inhabitants of the finished building; on any hazards they present to the builders; and on how their manufacture may have affected the environment. We have classified them into 'general construction' and 'insulation' categories for convenience, though in practice there is some overlap.

General construction

Concrete

Both the aggregate and cement components of concrete, from the orthodox builder's point of view a cheap material and one which gets used all too liberally, are negative to the environment. Everywhere we see vast excavations for aggregate.

To make cement, limestone, gypsum and clay are quarried and burnt, to the detriment of the environment, and consuming massive amounts of energy. Lime and cement are dangerous to handle, liable to damage lungs, skin and eyes. Concrete is not suspected of being hazardous to the house-occupant, but we think that it is good to cut down on it, even though it will still play an essential part in almost every building job. When it has to be strong, it is customary to increase the cement ratio. It is important for the self-builder to realize that to develop maximum strength, it is not a good idea to chuck in more cement. A rich but sloppy concrete may be easy to pour but it is expensive and will crack on setting. More important is to control water content to the least amount that will make the mix workable, and to tamp it well. Precious cement is saved.

As well as site-cast, concrete is available factory-cast into building blocks. Very often expanded aggregates or foaming agents are introduced for lightweight insulating blocks. They are almost universally used for the inner skins of brick cavity walls and for partitions in wet-and-heavy construction. They take plaster well and are cheap and quick to build with. They suffer the environmental drawbacks of all cement products in their making.

Bitumens

One thinks of the way tarmacadam has transformed road making by defeating the menace of mud and appreciates the value to us of bituminous products. They are useful in building too, beyond the waterproofing of our site roads and car parks.

In the old days of the production of town gas from coal, tar was an important by-product. It is now fairly rare, but comparable with it is the lowest fraction of the refining of crude oil, pitch, and at the moment it looks as though we shall have plenty of it into the mid to distant future. In a natural form, bitumen and sand are mined as rock asphalt and there is a lot of that about too. We use bituminous emulsion damp-proof membranes, pitch-polymer damp-proof courses and mastic sealants.

As for the environmental effects of the production of bitumens, we may not be very keen on oil refineries in our landscape, but while we place such value on the motor vehicle we must have them. It is certainly less objectionable to use all the fractions produced as by-products of fuel, ranging from margarine to vaseline to our indispensable roofing felt, than to let them go to waste.

All tar products are said to be carcinogenic to some degree, but the degree of exposure to them in building work is so small that the risk is negligible.

Clayware

Burnt clay building products have been with us for a long time, have proved their worth and not many people would find any fault with them. The basic materials are natural, not synthetic. Their main environmental drawbacks are that they require a large amount of energy to fire them and that they bring in their train the requirement for a lot of environmentally undesirable cement products—in the necessary concrete foundations and in the mortar used for bedding.

It is also customary when building 'wet and heavy' to plaster the walls and have concrete floors, which require cement and sand screeds. These materials are also the components of mortar, the essential partner of brick. It seems a paradox but is an important truth that mortar is not there to stick bricks together; its purpose is to hold them apart. If bricks were absolutely perfectly shaped, they could be fitted together dry and still perform their load-bearing and weatherproofing functions. But they are not, and would wobble about and be overstressed where they touched, if laid dry. Water would run through the joints. Bedded in mortar, these problems are overcome, but the mortar has to be strong enough only to stay put, and if made too rich in cement will shrink and crack, possibly taking the bricks with it. Weathertightness, which depends on the sponge-like, universal porosity of brick and joint, will be lost.

So once more, little is good: cut down the cement content to the minimum. A more workable, 'weaker' and better mortar can be made by incorporating lime. Besides performing better in use, a brick wall in lime mortar will be easier to dismantle and the bricks will be easier to clean for re-use.

The traditional image of 'house' in many areas of Britain is clay brick and tile. The material is abundant and used to be locally obtained from small clay pits, soon reabsorbed by the natural landscape. Nowadays industry dictates vast centralized plants with a big visual impact and big energy requirements. No longer does a brick house automatically fit into its locale. Red brick is not as Green as it once was.

From the self-build point of view, brickwork is a possible and reasonable choice—the material performs many functions well over a long period.

There is also scope for using ceramics in lightweight construction. Clay products, such as tiles for external wall hanging and quarry tiles internally, go well with timber-frame, as do clay roofing tiles.

For below-ground drainage there is now a choice

between stoneware and PVC systems. The self-builder will find the latter much easier to handle and install, but environmentally clay pipes score rather better in that although they are heavier, the embodied energy in their manufacture is less than half that of the equivalent PVC.

Stone

Quarrying is destructive, needs costly equipment and is wasteful from the point of view of the amount of walling stone produced, so stone is expensive. But it lends itself to reclamation and has a long life, besides being very beautiful when in the hands of a skilled mason.

Artificial stone is actually less wasteful to produce and is thus environmentally preferable in that respect, and it is easier to lay in the wall; but we have perhaps been unfortunate in seeing nothing but horrible parodies of the real thing wherever we have seen it used.

Stone is by tradition environment-friendly, but it is worth taking care to prevent a build-up of radon gas when building with granite, which is slightly radioactive.

Glass

The self-builder will be using a lot of glass: in glazing for windows and conservatories and in fibrous form for insulation. It needs some practice in safe handling but is a marvellous material and does humankind a great deal of good by allowing light energy from the sun to pass through, while preventing long-wave heat energy from passing back, thus trapping heat indoors, in the greenhouse and in the solar collector.

The raw material, silica, from which it is basically made, is abundant and only a moderate amount of energy is used in manufacture—even less when it is recycled—so it is environmentally quite acceptable.

In glass fibre form it is hazardous to handle and we advocate using protective clothing—gloves, eyeglasses or disposable respirators—lest skin, eye or respiratory irritations occur. Care must also be taken to keep it out of domestic water tanks.

Timber

Everybody loves wood and there can be no doubt that in its natural state it is a human-friendly building material (we caution you later about the toxic hazards of some preservatives). Being renewable, it is intrinsically environment-friendly too, provided that it is not over exploited.

Obviously the Segal method of building we advocate employs a lot of timber, but curiously only marginally more than conventional construction. Most houses have floors, roofs, internal walls, windows and doors of wood. The environmental score for either is good if the right timber is used and the design extracts full value from it. The right timber has to be harvested from renewable, properly managed resources. Enormous quantities of all qualities of structural and finishing wood can be acceptably produced. It is not always easy to be sure you have found it. At last something is being done about this. Operating under the auspices of the International Forest Stewardship Council, founded in 1993, the Soil Association has introduced a certification system in the UK called WOODMARK.[10]

They say it provides independent assurance to consumers and timber traders that products sold under its Woodmark label have come from forests meeting defined environmental, social and economic criteria, and that the processing chain from the forest to the point of sale has been audited to verify the origins of the certified timber products. By linking timber producers, traders and consumers, the timber certification system supports those who are already managing timber resources responsibly and encourages others to improve forestry practices in order to be certified.

The Friends of the Earth *Good Wood Guide*,[11] and FoE's attempts to establish good practice in the timber trade, are both admirable. Self-builders should be among the first to support them. Brian, speaking personally, suffers remorse because he had no knowledge of tropical rainforest over-exploitation when he made his choice of Philippine hardwood for the Herefordshire house. It now seems that temperate hardwoods from sustainable sources are available to the earnest seeker where softwood is not desirable.

There is a good chance that with better management of our woodlands, a significant amount of locally grown structural timber could be available to self-builders in the near future.

Timber preservatives and paints

The use of timber brings up the question of preservatives. This is a vexed question. In one part of south-east England, building regulations require you to use a preservative against the longhorn woodboring beetle, but its use has become general. The London Hazards Centre, in their book *Toxic Treatments*,[12] suggest that there is collusion between the preservation industry and the timber industry. Everywhere, tremendous pressure is applied to make you feel it is a common-sense precaution to impregnate all your timber with water repellants, insecticides and fungicides. Some of these chemical treatments are highly poisonous and at the very least need careful handling, particularly when the material arrives fresh on site from the immersion tanks, dripping noxious liquid. The poison can enter your body through the skin, as well as being inhaled as vapour. Subsequently there is some hazard from evaporation of certain chemicals over a period of time if a large surface of treated timber is exposed and ventilation is not good. While acknowledging that there is a case for treating structural timber in some circumstances to ensure durability, we see no need to treat decorative timber with volatile preservatives.

The government-appointed Advisory Committee on Pesticides in 1988 ordered a review of Lindane used in DIY timber-treatment products, now known to be a cause of aplastic anaemia. Two other chemicals, PCP and Tributyltin Oxide, are also under reconsideration. PCP is a constituent of some Protim products, whose use in the United States is restricted to outdoor use, as dust from treated timber contains dioxins which remain toxic for years. Tributyltin Oxide, already banned as an anti-fouling paint on boats because it stops all marine life reproducing, is also under reconsideration as a timber treatment.

There are professional DIY timber-preservative products that are bat-friendly, most of them based on permethrin. It is safe to assume these are human-friendly as well.

We also need to be careful with paint. Lead in paint joins lead in pipework and all asbestos products as the most extreme hazards around in building. We are helped by a ruling that has made all commercially produced paint lead-free since 1987.

We urge caution and restraint in the application of all timber preservatives. As Seymour and Girardet point out,[13] Europe has many timber-framed houses that have survived from medieval times without a drop of it. As long as wood is kept dry and well ventilated, neither insect attack nor rot is likely. Such water resisters as are needed can be non-toxic formulations. The German firms Auro Paints and O.S. Color[14] supply non-toxic paints and preservatives that are available in Britain. But the key to timber preservation is good design. Any moisture entering the structure must be free to drain away and good ventilation must be provided everywhere.

Manufactured boards

Plasterboard

About the cheapest board material around is plasterboard and it is extremely useful. It is fairly heavy and limp and therefore an excellent sound insulator, especially if built up in thicknesses. It has good fire-resistance and casings to columns and beams and under floors and stairs can be readily formed with it. Its paper face takes decoration well.

Plasterboard shares the same environmental drawbacks in manufacture as lime and cement products, though it is not harmful to handle. Basically, we don't think you can do without it, but continental plasterboard is to be avoided because, as we mentioned in Chapter 12, it is slightly radioactive.

Asbestos and its substitutes

We *can* do without asbestos. Asbestos sheets are no longer available for fire-resisting linings or corrugated roofing. Alternatives have been developed, made largely from cement and various particles—some wood, some mineral—and often these are reinforced with glass fibre. They seem to be pretty good.

Fibre building boards

Such fibre boards as hardboard, insulation board and medium-density fibreboard are made of materials ranging from wood pulp to shredded sugar cane, some of them simply by felting and compressing the fibres. These are ecologically benign. Some are bound with synthetic resins, and these require more caution. You will have to inquire of the manufacturer which is which.

Chipboard

Much chipboard, composed of wood particles embedded in resin, has been bound by urea-formaldehyde and it is thought that it can give off poisonous gas throughout its life. Avoid chipboard for unventilated kitchen cabinets. Other chipboards that avoid this danger are available on the Continent and may be imported to Britain. Chipboard can give off toxic fumes when it burns, as can all timber, if there are high temperatures but complete combustion does not take place.

Plywood

Allowing that one should avoid breathing the smoke when it burns, plywood is an excellent material with many uses. But the plys are commonly of hardwood, mainly of tropical origin, so we must try and wean ourselves off them for as long as they are not from renewable sources. Temperate hardwood plys are available and are to be preferred, though they are more expensive on our markets. Also good are North American WBP Exterior Grade, which is softwood, and Canadian Douglas fir plywood.

Blockboard and laminboard

Blockboard and laminboard are almost as useful as plywood and the strips of timber forming the board core are usually of softwood.

Woodwool

Woodwool slabs are employed extensively in the Walter Segal method, for the core of internal partitions and external walls and for roof decks. The slabs are made of long softwood shavings coated with cement and entraining air. They are quite cheap and, like plasterboard, are good both for sound and fire insulation. In addition, they have quite good thermal resistance and moderate strength. A little disappointingly, from the ecological point of view they are not, as might be imagined, made from waste wood; the particular chemical composition of the fibres is critical in manufacture and is provided only by a particular species of conifer, some of which is specially grown for the purpose in Scottish plantations. Apart from their cement component, they are not dangerous to manufacture and they are benign in use.

Thermal insulation

The choice of the correct insulation material is surprisingly difficult. The great benefits to the environment of well-insulated buildings can be reduced if the production of the insulants is itself a harmful process, or if they create hazards in use or when they are finished with and have to be disposed of.

Seymour and Girardet approach the problem by simply applying the maxim that natural substances should always be preferred to synthetics. Goodbye, they say, to anything made of plastics or even reconstituted wood. It is an attractive principle, but difficult to follow absolutely.

Pat Borer has a more helpful approach, acknowledging that there are degrees of offensiveness in the products of the chemical industry and yet facing the fact that we cannot be as energy-efficient as we need to be without using some synthetics. He has prepared a table which lists in descending order the 'naturalness' of the range of insulating materials, and records against each one its thermal conductivity.

How to choose the Greenest insulation

The list opposite is in descending order of naturalness. The synthetic materials towards the bottom have the higher insulation value per unit-thickness, are more polluting and more expensive. Many of them are made with CFCs (though substitutes are being introduced) or with formaldehyde, which gives off toxic fumes, particularly when burnt.

At the upper end (apart from cork and rubber, which have special applications) the materials tend

TYPE OF INSULATION MATERIAL	CONDUCTIVITY (λ) measured in Watts per m² per degree Celsius ·01 ·02 ·03 ·04 ·05 ·06 ·07 ·08 ·09 ·10	
ORGANIC: DERIVED FROM NATURAL VEGETATION RENEWABLE SOURCE RECLAIMABLE ON DEMOLITION	CORK	slabs, tiles and granular fill
	EXPANDED RUBBER	pipe sections etc.
	WOOD FIBRE	insulation board
	WOOD WOOL	rigid slabs
	CELLULOSE	loose fill in shredded paper pellet form
INORGANIC: DERIVED FROM NATURALLY OCCURRING MINERALS NON-RENEWABLE BUT PLENTIFUL SOURCE RECLAIMABLE ON DEMOLITION	MINERAL FIBRES	rockwool, slagwool ε fibreglass; quilts ε rigid bats
	PERLITE ε VERMICULITE	loose fill, aggregate for concrete
	AERATED CONCRETE	air-entrained in-situ concrete ε blocks
	FOAMED GLASS	glass in cellular form
SYNTHETIC ORGANIC: DERIVED BY CHEMICAL INDUSTRY FROM FOSSILIZED VEGETATION DIFFICULT TO RECLAIM ON DEMOLITION (check for use of C.F.C.s or urea formaldehyde in manufacture)	EXPANDED POLYSTYRENE	'bead board' - usually white
	EXTRUDED EXPANDED POLYSTYRENE	semi-rigid smooth-skinned closed cell boards, usually tinted
	POLYURETHANE FOAM	closed cell semi-rigid boards ε foam fill
	PHENOL FORMALDEHYDE FOAM	ditto
	UREA FORMALDEHYDE FOAM	ditto
	POLYISOCYANURATE FOAM	ditto

to have lower insulation value, be less polluting and cheaper.

If you cannot find a natural, renewable material to suit your purpose, the mineral fibre insulants provide the best compromise. Among these are the familiar exfoliated rock, the 'rockwool' type and the spun-glass filaments, the 'fibreglass' type.

It appears from first glance that the best insulators cluster at the bottom of the table, but it turns out that they are also far higher priced, so that the good news is that a greater thickness of benign material from nearer the top of the table can give the same degree of insulation at lower cost. You end up paying the same for equivalent insulation—therefore you need only use the synthetic chemicals where there is no room for a bulkier, more benign material.

Pat Borer groups first the organic substances of natural origin from renewable sources. At the end of their useful life they are bio-degradable.

Cork

The only wholly unprocessed natural substance included, it is a lovely material, but because of the limits of its renewable sources, it is quite rightly expensive. The Dutch builders of MW2 at Maaspoort used it to augment the insulation value of their solid timber walls as they liked its human-friendly character so much. The traditional place for cork in this country, as floor tiles, does seem sensible as it is pleasantly warm underfoot in places like bathrooms.

Wood fibre

Good old-fashioned fibreboard (or softboard or insulation board) is comparatively cheap and nice to handle but needs to be used in considerable thickness to provide good insulation in its bitumen impregnated form (trade names Bitvent and Frenit). It makes a useful outside skin to retain the loose-fill insulation in a 'breathing' wall.

Woodwool

Already mentioned as a rigid building and decking board board, it also has moderate insulating value but usually needs to be combined with some lower conductivity material in a wall sandwich.

Cellulose

Tons of unsold newspapers get recycled into a fluffy cellulose fibre by Excel Industries in Wales who treat it with borax against fire and rodent attack and market it as Warmcell loose-fill insulation. Installers bring bales of the material and a blowing machine to site (or you can buy or hire your own), spread it between ceiling rafters and floor joists and inject it under pressure into wall cavities. It fills every crevice and effectively checks wind penetration.

It is an excellent material for use in conjunction with timber frame construction, in which it can be kept dry but is not suitable for brick cavity walls—when wet it turns into mud.

Environmentally it is a winner, being easy to make, composed of waste material, with low embodied energy and negligible toxicity.

Borer then lists inorganic mineral-derived materials. The sources are not renewable but are not yet scarce. The processed materials are generally reclaimable.

Mineral fibres

These are mainly extruded glass fibre—which can be in woven or wool form—and mineral wools, made from slag or rock. They can be obtained loose, in quilts or pressed and bound into rigid 'bats'. Damage to the environment cannot be considered severe, though complaints of skin and eye irritations arising from close contact with manufactured mineral fibres are well known to production workers. We too as self-builders have suffered this irritation and advocate protective clothing, complete with gloves, goggles and masks.

After installation, there is negligible hazard from these materials to house occupants, though it is important to keep it out of water tanks. Vapour barriers will help in preventing fibres drifting into the house in dry construction. Their performance as thermal insulators, in the right application, is very good. If it is true that their manufacture is now (after close monitoring subsequent to the asbestos fiasco) safe and environmentally undemanding, then these are the materials we should choose, second to the natural ones. They are, nevertheless, to be used in moderation and treated with caution.

Perlite and vermiculite

They are expanded vitreous rock in small globules or crumbs and are ecologically unobjectionable, except that they do not occur locally. They are quite useful as loose fill between joists and as lightweight aggregate in plaster to make it fairly insulating, and are inoffensive to humans. Lytag, an aggregate for concrete made from pulverized fuel ash, is comparable.

Foamed glass

This has the environmental character of other forms of glass but is disappointing as an insulator compared with glass fibre and is very expensive in use.

Pat Borer's last category is that of synthetic materials derived from non-renewable fossilized vegetation—

the raw material of the petrochemical industry and by definition finite. Many of them involve very dangerous manufacturing processes which are damaging to the atmosphere. They are very difficult to dispose of at the end of their useful life.

Expanded polystyrene (EPS)

This is also called polystyrene beadboard. It is the least objectionable environmentally of this group of synthetics. It is clean and easy to handle and seems to be harmless to humans until it burns, when it produces toxic fumes. Being so light and cheap, it is attractive to use, though it is more expensive than the environmentally preferable (but nastier to handle) mineral and glass fibre products.

Extruded expanded polystyrene board

This is the smooth-faced, closed-cell material which has the advantage of very low moisture absorption and firm texture, making it very attractive for the upside-down roof layer—say, above bituminous felt and below gravel or grass sods. It is disproportionately expensive for its higher degree of thermal resistivity, but what has ruled it out from an ecological point of view until recently has been the disastrous environmental effect of the use of CFCs in its manufacture. The HCFCs now used are only less harmful to a certain degree.

Polyurethane foam

Available as foam and in boards, it is an efficient insulant but expensive. It is difficult to dispose of, and up till now has been made with CFCs.

Phenol formaldehyde foam and urea formaldehyde foam

These foams, popular with the chemical industry, are not human-friendly. Formaldehyde, a highly poisonous gas, is released very slowly over a long period, and unless ventilation is very good it can build up to hazardous proportions. It is not a good idea to fill the cavity in external walls with it. Fortunately, it is expensive in relation to its insulation value.

Polyisocyanurate foam

This, like the also highly efficient polyurethane foam, is not cheap for what it does, has been made with CFCs and is hard to dispose of.

It is easy to fall for the blandishments of the advertisers and in the field of insulation they can be very persuasive. In this important matter of insulation you need to be on your guard. The simplest and cheapest may well be the best, especially if the origin of the material is local.

Much more information on eco-materials and energy conservation has become available to self-builders since the first edition of this book was published.

Especially useful for those building in timber frame is the twenty page green guide to building materials in Pat Borer and Cindy Harris's new book *Out of the Woods*.[15]

The Association for Environment-Conscious Building[16] was launched in 1989 and has since published three editions of its products and services directory 'Greener Building' as well as a quarterly Journal 'Building for a Future'. It is particularly informative about good forestry products (it endorses the Soil Association Woodmark scheme), on the reed-bed sewage disposal method, on energy conservation and on alternatives to environmentally undesirable plastics products.

There are a number of completed buildings which are themselves ecological showcases of green building such as the new extension to the Horniman Museum in south-east London, by Architype, the Scotton Manor Visitor Centre in mid-Pembrokeshire by Peter Holden Architects, the Bishops Wood Environmental Centre at Stourport by the Hereford and Worcester County Council Architects Dept. and of course the many buildings and demonstration structures at the Centre for Alternative Technology at Machynlleth by Pat Borer.

Chapter 20

Permissions

We discuss planning permission, which is mainly to do with land use and appearance; building control, which ensures compliance with the building regulations; and other approvals required by the service suppliers and the financial institutions.

Satisfying official constraints and obtaining mandatory approvals

While you are engaged in finding a site, obtaining promises of finance and developing the design and ideas for construction, you will also need to have discussions with the various authorities dealing with planning permission, building regulations approval and service supplies and with the lending institutions. You will have to make formal applications to these bodies, accompanied by the relevant documents.

Planning permission

Planning permission is required to build almost all new buildings. Certain works known as 'permitted development' do not require consent. They include some domestic extensions, loft conversions, outbuildings, porches and the like. Refer to *Planning Permission: A Guide for Householders*, obtainable from your local planning department, and to the government's General Development Order,[1] the relevant part of which is reproduced on p.177. Most agricultural buildings are also exempt. In this connection it may be that the definition of what is and what is not an agricultural building is of importance if you are building in the countryside.

Planning law controls what uses land and buildings are put to: the size, position and form of new buildings, the internal arrangement, together with what buildings are made of and what they look like externally. Consideration will be given to the effect of your proposal on the owners of adjoining properties: will their rights to privacy and light be unreasonably curtailed? Will your proposed buildings be capable of being properly serviced with drains and water? Does your proposal deal adequately with the motor car; is there space to park off the highway, is your new driveway a danger to traffic? Is your proposal detrimental to the quality of the neighbourhood, is its scale similar, are the materials proposed sympathetic to the surroundings, is its form appropriate in its location? What provision is there for landscaping the spaces around the buildings?

Planning policies have been drawn up in each area to provide standards against which proposals are measured. These policies will show what uses are permitted in which areas, what intensity of development will be permitted in a particular location, expressed as a density measured in habitable rooms per acre, how many cars have to be provided for, what minimum distances are required between windows that face each other, and other things besides. Particular attention will be paid to the appearance of proposals that fall within or even near a conservation area—that is, an area designated as having a quality of environment that is worthy of being conserved and enhanced. You also need to obtain permission before you can demolish a building within a conservation area and you are required to give the planning authority six weeks' notice before felling, lopping or carrying out any other work to trees in a conservation area. In addition, other trees or groups of trees of particular quality may be subject to a Tree Preservation Order, in which case consent is again required to carry out any work to the trees.

Outline and full planning permissions

You may apply for outline permission, which establishes the basic principles only—the type of use and number of dwellings, for example—prior to full planning permission. An outline permission will refer to 'reserved matters' such as detailed design, the materials to be used, which will be subject to a subsequent detailed application. After a full permission has been granted, reserved matters, such as the type of brick to be used, may also be left for subsequent agreement by the planning authority. The advantage of applying for outline permission to begin with is that you can establish whether in principle the authority is likely to approve the development you have in mind without the time and expense of making a detailed application. This is particularly useful if you are proposing a potentially contentious scheme—to build on land which is not zoned for residential use, for instance. If considerations of this sort are not important, it generally saves time to make a detailed application straight away. Planning permission applies to the land to which it relates and not the person making the application. In this way sites are often sold with planning permission. You can make a planning application on land which is owned by someone else provided that they are notified. This would be the situation if you were thinking of buying a site without planning permission.

Planning permission generally expires after five years and unless it is either acted upon or renewed within that period, it will lapse. It should also be noted that renewing a permission need not necessarily be just a formality. Planning considerations may well have changed during the period in question to the extent that permission would no longer be granted for a particular development.

Planning permission is granted by the planning committee of the district or borough council or development corporation. The committee is composed of lay members of the council but they will most often follow the advice of the professional planning officers employed by the council. For this reason it is always advisable to consult those officers before submitting an application. They will advise you of the relevant planning policies and will tell you if there are likely to be any difficulties in approving your proposals. They are not generally in a position to give firm commitments and will say only something like, 'I would anticipate being able to recommend this proposal for approval', or, 'This scheme would not normally be expected to be approved.' The exception to this is that the officers will probably have 'delegated powers' to deal with certain detailed matters such as the approval of the colour of materials to be used, in which case they should be quite straightforward about what they will and what they will not accept.

The planning application

To make a planning application you need to submit:

- ▷ at least four copies of the application form;
- ▷ a certificate that states that you either own the land concerned or, if it is owned by someone else, have notified them;
- ▷ a location plan at a scale of 1:1250 showing the position of the site;
- ▷ a site plan usually at a scale of 1:200 showing the position of the proposed buildings and road access, together with the treatment of the area around the buildings;
- ▷ plans and elevations at a scale of at least 1:100 showing the materials to be used;
- ▷ a perspective view showing the new buildings in relation to the other existing buildings around (this is not vital but it is often a good idea to include it);
- ▷ last but not least, a cheque for the planning fee; this is currently £140 per new dwelling, so you can see that the initial cost can add up to thousands for a scheme of twenty or so houses. For this reason it is important to have established as far as possible that the proposals will be acceptable beforehand.

Planning applications should be dealt with within a period of eight weeks. In some of the larger city authorities it can take much longer, currently up to six months in some London boroughs, so it is important to get your application in as soon as you can.

If your application is turned down, you have the opportunity to appeal to the Secretary of State for the Environment against the decision of the local authority. You can also appeal if your application has not been resolved in one way or another at the end of the statutory eight-week period. The chances of success are slightly worse than evens and it takes at least six months. You need to fill in a form and make a statement of your case. You can opt to have the matter dealt with entirely on the basis of written submission or by a hearing before an inspector appointed by the DoE.

Obtaining planning permission is a vital and necessary step in any building project. There can be quite delicate negotiations to be undertaken, which may mean playing one point off against another—reducing the number of dwellings, for instance, but making them simpler and therefore cheaper to build, thus offsetting the increased cost of land. It is also the case that most planning officers will react against a scheme that is presented in an amateurish manner. For these reasons it is advisable to employ a professional person, generally an architect, in these matters.

Building regulations approval

All but the smallest outbuildings (detached outbuildings below 30m² (325 square feet) are exempt, for instance) are required to conform to the building regulations.

There is a choice of procedure under the regulations which is between what is known as the 'full plans route' and the 'building notice route'. The full plans procedure involves getting all the details of the proposed building down on paper and approved before you start. Under the building notice procedure all that you are required to do is give the building inspector a minimum of three days' notice of when you intend to start building and then all matters will be dealt with as they occur on site. This latter situation requires that you are very confident of your ability to conform to the regulations and is not recommended for self-builders. The only time it may prove useful is if you need to get a very quick start on site.

The regulations control most aspects of building and are divided up into parts that cover:

▷ means of escape in case of fire;
▷ materials and workmanship;
▷ structure;
▷ fire;
▷ site preparation including soil contamination and resistance to ground moisture;
▷ toxicity of cavity insulation;
▷ insulation against airborne and impact sound transmission;
▷ ventilation and condensation;
▷ hot-water storage and sanitary and washing facilities;
▷ drainage and waste disposal including waste pipes, cesspools and rainwater;
▷ heat-producing appliances;
▷ stairways, ramps and guards;
▷ conservation of fuel and power;
▷ access for disabled people.

The regulations themselves are framed in a relatively general way. For instance, it is a requirement to protect the building against dampness in the ground. There are, however, what are known as 'deemed to satisfy' clauses, which describe in detail how that can be achieved and which specify one or more types of construction that are approved in this situation and which are 'deemed to satisfy'. A complication is that the regulations refer in many instances to the hundreds of British Standards that apply to building and copies of the British Standards are expensive to buy; the full set of documents referred to in the regulations costs many hundreds if not thousands of pounds.

It is a good idea to go and see the building inspector before submitting an application (but remember they are usually in their office only up until ten o'clock or so in the morning and after four in the afternoon as they are out on site the rest of the day). They are very knowledgeable about what is in the regulations and will advise on points to be dealt with. Building inspectors will sometimes be sceptical of the abilities of non-professional builders to begin with, but will usually prove to be very helpful once they see that you mean to do things properly. They are also often very experienced in building generally and can be of great help to self-builders. One hears stories of building inspectors going way beyond the call of duty to help out. Our experience in Lewisham was no exception; the drainage inspector was found in his wellies one Saturday morning, helping to construct a particularly deep and difficult manhole on one of the self-build sites.

To make a building regulations application you need:

▷ a completed application form;
▷ a site plan showing the building in relation to the boundaries of the site; it may also be used to show the layout of the drainage system;
▷ plans of each floor level of the building at a scale of 1:100, 1:50 or 1:20;

- ▷ elevations showing the window sizes and the amount of window that opens for ventilation;
- ▷ details of the principal elements of the construction: roof, walls, floors, foundations, stairs, flues, drainage;
- ▷ specification of the materials to be used;
- ▷ calculations of the structure for any elements that are not entirely standard (the regulations give 'deemed to satisfy' requirements for many elements such as floor joists, rafters, purlins);
- ▷ calculations showing the thermal performance of the building. A revision of the Regulations in July 1995 radically changes the requirements with regard to the thermal performance of buildings. It is mandatory for all new dwellings to have an energy rating, calculated using the Standard Assessment Procedure (SAP). This is similar to current voluntary assessment schemes such as NHER and Starpoint. It is based on the projected total annual hot water and space heating costs and includes standing charges and the energy consumption of pumps and fans. The SAP rating varies from 1 for a very poor energy performance to 100 for a very energy efficient building. Most new houses built to the current Building Regulations standards achieve around 70. The Housing Corporation require that all new housing association housing has a SAP rating of at least 75.

The SAP rating is relatively complex to calculate and takes into account factors such as the type of heating system, insulation, size and type of glazing, orientation, ventilation system and type of hot water cylinder and heating controls. You will probably need assistance to calculate the SAP rating.[2]

This new system will encourage the design of more energy efficient buildings and will enable you to make more informed choices. There are some limitations to the SAP rating, however. It does not take into account the energy used for lighting and power or the CO_2 emissions of different fuels. It is more difficult to improve insulation standards than to upgrade a heating system and so one would expect insulation standards to have a higher weighting. Location is not taken into account which ignores the increased benefits of higher insulation standards the further north you are or the advantage of increased areas of south facing glazing in the generally sunnier, milder south of the country.

- ▷ calculations which show the position and size of 'unprotected areas' (these are the windows, doors, other openings and any areas of the walls that do not achieve the minimum degree of fire resistance that is required taking into account their distance from the 'relevant boundary') to ensure that there is not a danger of fire spreading from one building to another (this is a complex procedure and you will have to refer to the regulations to see exactly what is involved);
- ▷ and, of course, a fee; this is calculated on a sliding scale based on seventy per cent of the building cost, which will have to include a notional amount for the value of your free labour, and is payable in two parts, the first on submission of the plans, the other as soon as the site inspections are due.

After submitting the documents you may receive a letter asking lots of questions like, 'How do you propose to conform to clause 4.6 subsection (II)?' Don't worry. Nobody can get it all right first time around. Answer the outstanding points and approval of the plans should follow within the statutory period of five weeks. Some of the large city authorities are not able to approve within this period. In this case there is little you can do except to grant an extension of time, as not to do so will result in your being issued with a 'deemed refusal'.

The approval will be accompanied by a handful of cards to be dispatched to the local authority at certain key stages in the construction process; for instance, when the foundations have been excavated. The building inspector will visit at these stages and may make random visits in between.

Preparing the drawings, specifications and calculations necessary for a building regulations submission and dealing with the queries that arise are complicated matters and are generally best dealt with by a professional person.

Other approvals

These might include:

- ▷ Registration of your self-build home or group development for cover under a structural insurance scheme. This would cover you against structural defects that arise during the first tern years of the life of the building. The most widely known is the Warranty issued by the National House Builders Council (NHBC). This is only available to registered builders and developers, not generally to self-builders. The NHBC will consider registering a self-build group if it can prove that they have professional builders amongst the membership and/or that they are under the direction of an experienced contracts manager. Each case is considered on its merits.

However, ten year cover is generally available to self-builders under the 'Custom Build' scheme run by Zurich Municipal. This replaces the 'Foundation 15' scheme run by Municipal Mutual who have been taken over by Zurich Municipal. If you are developing a self-build project with a housing association, they may be members of the Housing Association Property Mutual (HAPM). Those associations that are members can get cover for any self-build projects that they develop. The general procedure with these schemes is that you submit full details including drawings, specifications and calculations. They will be evaluated for any risky or unsatisfactory construction by a surveyor who will comment on any amendments that are required. When the job is under construction, their surveyor will make periodic visits to satisfy himself that the construction is being carried out in accordance with the design. There is quite a lot of additional paperwork involved but you may find the reassurance worthwhile.

▷ approval from the funding authority. They will be looking to see that the design and layout that you propose has a value when complete that is large enough to ensure that their loan will be secure. They will need a site plan and plans and elevations of the dwellings.

▷ agreement to your proposals from the Electricity Board, Gas Board, Water Board and British Telecom. First, send them a location plan of the site and ask them to send you details of their existing services adjacent to the site. You will then be able to work out the best route for your new service. You will then have to request a quotation for supply, enclosing your proposals.

Electricity

Generally speaking the Electricity Board will lay their main cables about 450 mm (1′6″) deep, in a trench dug by them. You will need to provide a duct into the building to the 'intake position' where the cable comes into the meter for the individual house supply. This duct will be 100 mm (4″) in diameter and will have to have an easy bend as it comes up from the ground and be provided with a draw rope to pull the cable through later. You will also need to provide a duct where the cables have to pass under any road that you may be constructing. The Electricity Board will provide a domestic service for a charge, currently between £170 and £250, unless there are special circumstances—say, if the cable has to come on poles from six fields away. The Electricity Board will not lay a cable while the site is still liable to be excavated for any reason and neither will they lay a cable into a building until it is secure and weatherproof. For this reason you will probably need to get them to provide a temporary builder's supply whilst you are building. This can cost up to several hundreds of pounds and requires a secure, weatherproof, masonry enclosure. You may be able to come to an arrangement with a helpful neighbour to take power from their meter and pay them back for it. Before the Electricity Board will connect your supply and install a meter, you will have to submit a signed test certificate certifying that the installation has been tested and that it conforms to the current edition of the Institute of Electrical Engineers (IEE) regulations. You may want to get a qualified electrician to test the installation to be sure that it is properly installed.

Gas

The Gas Board similarly lay their own pipes about 450 mm (1′6″) deep, in a trench dug by them. The Gas Board have standard requirements about how the pipe enters the building so that it can be replaced if necessary and they will provide details of these arrangements. If you are likely to be using a substantial amount of gas over the years—installing gas-fired central heating, for instance—then the Gas Board will put in a new domestic supply for no charge. There is a small charge if you want the Gas Board to test the installation of pipework in the house before they connect the supply. The installation will have to conform to the Gas Regulations and it should be possible to obtain a copy of these from the Gas Board. They cover such matters as the position permitted for the installation of gas flues and the ventilation requirement for gas appliances. The new regulations require gas installations to be carried out by a 'CORGI' registered fitter.

Both the Gas and Electricity Boards are encouraging people to install all new meters in a position where they can be read from the outside of the building. Both supply ugly external meter boxes free of charge for building into an external wall for this purpose. You can devise your own meter box but remember that a gas meter must not be in the same enclosure as an electric meter and that a gas meter enclosure must be ventilated for reasons of safety.

Water

The Water Board will bring a new service pipe as far as the boundary of your property, where it will be terminated with a stop valve about 750mm (2'6") below ground level. You will have to excavate and lay your own supply pipe from this point at least as deep, to avoid it freezing in winter. It is a good idea to save digging by laying the water pipe in the same trench as the drainage if this can be arranged. The water supply industry now charges a substantial amount for providing a service connection—maybe £1200 to £1500 including infrastructure charges for water and sewerage, even if the water main is immediately outside the site. Every house has to have its own separate supply pipe to the boundary. This can get expensive if you have a dozen houses or more. In this case it may be a good idea to get the Water Board to put a 'mains extension' in on the site, with a short supply pipe to each house. This may cost a few thousand pounds but could be cost-effective and save work.

Telephone

There are now two principal networks and there may be more in the future. Everywhere will be served by British Telecom but you may wish to check if Mercury have laid ducts in your area.

If you are in an area with an underground telephone service British Telecom will supply you free of charge with sufficient ducting to get from each house to the 'joint box', where they connect into the main cable. You will have to lay this duct with a draw rope in it about 450mm (1'6") deep. If you are in an area with an overhead service, you do not need to make any particular arrangements for the telephone installation. British Telecom lay their cable and charge when the telephone itself is installed.

Cable TV

Many cities now have ducts for cable TV laid in the streets. You may wish to take advantage of this service in which case you may want to incorporate a duct into the house. Contact your local cable TV company for their requirements.

With an approved design and arrangements made for the supply of services you are at long last ready to build.

GDO 1995: Development within the Curtilage of a Dwellinghouse — Permitted development:

E. The provision within the curtilage of a dwellinghouse of any building or enclosure, swimming or other pool required for a person incidental to the enjoymernt of the dwellinghouse, or the maintenance, improvement or other alteration of such a building or enclosure.

Development not permitted: E.1 Development is not permitted by Class E if—
(a) it relates to a dwelling or a satellite antenna;
(b) any part of the building or enclosure to be constructed or provided would be nearer to any highway which bounds the curtilage than—(i) the part of the original dwellinghouse nearest to that highway, or (ii) 20 metres, whichever is nearest to the highway;
(c) where the building to be constructed or provided would have a cubic content greater than 10 cubic metres, any part of it would be within 5 metres of any part of the dwellinghouse;
(d) the height of that building would exceed—(i) 4 metres, in the case of a building with a ridged roof; or (ii) 3 metres, in any other case;
(e) the total area of ground covered by buildings or enclosures within the curtilage (other than the original dwellinghouse) would exceed 50% of the total area of the curtilage (excluding the ground area of the original dwellinghouse); or
(f) in the case of any article 1(5) land or land within the curtilage of a listed building, it would consist of the provision, alteration or improvement of a building with a cubic content greater than 10 cubic metres.

Interpretation of Class E. E.2 For the purposes of Class E 'purpose incidental to the enjoyment of the dwellinghouse' includes the keeping of poultry, bees, pet animals, birds or other livestock for the domestic needs or personal enjoyment of the occupants of the dwellinghouse.

Chapter 21

Bringing the strands together

At long last, after all the weeks and months of the planning stages with the uncertainties and possible frustrations involved, you are now in a position to start the practical work. This is the enjoyable part, seeing the results of all your efforts so far taking shape as you work. Before you can start, you will need to bring all the strands of the planning process together and plan the project on site.

Project planning

A successful building operation depends on good organization. Your own morale, and the morale of a group, will be able to stand up to the long weeks of hard work ahead if things are properly planned, if materials and the tools for the job are to hand when they are needed, money is available to pay for things and people are clear about what they have to do.

Programme

You should set out a programme for the work, with target dates for completing each stage of the building. It need not be a complicated, critical path diagram. It simply needs to list the operations that have to be done and the order in which they need to be done and give a forecast of how long each of them will take. Do not be over optimistic in your estimates of time; things have a habit of taking longer than you think. This programme must show when you need to order the various materials so that they are on site when they are needed. It will enable you to give notice of when subcontractors, statutory authorities and the building inspector will be needed on site. You will be able to measure actual against planned progress and be able to see if some operations are behind target and others can be carried out ahead of programme so that the overall construction period is not extended, with the additional financing costs that this would incur.

Working methods

You will have to decide on the working arrangements on site. Are you going to have people nominated to be in charge? Are you going to work in teams? Are you going to build with a sequence of building trades? Are people with particular trade skills going to be responsible for those parts of the work? If you are building using the Segal method in particular, there is a balance between group working and individual working which will need to be decided on. This aspect is discussed for a Segal building programme in the next chapter.

Finance

You will have to finalize a schedule of payments from your source of funding. You may be required to prepare a cash flow chart showing the anticipated expenditure month by month. Agree the procedure for obtaining certificates and presenting them for payment.

Suppliers

You will need to organize obtaining materials. You should by now have obtained competitive quotations for the supply of materials. You will need to open accounts with suppliers and negotiate a discounted price that is based on the size of the order. If you are building a number of houses, this could be as much as £500,000. Merchants will do a lot to secure an order of this size. Also important in this discussion

SELF BUILD **BUILDING PROGRAMME**

Aug-90 REVISION

WEEKS	OPERATIONS	MATERIALS	EXTERNAL CONTRACTS	NOTES
1		Fencing, cement, ballast	Subcontract for road base, kerbs,	Site accommodation in garage.
2	Establish boundaries	Blocks, sand	main drains & service ducts	Constructed in Segal method
3	Construct temp. meter enclosure	Timber, panels		as a training exercise
4	Establish site accommodation			
5				
6			Temporary electricity & water supply	Site access and temporary services
7	Setting out	Cement, ballast		from this point
8	Foundations			Divide into 5 teams of 5 households
9			Building inspector	
10		Shingle		
11	Oversite	Structural timber & bolts		
12	Framing		Building inspector	
13		Lead		
14			Erect frames	
15	Joists and bracing			
16				
17		2nd fix timber		
18	Fascias			
19		Sterling board		
20	Roof deck			
21		Timber stain	Roofing subcontract	
22	Roof capping			Under cover from this point
23		Pipe & fittings		
24	1st fix plumbing			
25		Insulation		
26	Fix floor deck			
27		Board materials		
28	Prepare battens & plasterboard			
29	Fix external walls			
30				
31				
32				
33		Joinery timber		
34	Fix door & window frames			
35				
36				
37		Doors & ironmongery		
38	Fit external doors		Measure & make windows &	
39		Windows & double glazing	double glazing units	
40	Fit windows			
41		Cable		Weatherproof & secure
42	1st fix wiring			from this point
43	Fix internal walls			Divide into single households
44				at this point
45				
46				
47	Fix internal door frames			
48		Boiler & radiators		
49	2nd fix plumbing			
50				
51				
52	Fix ceilings		Test gas & connect	
53				
54		Electrical accessories		
55	2nd fix electrics			
56		water pipe, drains, ducts		
57	House drainage, phone duct		Test electrics & connect	
58	& water service pipe			
59				
60		Kitchen units & sanitary ware		
61	Fix kitchen & bathroom		Test drains & water & connect	Heating on from this point
62				
63				
64				
65	Stair			
66				
67				
68	Fit internal doors			
69				
70	Fix shelves & cupboards			
71				
72	Fix verandah & steps			
73				
74		Floor finishes		
75	Floor finishes	Paving		
76	Path & carspace			
77		Fencing	Connect phones	
78	Plot fencing		Subcontract for road finish	

is the question of prompt payment of accounts, which can be worth another two and a half per cent. Just as important, and not always given enough weight, is the question of service. Will they deliver at weekends? Do they have trucks with a crane offload facility? Will their accounts department be sure to identify which house a particular invoice relates to so that you can sort out the final account at the end? You will probably need:

▷ a timber merchant;
▷ a general builder's merchant, who will often also supply the timber;
▷ a plumber's merchant;
▷ an electrical supplier;
▷ a window supplier and glass merchant.

Get recommendations if you can and check with their customers. You may decide that you can obtain some materials direct from the factory at a better price. These may include:

▷ aggregates;
▷ fixings;
▷ paint and timber stain.

You will need to set up a procedure for placing orders. Some groups nominate one person to be responsible for this. It could be a group member, or the foreman or book-keeper if you have one. Make sure that all orders are confirmed in writing and keep a copy. This can save a great deal of confusion later. You may decide to have a simple order form that clearly identifies which house the order relates to. Agree a programme of deliveries with the suppliers and set up a procedure for receiving deliveries. Make sure the delivery notes are marked up with shortages and damages and returned to the person responsible for ordering.

Plant and tools

It is nearly always worth buying plant and tools and selling them on completion rather than hiring them. This may represent too big an investment, in which case you will have to hire what you cannot afford to buy. It is worth buying good-quality tools, as they last longer and are easier to use. If you do hire, make sure that they are not sitting around the site during the week unused and clocking up hire charges. A powerful electric drill and saw are essential tools and should be 110-volt equipment for use in the open. You will have to hire a transformer until you are working under cover. The new battery-powered drills that are now on the market are really good, powerful and convenient tools and they can operate as a power screwdriver as well; a really good investment at £100 or so. Also good if you can afford it is a radial-arm saw. This really adaptable woodworking tool can rip-saw and cross-cut at any angle and can make rebates and grooves with great accuracy; another good investment if you can afford it at £350 or so. A good strong yet light ladder is essential and well worth buying to do the maintenance from in the years ahead.

Subcontracts

You will also have to place orders for parts of the work to be subcontracted. These may include brickwork and plasterwork if you do not have skilled members and possibly the roofing, main drainage and road works. Working on pitched roofs is dangerous for the inexperienced and laying most flat-roofed materials requires special equipment, such as a tar boiler. Drain laying is not easy for the inexperienced and is again dangerous if you are working in deep trenches. Road laying is not so much a problem if you hire an excavator with its driver, but some groups reckon that it is worth getting the road base (but not the finish, which can get ruined during the course of construction) constructed by a contractor as quickly as possible so that you can get materials delivered to the house plots easily. Obtain competitive quotations and give a written order, setting out exactly what is to be done, when it is to be done and how much it is to be done for. Obtain recommendations where you can and double check.

Services

You should have already obtained quotations for supply but you will need to liaise with the statutory authorities to verify their requirements and agree a programme for their work. Place written orders and organize any temporary supplies that may be required.

Building control

You should have received a set of cards with your building regulations approval for notifications of the building control authority at certain key stages of the construction—to inspect the foundation trenches before placing concrete, for instance. Nevertheless, it is a good idea to contact the building control department at this time to agree procedures.

Safety

Building sites are potentially dangerous places. Some of the commonest causes of accidents are falling from an improvised working platform, the collapse of inadequately shored excavations, things dropped from a height on people below. Keep the site tidy and dispose of rubbish as you go. Make sure that you wear strong boots that will stop a nail from going through your foot in a way that trainers will not. Wear a mask and goggles against dust, a hard hat against things falling on your head and gloves against the toxic chemicals used to treat timber. Make sure there is a first-aid box. On a larger site appoint a safety officer to ensure that these precautions are taken. The factory inspector can inspect your site like any other. Remember that it is when a serious accident has already taken place that the questions get asked.

Insurance

You should have already obtained quotations for insurances. You will need the following cover:

- ▷ 'contractor's all risks', which insures the building, tools and materials against fire, theft and damage;
- ▷ 'employer's liability', which covers the members of a housing association or any labour-only subcontractors;
- ▷ 'public liability', which insures you against any death, injury or damage arising out of the works.

You may also decide to have health, personal accident and life cover in case unforeseen circumstances delay completion while debt charges are still mounting. You may also be required by the funding institutions to have fidelity guarantee cover. The Norwich Union offers a complete insurance package for self-builders.

Book-keeping

Set up a simple book-keeping system so that you can keep track of the money. If there is more than one house, it must identify which building the expenditure relates to. You will have to keep VAT records and set up a system for dealing with petty cash items. Make arrangements for having the accounts audited. Decide if you are going to make a group member responsible for keeping the books or whether you want to employ a part-time book-keeper.

Supervision

The architect will visit the site from time to time and ensure that the construction is satisfactory from a technical point of view. He or she should be able to offer advice and help sort out problems. The architect will not be available at all times that people are working on site and for this reason you may decide to appoint someone who is technically competent to direct operations. The funding institutions may insist on such a person. It could be a group member if they have the relevant experience or you might appoint a clerk of works or site agent. There are agencies that supply staff to the building trade who have such people on their books. You need someone who is experienced and if the Segal method is chosen, it is an advantage if there is someone available who has knowledge of Segal construction and experience in a training role. They will be working part-time and at the weekend. In any case, agree a job description, working hours and conditions of employment. Arrange interviews and appoint as early as possible so that they will be able to assist you in the project planning stages.

Training

You may decide to arrange for some training before going on site. A number of adult education institutes and technical colleges around the country run evening classes in DIY skills. Some self-build groups have organized their own specialized course of this kind. A series of, say, twelve evenings might cover going through the building process with slides, a detailed examination of the documentation, using hand and power tools to gain confidence in their use before getting on site, and learning the basic techniques of plumbing and wiring. The Centre for Alternative Technology[1] has run a course based on Segal-type construction over a long weekend for many years. These courses combine theoretical work with a chance for some practical 'hands on' experience. You cannot beat learning by doing. In our experience the things that people seem to find hardest are the plumbing and the drain laying. (Both operations, however, are easier then they used to be with the introduction of plastic pipework and drainage systems.) In a group, people learn from one another to a great extent and you may decide that formal training is not necessary. What is more important is that people have the right attitude of mind; confident without being over confident, careful, methodical, able to think things through in advance and able to plan ahead. A successful building project of any kind relies on planning ahead.

Work on site

Once the preparations are made, you start the actual work. We do not describe this here as there is such a variety of building techniques, information on which can be obtained from the large number of books available about conventional building methods. (The next part of the book, however, describes the construction of a Segal building in detail.) One of the most important things is to keep the flow of materials and money to pay for them in step with the progress of the work on site.

Completion

When the house is finished you will have to:

▷ arrange for services to be connected;
▷ obtain a certificate of practical completion from the architect that the house has been constructed in accordance with the drawings and specifications;
▷ arrange house insurance.

If you are building as a group, you will probably have an arrangement whereby each house can be occupied under licence as it is finished. When the last house is completed you will need to:

▷ arrange for the legal formalities of transfer of ownership to be drawn up;
▷ prepare a final account and settle all outstanding bills;
▷ obtain long-term mortgage finance and pay off the development loan;
▷ obtain a final audit of the housing association books and wind up the association, unless you are organized as a co-operative, in which case the co-op will continue to exist to hold the group mortgage and charge rent to the self-builders.
▷ set up an on-going management structure if appropriate, to be responsible for the common parts.

Then you can organize the party!

Maintenance

You will have to take time out every once in a while to carry out the maintenance. In particular external timber will have to be painted or stained between every three to five years, depending on its exposure. Regular maintenance is most important; given regular attention a building will last indefinitely.

Part Four

The Segal method

In earlier Chapters (2, 5, 7, 18 and 19) we have made frequent reference to Walter Segal. Here we explain why we advocate his method and describe the construction of a Segal-method building in detail. We believe this method has much to offer self-builders and there is little information available elsewhere. The documentation for a Segal building is examined, together with alternative working methods and the tools required.

Each part of the construction is documented and we give general step-by-step instructions of how to construct each element. These instructions are of a general nature here so that they will be relevant to a range of different situations. Because the Segal method is an approach to how to build and not a closed system of construction, a great many variations are possible within the basic principles—no two Segal buildings need to be the same, so it may be necessary to prepare instructions specific to each situation.

Those operations involved in building a Segal house which are the same as for building any other form of house—the electrical installation, for example—are not described in detail here and you should refer to a standard work on these aspects of house building.

Chapter 22

An introduction to the Segal method

The basic features of the Segal method and how it has developed as a way of building with real advantages for the self-builder, and for other applications. We describe the advantages and limitations of this form of construction.

Walter Segal's insight was to see that the proven advantages of truly traditional timber building could be retained using the materials available today if the construction was properly detailed. Not only did he devise techniques that discouraged water from getting into the structure but he ensured that if it did get in, it could always get out again, so that the well-ventilated timber would quickly dry out without harm. He proved the value of this approach in 1962 with a little building in his garden. The main task was to rebuild (in brick) his house in Highgate, north London. Meanwhile, his sizeable family had to be temporarily accommodated, so he decided to build, very cheaply and quickly, a temporary house on his back lawn. He made this ingenious, compact and convenient little house a test bed for his idea that time, trouble and money could be saved if the materials used were readily available on the market and were incorporated into the building as far as possible in their 'as manufactured' state, without the subsequent processing that adds cost to prefabricated components.

He therefore devised a dimensional grid for the timber frame within which these straightforward materials in their basic uncut sizes could be combined, using simple, dry-jointing methods such as bolts and screws. The result was dramatically cheap—it cost £835, including two carpenters' labour (adjusted to 1988 prices, the equivalent of £6,500). It was equally successful as a demonstration of his method of sound, practical detailing—the house still stands over thirty years later in the garden of the rebuilt house.

News of it leaked out, and a succession of private clients in England and Ireland, impressed by the economy and practicality of this approach, commissioned Segal to build about two dozen or so such houses over the next fifteen years. Considerable interest has been aroused abroad; a student's residence based on Walter Segal's ideas was designed and built by the students at Stuttgart University in 1983 and a Segal house was even designed and built in Australia after two long telephone calls to Segal in 1980.

The Segal method and self-build

It was in 1971 that one of Segal's private clients, a school teacher, decided after watching the carpenter recommended by Segal start work on his new house that the work looked relatively straightforward and he could do it himself, thus saving a lot of expense. He telephoned the architect that evening and told him that he had asked the carpenter to leave the job and was that all right. Subsequently other individual clients built their own houses in this manner.

Once it had been demonstrated that his method of building could be mastered by someone without any previous building experience, Walter Segal was keen to exploit the potential for small groups of families building their own economical and practical houses designed for their own particular needs. He was able to see this come to fruition with the self-build schemes in Lewisham described in Chapter 7. The coming together there of council-owned sites, a suitable financial arrangement and an appropriate way of building enabled ordinary people with relatively low incomes and, most significantly, no particular building experience or skills to design and then build their own houses. Experience has shown that anyone with sufficient commitment can build a house for

Interior of Coin Street

themselves in this way; houses have been successfully completed by a single mother and by a number of couples nearing retirement.

Since Walter Segal died in 1985 a number of buildings have been built by contractors at a speed and a cost that demonstrate the inherent economy of the approach. The Coin Street Design Centre was built for The Coin Street Community Builders, which is a community group which came together to develop a large inner-city site in London for co-operative housing. The building provides meeting rooms and office spaces, together with the associated storage, kitchen and toilet accommodation. The building, which has an area of 135 m² (1440 square feet), was constructed in just eight weeks in the early part of 1988. It was built by a small team of builders, two of whom had built their own houses on the Lewisham self-build scheme and who are an example of the way that the experience gained through self-build can often open up new horizons. The cost, including labour, was less than £50,000 at the time.

A meeting hall was built in Camden, North London, for the tenants of a large council estate completed in the 1970s but not provided with any community facilities. The hall was built for the local authority as part of a small development of four houses for homeless households adjacent to the estate. It is used by a playgroup, an old people's lunch club, the tenants' association and others. There is a hall with the associated kitchen, storage and toilet facilities. The building is 105 m² (1130 square feet) and cost £44,000 to build at the end of 1987. The client in this case was a local authority housing department and regrettably the users of the building did not have any say in its design. The contractor was a conventional commercial building company, albeit with one director with an interest in timber construction. They completed the development of four houses and the community building in eight weeks. Impressed by the speed and economy of construction, they have gone on to build other buildings using the Segal method, including a museum building in Kent which is three

Calthorpe Institute Kings Cross: A community complex built in the Segal method of the 1990s

times the size of this building. The method has been developed to incorporate energy saving and ecological design principles.

The Segal method is an approach that suggests how to build rather than a system of building. It is an attitude of mind based on a rigorous simplification of the whole building process, including design and documentation as well as the actual processes on site. It includes a radical rethinking of common problems in building construction, such as control of condensation in the thickness of the construction, how to produce a flat roof that does not fail and so on. The notion of a way of building based on using mass-produced building products which is yet simple so that ordinary people can design and then build their own houses suggests a vernacular form of building appropriate for our times. It is not imposed for stylistic reasons but rather stems from the basic products and skills of modern industry, understood by and under the control of ordinary people. It suggests a style of building more appropriate to the twentieth century than the pseudo-vernacular product of a wholesale rejection of the reality of today for some romantic vision of the past.

The main advantages of the Segal approach are:

- ▷ The construction process is *simple*: the basic carpentry skills involved can be quickly mastered by someone without any previous building experience.
- ▷ The construction method is *economical* for two main reasons: the cost of foundations is much less than most buildings and the need to employ costly subcontractors is all but eliminated because the building method does not require any input of skilled labour beyond possibly a roofer.
- ▷ The construction is relatively *quick*: site work is reduced by using building materials in their basic stock size as far as possible—straight off the lorry and into the building.

- ▷ The building is supported above the ground on posts, which reduces the extent of foundations necessary and removes the need to level the site. This makes the *ground works manageable* for self-builders working without heavy machinery.
- ▷ The construction is *economic on difficult sites* with steep slopes or poor ground conditions. Trees can be kept undisturbed right up to the building.
- ▷ Because it is possible for a family to master all the aspects of building their own house from start to finish, it is possible to organize a group so that *each self-builder is individually responsible for building his or her own house*, coming together only to carry out the common parts such as roads and drains in a co-operative manner. This more individual building process means that you do not have to implement the onerous regime of strict working regulations with set working hours and a system of fines to enforce them, which is the normal way of organizing a self-build group. This in turn avoids many of the pressures that self-builders are under. Women and children can take part rather than be excluded as is usual. Each household can proceed at a pace that suits their particular lifestyle and other commitments and can achieve a level of workmanship that suits them.
- ▷ Another aspect of this is that each house in a group scheme can be of an *individual design* without causing difficult issues to do with some members benefiting more than others from the group work effort. The resultant variety (within the discipline of the method) seems to us to be entirely appropriate to a democratic way of life and preferable to the sterile uniformity of so much modern housing.
- ▷ Because the walls are formed of panels of standard size arranged on a modular grid and do not support the weight of the building, it is quite easy to move them and rearrange the layout of the building. This is in contrast to typical masonry construction, which, because the walls hold up the building, is rather difficult to alter. A Segal house is very *flexible and adaptable* and can be altered to suit the changing needs of the household as children come and go, and grannies too. One couple in a Segal house had a party. Walter recalled that, 'All the wife's family from the Auvergne were coming. They could have hired a room of course, but instead took out all the partitions in their house, stood them against the wall and had the whole space free. Filled with tables and so forth, the whole family could celebrate the family day at home—and afterwards it was put back in two hours.'
- ▷ It is also possible for the *layout of the house to develop as you build*: to decide exactly where to place the windows only when you can stand in the building and appreciate the particular view from that position, to determine the precise size of the rooms when you can sense the volume of the room full size on the building site. In this way the design can develop in a natural manner and be controlled fully by someone who is not used to reading drawings. It is always difficult to visualize what spaces will be like when built. Self-builders have used the freedom that the Segal method offers extensively.
- ▷ It is *easy to extend* a house of this kind. Two of the houses in Lewisham have been extended. One had an extra bedroom constructed for just £1,200, something like twenty per cent of the cost of a brick-built extension of similar dimensions, when a new baby arrived. The new room was fully enclosed after three weekends' work!

There are limitations:

- ▷ This form of building is particularly suited to building detached houses. It is possible to build in terraces but it leads to complications and additional expense. You lose the advantage of people being able to build independently of their neighbours. There are real advantages in having a detached house, such as the absence of nuisance from noise made by your neighbours, but there may be situations where high land costs or the size and shape of the site demand a terraced arrangement.
- ▷ The lightweight nature of the buildings means that sound insulation is generally not as good between rooms as in a house built with thick, solid walls. There are always ways of improving the situation by constructing built-in wardrobes between bedrooms and having a sufficiently spread-out plan so that rooms which are sensitive need not be next to one another.

And there are prejudices:

- ▷ in Britain against timber construction. This is in spite of the fact that many of the oldest buildings around are made of wood and that the image of a half-timbered black and white cottage is still a very potent one. This prejudice used to be underpinned by the building societies, who were reluctant to lend on timber houses. This situation has changed, possibly because the advantages of speed of erection and economy have seen a large proportion of new houses built with a timber structure. In the richest countries, in North America and Scandinavia, the houses of rich and poor alike are built of wood.

▷ against the unusual panelled appearance—like prefabs, say the critics. The point is that the Segal method is an approach, not an aesthetic. The particular look which comes from the use of standard, low-maintenance cladding panels does provide a strong visual character which has much in common with the half-timbered look of medieval buildings and which serves to provide a visual framework within which the variety of the different houses can be expressed without losing a sense of order. There are many other types of cladding that would retain the basic idea of lightweight, dry construction while giving a different image—tile hanging or weatherboarding, for instance.

▷ against flat roofs. Again, this is not an intrinsic part of the method and plenty of Segal buildings have been built with conventional pitched roofs. There is good reason to be nervous about flat-roof construction, as it has caused many failures in the recent past. Walter Segal, however, devised a method of flat-roof construction that works and has encouraged his self-build clients to use it so that they have been able to reap the benefits of easy and economical construction.

As the number of Segal-method buildings grows and the variety in their design increases, both in terms of appearance and in the ways very high standards of performance are achieved, so the grounds for hostile prejudice recede.

Developments in Segal-method construction

'Classic' Segal method building has developed in a number of ways over the years. Building standards and expectations are higher (the standard for the thermal insulation of walls is more than twice as high as when the Lewisham houses were built for example). The form and appearance of the buildings has tended to be more conventional. Practitioners working with the Segal method have embraced the environmental agenda in its many aspects including energy conservation, healthy building practice and sustainability all of which have an impact on construction techniques.

The consequences of this trend include more complicated construction, higher costs and a degree of convergence with the best conventional timber practice. There is a danger that the inherent simplicity, economy and comprehensibility of the original idea may seem to be compromised and it may become less accessible to lay self-builders.

In fact the elaboration is fairly superficial. The fundamental concept of a post and beam frame standing on minimal foundations and clad with readily available sheet materials put together 'dry' stands unchanged. While not as dramatically cheap to build as Walter's buildings of the 1970s, the method as it is now applied to the higher environmental standards of today and tomorrow still gives extremely good value for money and effort.

There have been two principal directions to the development work. The first has been aimed at bringing Segal method into the main stream by making its appearance more acceptable to the public and to the planning and housing officials who control the housing process.

The second has been to bring it to the forefront of 'green' building practice. Energy consumption in use has been reduced and materials with a low 'embodied-energy' count have been preferred, obtained where possible from local sources. Design has been responsive to the building's setting and concerned to promote a healthy internal environment.

Control over the design and management by the occupants is an essential feature of any form of housing sustainable in the long term.

Recent self-build projects such as the Brighton Diggers outlined in Chapter 11 combine this control with an environmentally sensitive approach to building. It is one of the few examples of truly sustainable housebuilding in Britain, where the occupant has usually had so limited a role.

The tendency away from the 'classic' Segal house with its flat roof and repetitive rhythm of battens on the panelled walls reflects a wish to distance the image of Segal building from the somewhat pejorative association with prefabs. Many current Segal-method schemes have conventional steeply pitched tiled roofs, others have low-pitched roofs—literally green and grass-grown. Walls are finished in different patterns of timber boarding and rendered surfaces, often in combination with panelling. Tile hung walls are also possible, conforming well to the dry-construction principle.

The main reason for changing from the 'classic' Segal sandwich wall as used by Ken Atkins in the early Lewisham days is the increased insulation now demanded. Walter Segal believed in putting on more clothes if you were cold and his response to the requirement for more insulation was grudging at first. The thickness of the plasterboard in the external walls of the first Lewisham scheme was increased from half an inch to three quarters of an inch in order to achieve the then required U value of 1.0 W/sq m/deg C. An additional layer of foam plastic insulation was then incorporated into the sandwich build up for the second Lewisham scheme to achieve a U value of 0.6W/sq m/deg C. The first attempts to construct a low-energy Segal house, the demonstration houses at the Centre for Alternative Technology and the Glasgow Garden Festival dispensed with woodwool altogether and used the more effective insulation properties of foam plastic (this was before the realization that the CFCs used in its manufacture were so undesirable). The core of the external wall was also increased from 50mm to 100mm thick. This led to a change in the tartan grid into which a 100mm zone was introduced for the external skin of the building, and had the advantage of accommodating a 100mm square post—a more suitable structural section for the main frame. The U value was improved to 0.23W/sq m/deg C.

The foam plastic was not as strong as woodwool and had to be stiffened with timber studs. In theory, they were in themselves capable of holding up the building and called into question the necessity for the main frame itself. But the post and beam frame is still thoroughly worthwhile, for its clarity; for the flexibility in design it permits—doors and windows can be placed freely within it; for the advantage of a building sequence that allows preparatory work to be done at ground level and hoisted complete; for the early stage the roof goes on, for the ease and safety of making subsequent alterations and most emphatically for the minimal amount of site work necessary in levelling the ground and digging foundations.

The next step in the evolution of the external wall was to dispense with the dubious rigid foam plastics to 'greener' (and cheaper) insulation materials. Many of these, such as the quite ecologically friendly mineral wool, are not rigid and can only be used when studs are introduced into the framing. The outstandingly good loose-fill insulation that has emerged in recent years for use in conjunction with timber framing, is cellulose fibre. It is produced from all those returned unsold newspapers from your newsagent. The manufacturer, Excell Industries, pulverizes the waste paper, treats it to make it fire resisting and unattractive to rodents and markets it as Warmcell. A negligible amount of energy is used in making it, and it is unpolluting. The chemical it is treated with is borax, a naturally occurring salt crystal—a well known mild substance long used for its cleaning and antiseptic properties. Because it will not support flame, and excludes air from the sides of the timbers it is in contact with, Warmcell actually makes the structure safer in the case of fire attack.

When using loose-fill insulation in a double-skin wall framed with timber studs, it becomes less important to stick to the characteristic Walter Segal 'tartan' grid. The more conventional 'centre-line' grid has been commonly used (an explanation of the advantages and disadvantages are discussed in the next chapter). A typical U value achievable with 150mm of Warmcell in the wall is 0.22W/sq m/deg C.

There have been ingenious solutions to the apparent problem of building bulky, well-insulated walls into the lean Segal system without adding a lot of additional heavy timber framing. To save material and avoid 'cold-bridging', devices like split (or spaced) studs and double layer constructions of vertical studs and horizontal rails have been used.

The 'breathing wall' as a preferable option to impervious vapour barriers has been perfected as a way of countering the harmful effects of condensation (and is fully described in the next chapter).

The quest for high levels of insulation in walls has been a complex affair. Along the way, most aspects of

Hand-made clay tiles and weatherboarding on a Segal-method frame

the construction—the planning grid, the structural frame, the employment of secondary stud framing and the appearance, internal and external, have been affected.

Higher insulation standards have had implications for window design as well as walls. At the beginning, some self-builders made very cheap, simple single-glazed sliding windows to Walter Segal's design, but it soon became the habit to buy in ready made windows that had a tighter fit and saved time, though costing more. Recently, very high performance timber windows with good storm sealing and factory made double (or even triple) glazing have become usual (at still higher initial cost, naturally).

Another concern that has emerged quite recently is the amount of embodied energy in the building's construction; that is the amount of fossil fuels (coal, oil, gas) used to produce, process, transport and erect the materials from which the house is constituted.

It is measurable in kilowatt hours, and for a typical orthodox masonry house is approximately 115,000 kWh. By using the Segal method of framing with locally grown timber, and with Warmcell insulation and the avoidance of too many metal and plastic components, this figure has been reduced by a half. The benefit extends to a very real reduction in the amount of CO_2 released into the atmosphere.

Furthermore, the self-build process is not very energy intensive, relying predominately on the use of hand tools rather than heavy machinery.

Chapter 23

Designing a Segal-method house

A path 'Out of the Woods'

For this chapter of our book we will explore with you a path through the bewildering array of choices offered to the timber frame designer.

A major step forward in making the best practice in environmentally sound building accessible to all was the publication in 1994 of a book about environmental timber-frame design for self-build by Pat Borer and Cindy Harris. The authors of *Out of the Woods*[1] have been engaged in the development of the Segal-method at the Centre for Alternative Technology (and on many projects elsewhere) for many years, and it was Pat Borer's presence as staff architect at the CAT during the last decade of Walter Segal's life that led to the programme of self-build courses that continues to this day (see p.24).

Their theme is that at this advanced stage of development, there is a wide range of choice in Segal-method construction. While some building procedures are still common to all Segal method projects, the majority depend on what particular result you want and on how you choose to go about achieving it.

'For every built design,' they say, 'there are a thousand unbuilt solutions that might have been equally satisfactory' (and that is not to mention that thousands more that wouldn't have been quite as good).

Initially, faced as it were with a blank piece of paper, the range of options may seem utterly bewildering. Borer and Harris give sound guidance on finding the right way to apply the Segal method to your particular project.

One of the delights of self-build is the making of choices that perfectly suit each individual's requirements and preferences. As we advocated studying *A Pattern Language*[2] to trace your way through the related series of design decisions that lead you to generate a building that is alive (p.25) so we now suggest you use *Out of the Woods* to help you develop your timber frame structure.

Because we cannot go down every by-way and consider the results of every alternative we shall invent an imaginary, but fairly typical, example and go through the thinking process that this one building entails. Other choices open to you will produce different solutions when you come to build for yourself, but the process will be broadly similar. This is just to illustrate how you think your way through the building before actually doing it.

For the purpose of the exercise let us make some assumptions, that:

- ▷ you have a sloping site,
- ▷ you have decided upon a two-storey pitched roof house,
- ▷ a tiled roof is a requirement,
- ▷ you need a high level of insulation,
- ▷ you want to keep open the options of having interchangeable modular panelled walls for flexibility (on a tartan grid) or, for the sake of speed and economy, forgoing relocatable walls and having fixed walls (on a centreline grid),
- ▷ you are aiming to make as many environmentally correct choices you can, both in the origin and and performance of the materials you use.

Sketch plan

The first thing to do with your blank paper is to sketch out your broad ideas. You begin by making loose diagrams showing general relationships of the spaces required. You need to develop these until

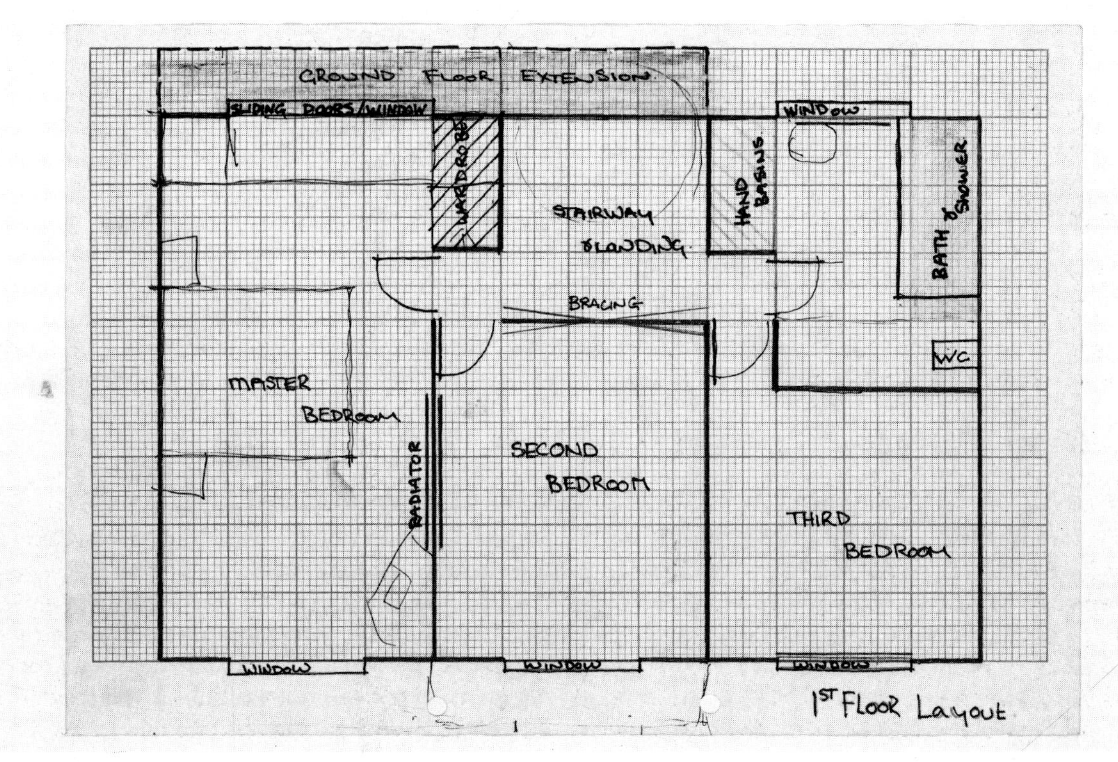

A self-builder's plan

you are able to arrive at a rough set of plans, the most illustrative being the floor plan, or layout plan. This can be drawn up on squared paper, relating the various room sizes by varying the number of modular increments forming them. The decision made at the outset as to what module to use is crucial, but once decided upon, makes detail designing and building a great deal easier.

Once the basic layout has been decided, the arrangement of the structure can be worked out. It is a post-and-beam timber frame with the loads carried by columns spaced up to 4m (13′) apart, connected together with beams. It has much in common with the medieval techniques of timber construction we see in the black and white half-timbered buildings of Britain. The difference is that one is using modern materials and jointing methods; building board panels for the infill making the walls instead of wattle daubed with a mixture of straw, clay and cow dung; and galvanized steel bolts in place of the elaborate carpentry joints of medieval times. This post-and-beam frame has two main advantages. First, the loads are concentrated to a limited number of columns—about twelve, say, for a typical three-bedroom house, each of which has a small, isolated foundation. This is considerably less than is required for a house made of load-bearing wall panels, each of which has to be supported along its whole length by a strip foundation. The second main advantage is that you have great freedom of layout, which can be changed very easily in the future, unlike the situation in a house where the walls themselves are holding up the building. The particular sort of frame used in a Segal-type building is very adaptable in that it allows the structural arrangement to be worked out around the desired plan, unlike the situation with most frames, which tend to force the plan into a shape demanded by the structure.

The principal frame lines are established. Column positions are determined bearing in mind the fact that

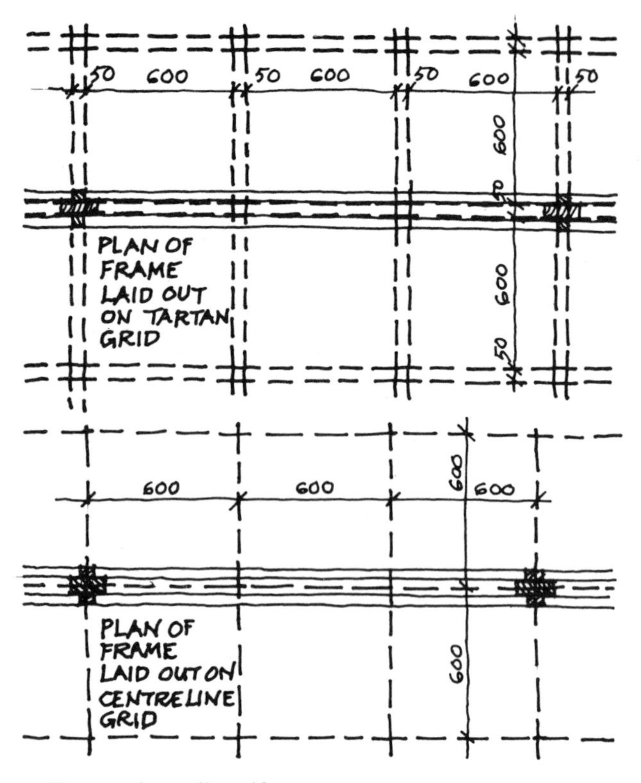

Tartan and centreline grids

the beams will span up to six modules without difficulty. Beyond this the deflection becomes a limiting factor. The floor joists will span an equal distance without trouble. This gives a maximum economic spacing of posts of 3.85m (12′6″) in each direction. This spacing will accommodate most domestic-size rooms. Laminated timber beams or composite trusses can be used to create greater spans if necessary—in a community building, for instance. The position of bracing against wind forces has to be established. Cantilevers can be employed. They have the property of reducing the effects of bending and deflection in the centre of the main spans of the beams. This can be useful although it does tend to mean that you have freestanding columns within the building, which may be inconvenient. Other devices can be employed to fit the structure to the plan, such as supporting areas of the floor on hangers from the roof. When the most economic structural arrangement has been devised, it may be necessary at this point to modify the basic plan in some way to make it fit the structure neatly. Once you have settled the overall structural arrangement, it will be necessary to position the columns relative to the grid and determine where you may need to joint beams and joists in their length in case the members are not obtainable in one length of timber. The whole process of design is one of getting all the aspects, structural as well as functional, into a balance, and while you are devising a plan arrangement you should always have the other structural and cost implications of what you are doing at the back of your mind. This juggling of the variables involved in design comes with experience and explains why it is desirable as an amateur designer to have an experienced person involved.

Layout drawings

Once the structural arrangement has been decided, it can be co-ordinated with the plan arrangement of the layout drawings. These are simple diagrams that show the position of the walls, windows and doors relative to the modular grid and the columns. (See the set reproduced on page 210.) Experience shows that once the structure of the building is erected, the modular nature of the layout becomes clear, and it is not very often that one needs to refer to the drawings in adding the later elements.

Planning grids

To go further with the integration of our structure and our layout plan we must now make our choice of modular dimensions and planning grids.

The module we advocate here is 600mm (2′0″) square. This is small enough to generate a good range of room sizes and large enough to involve very little cutting of panel materials. Many are made 600mm wide; most are 1200mm wide. The standard size material common in the building industry is the old 8′×4′ sheet (2400mm × 600mm), be it plywood, particle board or plasterboard.

The modular units will fit into a broader planning grid pattern. There are two ‘families’ of grids. ‘Tartan’ grids are based on the module as the main incremental dimension, with a narrower measurement of free space separating each module from the next. Walter Segal employed the tartan grid, choosing 50mm (2″) for the narrower bands between the 600 mm panels and used these spaces as the zones accomodating the thickness of a panel and the joints between panels. He designed timber frame structures with members only 50 mm (2″) thick so that panels fitted freely between framing members. 'Centreline' grids, instead of creating zones into which modular components fit, are lines drawn at regular modular intervals throughout, and members are related to the grid by measurements to their centre lines.

Tartan grid

The advantages of the tartan grid are listed by Borer and Harris as:
▷ walls and ceiling panels can be loose fit and reusable;
▷ materials cutting is minimized;
▷ the gaps between panels can be used for services.
And the disadvantages:
▷ column width is limited to the smaller tartan dimension;
▷ gaps between panels can lead to poor thermal and acoustic performance;
▷ quite a lot of material is required to cover gaps.
It could be added that where external walls fall on the tartan they are either limited to a thickness of 50 mm (2″) which is too thin for today's standards of thermal insulation to be achieved, or 650 mm (2′2″) which is extravagantly thick.

However, as external walls usually fall on the very outside of the grid pattern it is usually possible to add non-modular amounts of thickness without disturbing the system. There remains the difficulty of forming external and re-entrant corners off-grid, which needs some ingenuity to resolve—there is some loss of the stark simplicity of the Segal original.

Centreline grid

Of the centreline grid, which is the one usually used in conventional timber stud frame construction, Borer and Harris make these observations:
Advantages:
▷ wall and ceiling panels can be fixed direct to studs/joists;
▷ loose-fill insulation can be used;

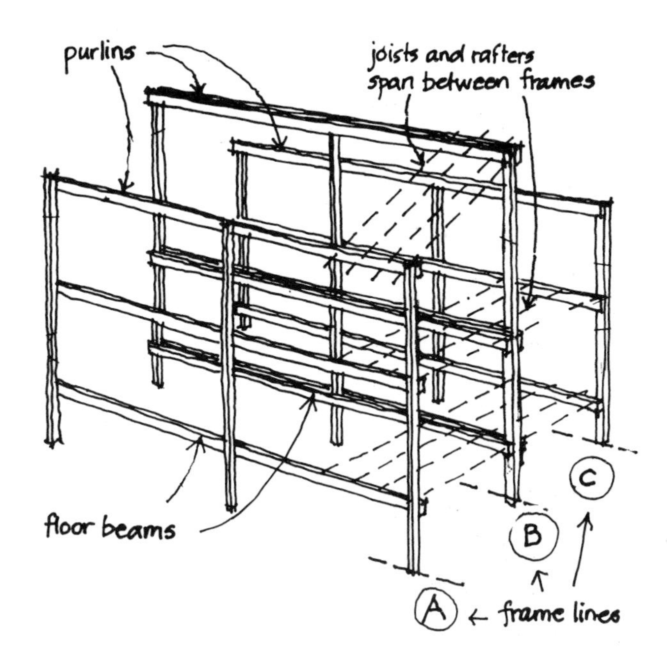

Purlin frames

▷ sheet materials can be used structurally.
Disadvantages:
▷ materials have to be cut on corners and junctions of external and internal walls;
▷ windows and doors adjacent to posts have to be a non-modular size;
▷ there is no 'services void' between wall panels;
▷ the structure is not visible which can result in featureless rooms.

The conventional practice that this system implies, of nailing the edges of panel materials to studs rather than clipping them in behind coverstrips is quicker to do, but renders the panels unreusable as the nails cannot be withdrawn without damaging the material. Insulation value of the wall is increased, universal flexibility is diminished.

So, for our example, we choose a tartan grid and keep the flexibility option open.

Having determined the grid it will help with laying out the plan in the preparation of the construction documents and also will help during construction.

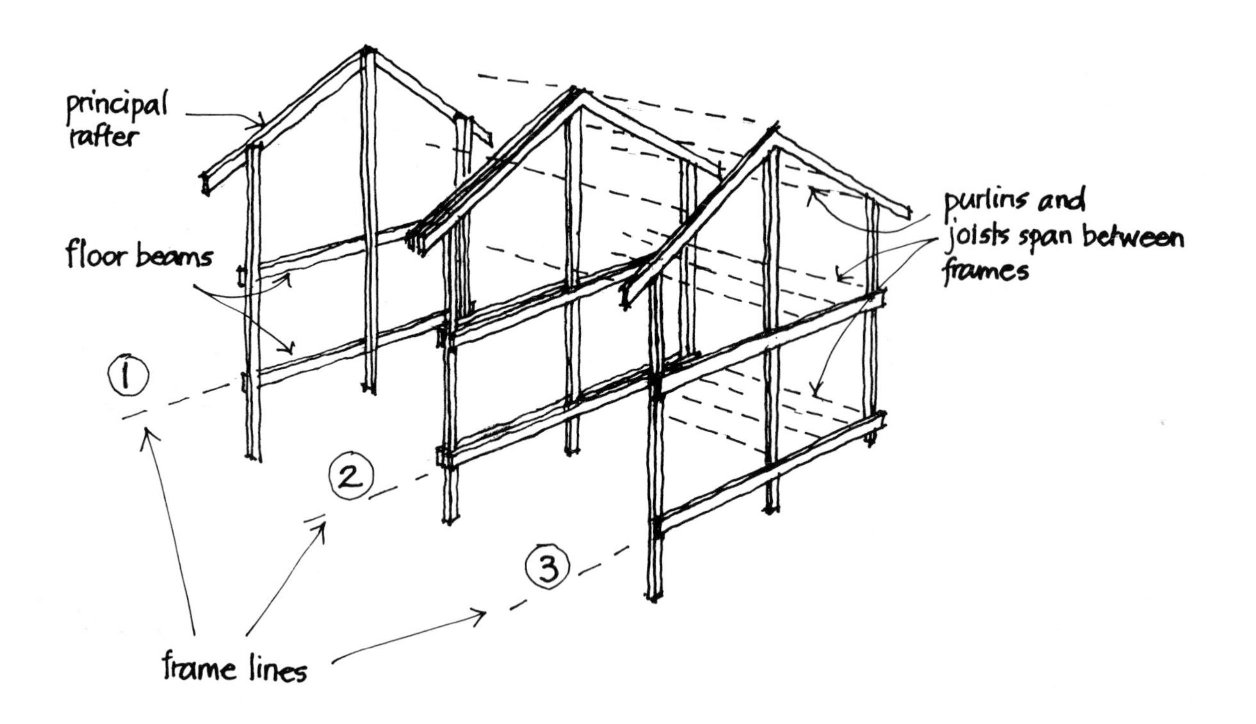

Principal rafter frames

Grid references

Cross-referencing the grid lines by numbers and letters enables us to identify and locate without ambiguity any member anywhere in the building. It is the modern equivalent of the medieval carpenters' mark, identifying each joint in the cruck barn or great aisled hall.

The main frame lines can now be superimposed on the diagrammatic modular layout plan.

Frame types

We talk about establishing the frame lines, but it can be a bit puzzling to know in which direction they should run—along or across the building. With a flat roof it doesn't matter, it probably comes down to what is convenient on a particular site, but with pitched roofs there are other implications. Borer and Harris give names to the types that occur in Segal-method building.

Pitched roof: purlin frame

Purlins are the horizontal roof beams that support sloping rafters (that are notched over them with joints charmingly called birdsmouths—but which are notoriously tricky to get absolutely right!). Purlin frames therefore run along the length of the building parallel with the roof edge.

Purlin frames being rectangular are easy to set out and make, but if the building is long and narrow they are unwieldy to erect.

Pitched roof: principal rafter frame

These go across the narrow width of the building at right angles to the ridge. The tops of the frames therefore have to take up the slope of the roof and those members are called principal rafters. They are

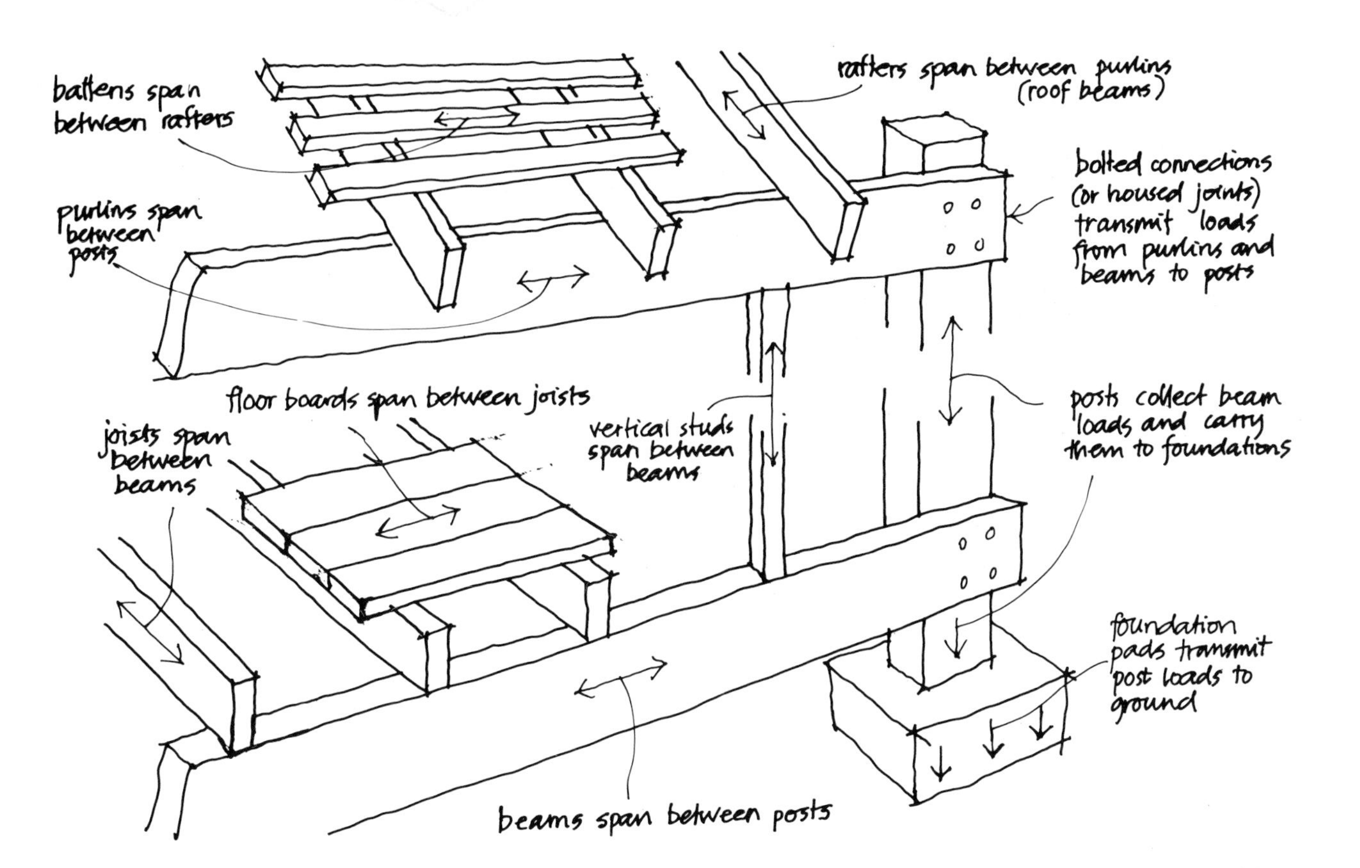
battens span between rafters
rafters span between purlins (roof beams)
purlins span between posts
bolted connections (or housed joints) transmit loads from purlins and beams to posts
floor boards span between joists
joists span between beams
vertical studs span between beams
posts collect beam loads and carry them to foundations
foundation pads transmit post loads to ground
beams span between posts

more demanding to make, but at least the difficult angled cuts are made while the frame is on the ground.

Framing scheme

Our for-example house is not long and narrow so we select purlin frames to go along the length of the building. We can do it with three frames—two lower ones in the plane of the outside walls and a higher one set in the centre under the ridge. They will support rafters sloping from the centre to the outside frames.

Framing members

Beams, Cantilevers and Joists

After positioning our posts and frame lines in relation to the grid, Borer and Harris now invite us to consider the beams.

All horizontal load bearing members come in the family 'beams', be they main beams, purlins [at roof level], joists [smaller members spanning between beams] or cantilevers [beams that overhang their supports]. They can be 'simply supported'—spanning gaps in the structure and only held up at each end, or 'continuous'—having several supports along their length. The advantage of 'continuity' is that the bending at mid-span is counteracted to some extent by reverse bending over the supports. This greatly reduces deflection, the tendency to droop that is the controlling factor in the design of simply supported beams.

The main framing beams at floor level can, in our case, be 'continuous'. Any cantilevers at the ends must needs be continuous (they cannot just be stuck on the face of the posts) and can be helpful in relieving stress in the adjacent span, as medieval builders found when building their overhanging jetties.

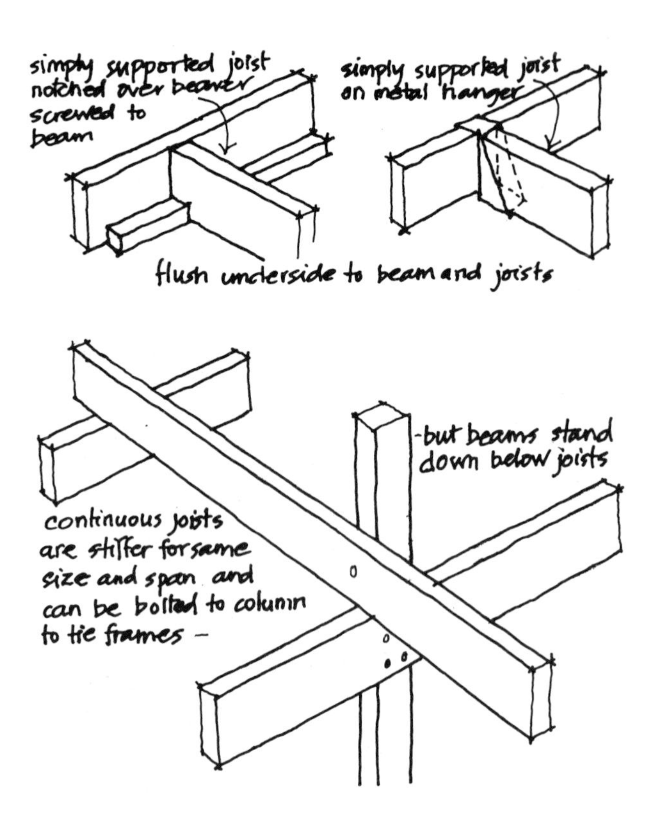

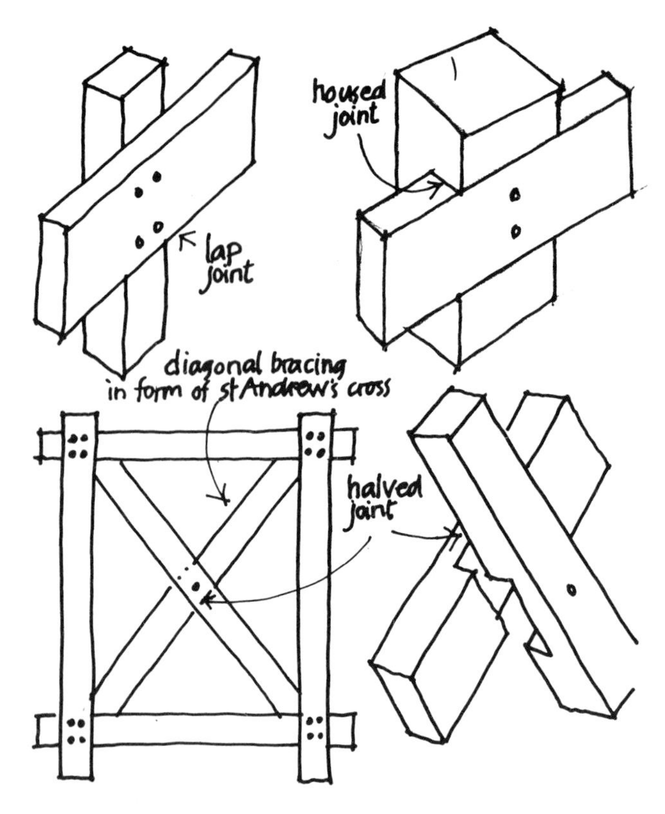

Long overhangs can be further supported by knee braces.

Advantages of cantilevers:
▷ fewer posts in the main frames;
▷ smaller beam sizes;
▷ the overhanging building shelters the feet of the posts;
▷ the triangulation of the brace can provide racking resistance to withstand wind pressure.
Disadvantages:
▷ frames can be unwieldy to raise as there are fewer posts to get people around;
▷ overhangs will always sag a little as timber dries out and slack in the joists is taken up.

The floor joists spanning between frames, can also be continuous with the same strength advantage, but on a fairly level site, the fact that they lie across the top of the beams may make the finished floor unacceptably high. If joists are simply supported between beams and flush with them, with ends supported on ledges or metal joist hangers nailed to the beam faces they will have to be correspondingly bigger to take their allotted load.

On our imaginary sloping site the height factor is probably unimportant so we can afford to choose continuous floor joists. These are easier to install and stiffen the building better as they can be fixed to the posts they pass by.

At first floor level the main frame beams will again be continuous, possibly with cantilevers, but the choice of first floor joists depends on whether you demand an uninterrupted flat ceiling throughout the house and thus have one standard partition height everywhere (as Walter Segal did in the Lewisham designs) or whether you can tolerate beams below the ceiling.

If the extra complication of fitting some walls under or around beams doesn't bother you, you can have 'continuous' first floor joists as for the ground floor, with the same advantages. Should you go for the fully interchangeable relocatable wall panels you will use simply supported joists the same depth as the beams. These may be bigger (or of higher quality timber) than the ground floor joists.

Purlins and rafters

At roof level the sloping rafters and the purlins they rest on can be 'continuous' as downstanding beams are not usually inconvenient in the roof space.

Next, Borer and Harris come to consider how you fix the frame together—the joints between timber framing members.

Joints

Lapped bolted frame-joint

This meccano-style joint is the characteristic one in Walter Segal's own buildings. It is very simple, though not quite as uncomplicated in wood as in metal. Because timber is soft and steel bolts are hard it is usually necessary to have a cluster of bolts at each joint, sized to transfer the loads to the timber without rupturing it. Nor can tightening the nuts be relied on to stiffen the joint—the wood fibres are too easily crushed. As the wood shrinks during its life the nuts slacken off anyway. The bolts have to be regarded just as pegs, with the nuts stopping them coming out.

The joints are simple to make, though proper care has to be taken to drill accurately, and need to be precisely calculated as the size of beam and post will sometimes be dictated by the number and pattern of bolts needed.

Housed joint

The posts we have chosen are only 50mm thick, but if in our example we had square or cruciform section posts we could house the beams into the posts. The utter simplicity and the speed of making the lap joint is lost, but the advantages are great, as the area of timber to timber in direct contact is ample to transmit compression loads. Just small bolts or even screws can hold everything in position.

Half-lap joint

Material is taken out of each of the members passing each other. This joint is used where diagonal braces cross each other.

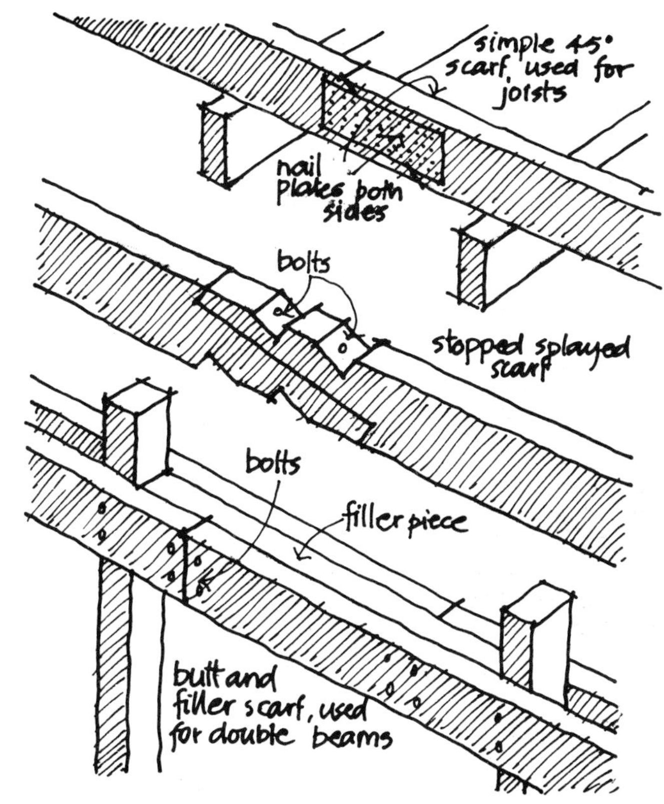

Scarfed joints

Scarf joint

This is the only other special joint employed in Segal-method building. It is used for joining timbers end on to form longer pieces than can be sawn from the log.

There are several forms. When the joint has to be in one plane it is cut to one of the profiles shown, and bolted. It is even simpler with double members to insert a block between them.

Frame wind bracing

As well as the force of gravity imposing downward loads on the building, the wind blows at it sideways from all directions. These forces have to be resisted in a frame building by introducing members that counter the tendency of rectangular frames to lozenge. It can be done in three ways:

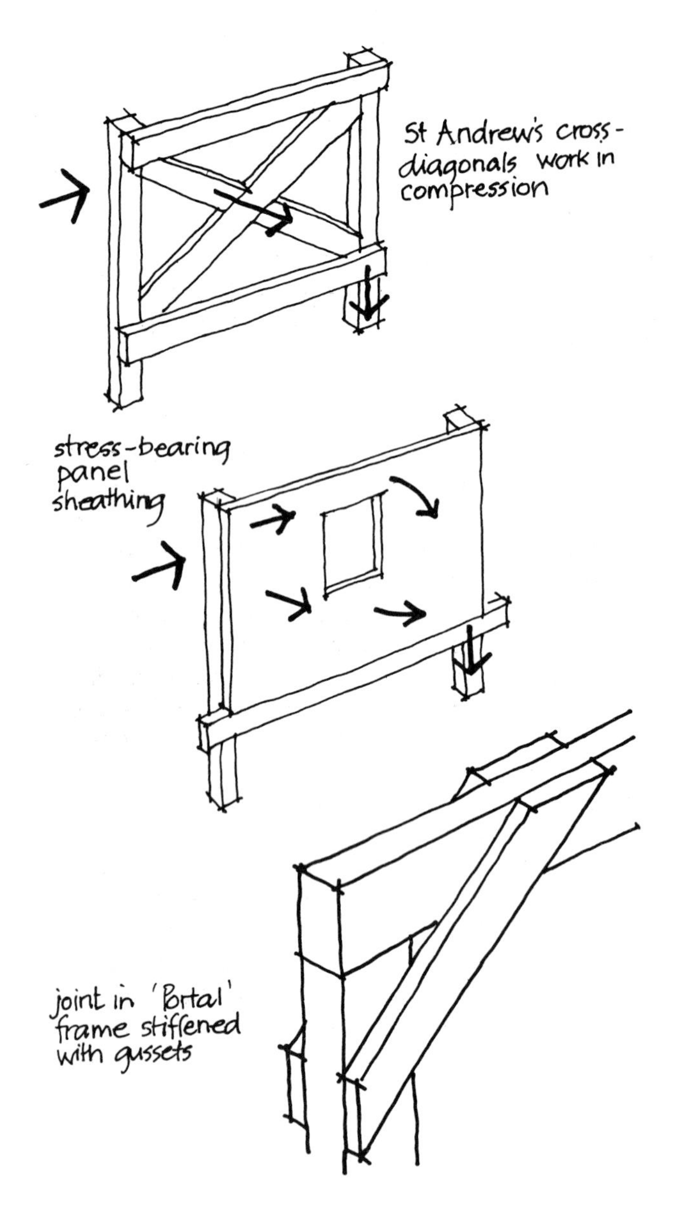

Wind bracing

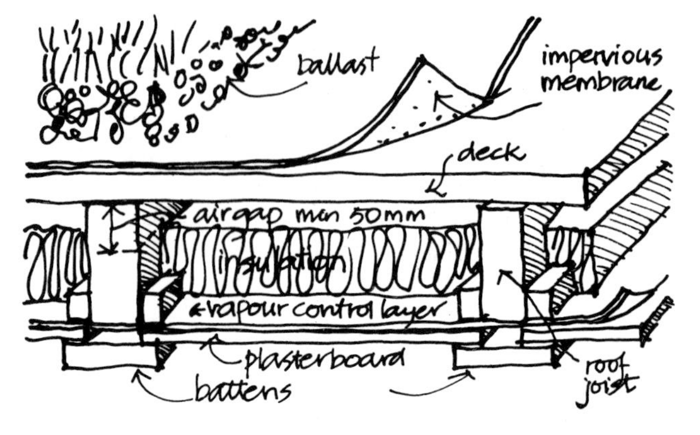

Ventilated cold roof

Portal frames

By having stiff joints at the corners (technically known as a Portal frame). It is fairly difficult to make them strong enough—the self-builder will probably do it with plywood gussets across the corners. These tend to be ungainly where they show in a house.

Diagonal braces

By converting the rectangles, which can deform, into triangles which cannot. (However loose the joints, you can only distort a triangle by breaking one of the sides—a very useful characteristic widely exploited in all structural engineering.) You do it by introducing diagonal braces into the frames—usually crossed so that forces from either direction can be taken up in compression (tension joints are difficult to make in wood).

Diaphragms

By fixing a diaphragm of sheathing material across a whole frame. This is cheap (the sheathing can form one layer of the wall construction) and strong. Such walls should be marked during construction as they must not subsequently be removed without recalculating the racking resistance. Diagonal braces show their purpose very clearly.

The diaphragm principle admirably suits the bracing that is required in floor and roof planes, elements you are not likely to switch around later.

If in our imaginary example we choose diagonal bracing we have to find locations for 'St Andrews crosses' in line with and across the main frames.

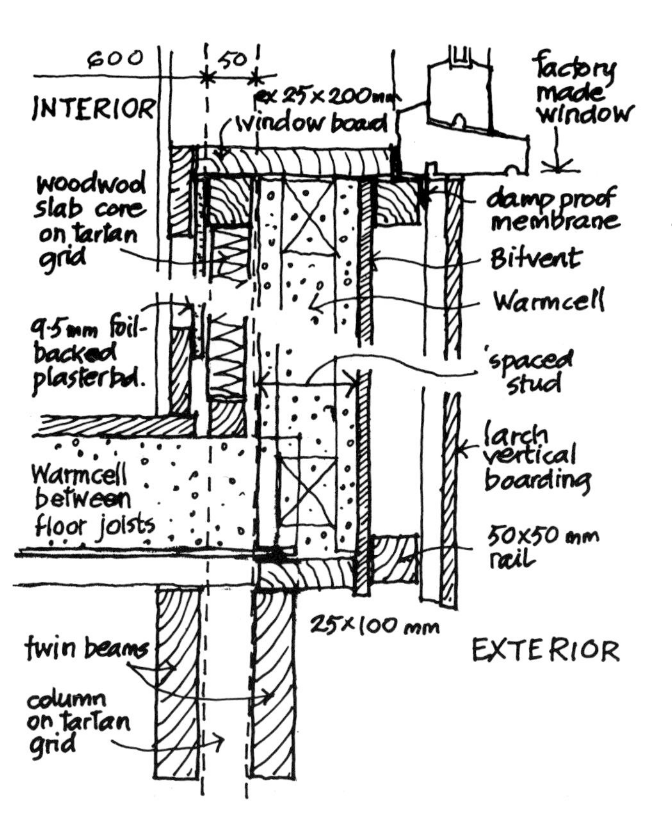

Section, suspended floor and wall

External skin

Roof

Here the range of choice gets significantly wider.

Ventilated cold roof

You can opt for Walter Segal's original cheap, superbly simple, level flat roof. It has the advantage over a sloping roof that all walls fitted up to it can be a standard height. Made to his 'loose laid' principle it does not fail (as flat roofs used to do when they were stuck down firmly to a deck and subsequently split or cracked). A 'table-cloth' impermeable membrane is draped across the roof, and only clipped at the edges. It is ballasted with shingle against wind uplift. Insulation is under the membrane and above the ceiling, which means you must have a vapour barrier and a ventilated space to avoid condensation problems.

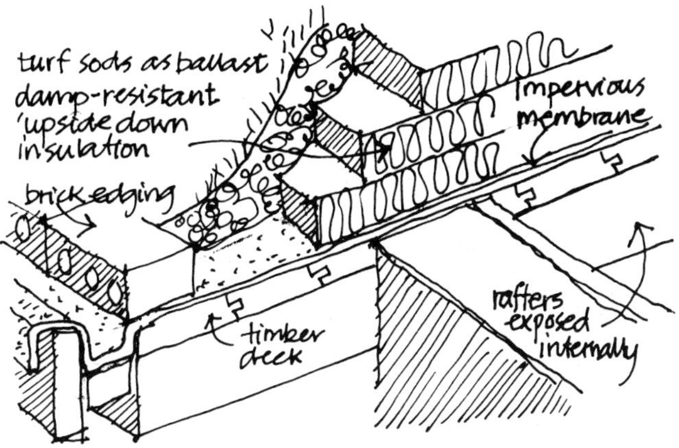

Unventilated warm roof

Borer and Harris class this as a 'ventilated cold roof' (in that the roof construction is outside the insulation). Somewhat like it is the low-pitch membrane-covered roof. It is an attractive idea to ballast such a roof with turf (as we have both done, Jon on his recent house and Brian on his house and workshops, see pp. 39 to 41 and pp. 63 to 65).

Incline the pitch steeper, and instead of relying on a membrane you can use the age-old fishscale principle and shed the rain from lap to lap of slates, tiles or shingles.

Our 'imaginary' pitched tiled roof house, if it is to have a warm attic space, still comes into the family 'ventilated cold roof', that is, part of the roof structure itself is out in the cold and therefore has to be kept ventilated to avoid condensation. There has to be a ceiling, a vapour check layer, a sufficient depth of loose fill insulation, say 150mm (6″), a ventilation space of at least 50mm (2″), an underfelt (which can be microporous, ie. it will shed any water or snow driven under the tiles but will allow vapour to disperse through it), then tiling battens nailed to the rafters and tiles. For the main area of the roof the tiles can be merely 'hooked' over the battens, but at all edges they must be nailed or clipped down to withstand wind uplift.

The advantages are a breathing construction and 'benign' insulation (good old Warmcell again). The disadvantages are that good attention has to be paid to ventilation (for instance the Warmcell has to be injected *after* the roof is made waterproof but must somehow be kept 50mm (2″) below the underfelt. It is done by stapling scrim between the rafters before hand at the appropriate level, a fussy job). Another disadvantage is the need for abnormally deep rafters, to give space for the insulation fill (a convenient product is now available, a factory made composite rafter with small softwood members at top and bottom and a tempered hardboard web, much cheaper and lighter than a solid timber rafter with the same dimensions would be).

Unventilated warm roof

The other family of roof construction identified by Borer and Harris is 'unventilated warm roof' (as at Romilly). If you have insulation below the membrane you must have a ceiling to support it, but if you prefer to show the roof structure internally and omit the ceiling, you can have the so called upside down roof. The principle here is that the structure is kept warm and therefore free from condensation by putting the insulation on the outside. Nobody has yet found a way of doing this in thoroughly environmentally benign materials. For instance although the turf roof as used by Brian at Romilly performs admirably, it relies on closed cell rigid expanded polystyrene insulation and a high-embodied-energy tough water-proof membrane, both products of the petrochemical industry.

Spaced stud on a tartan grid

It is timely to repeat here that there are always choices and not one of them tracks an absolutely clear path to economical, ecologically correct, easy to build, beautiful houses.

Floors

Suspended floor

Here the choice is not so wide. With a sloping site and a post and beam frame a suspended ground floor is obviously indicated (that is a floor clear of the earth, not resting on it). Only if there is a strict height restriction or if particularly heavy loads are going to be imposed, such as a garage for a car, do you have the tedious and heavy task of making a solid concrete floor on a firm foundation.

Our suspended floor needs good insulation as the cold winds can blow underneath the house. There is plenty of space between the joists to accommodate loose fill insulation (such as Warmcell) and the wooden floorboards provide enough of a vapour check to ensure there is no interstitial condensation. You can either have solid tongued and grooved wooden boards which present a nice finish in themselves or use a manufactured board (plywood or oriented strand board) and have it as a subfloor for carpet, lino or cork tiles.

To contain the floor insulation you can use panels dropped in between the joists and resting on timber fillets. Fibre cement sheets (Minarit) are often used —the tartan grid layout ensures that they fit without cutting.

Solid ground floor

If you have to keep the building low and/or if you have heavy loads such as a car or workshop machinery to accommodate indoors you cannot so easily have a suspended timber floor, and will instead cast a concrete slab on the ground, duly excavated and consolidated with hardcore fill.

You can still put post and beam framing on it but without the special benefit of cheap foundations. You must work accurately in the substructure because you will have to fix the posts to the concrete in the absence of restraint from floor beams.

Expensive rigid insulation will have to go under the slab for an insulated floor.

External walls

Out of the Woods examines and evaluates a number of ways of building walls that meet current standards and go far beyond them.

Spaced panel wall

It is possible to build two simple Walter Segal type walls spaced apart and fill the void with loose fill insulation. It is low cost and environmentally benign with good sound resistance but it makes a very heavy and thick construction.

Spaced stud wall on a tartan grid

These are made with a light timber framework separating the inside and outside skins.

The interior side can be plasterboard on a woodwool core as in the early Segal designs, but now with loose fill insulation injected into the void behind an outer sheathing of bitumenised fibreboard. Spaced off this is a rainshield of weather-boarding or panels of Glasal, Minarit or what-you-like. The woodwool is relatively heavy, rigid and a good sound insulator.

Spaced studs and studs with counterbattens on a centreline grid

A disadvantage of combining timber studs and woodwool slabs is that there is a degree of redundancy. The woodwool slabs can be omitted. Two thin skins of material are held apart by 'latticed' timber studs, or studs and counterbattens, and the wide cavity thus formed is filled with loose insulation (Warmcell). It can be considered a more practical operation to pin or screw the skins to the framing instead of clamping them with cover battens. This system is best arranged on a centreline grid when the sheets in their manufactured widths can be closely butted over the studs. It is speedy and cheap but not relocatable—you damage the materials if you take them down again (the close butting means there is no leeway for ill fitting sheets—accurate working is important).

You may think we are labouring a point of detail here, but it may prove to be a significant one in terms of the rest of the design—the choice of grid for instance affects all the parts. It may have a profound effect on the appearance of the building, inside and out. The balance between the pros and cons may appear to be fine, but the judgement has to be carefully decided.

Should buildings breathe?

Whatever build-up of wall materials you use, if it is to be well insulated you must guard against the harmful effects of condensation. If the dew-point (the point at which cooling moist air deposits its vapour as droplets of water) occurs in the body of the wall, the dampness destroys the insulation value and promotes rot. The phenomenon is called interstitial condensation. There are two ways of preventing it.

Vapour barrier

The first is by providing an impermeable vapour check such as polythene sheet behind the internal surface of the wall. Because this surface is warm there will be no condensation of the high humidity air within the room. Outside the impermeable layer the humidity will be at the generally lower outdoor level.

This has become the conventional way of dealing with the problem. The disadvantage is that the interior atmospheric environment is isolated from the outside. It is not pleasant to be living 'inside a plastic bag'. More important, it is difficult to carry out the work effectively, for a completely sealed skin has to be achieved. You can imagine the difficulty of dealing with the places where pipes and wires emerge. The polythene can quite easily be damaged during subsequent operations.

Breathing construction

The second and preferable way is the so-called

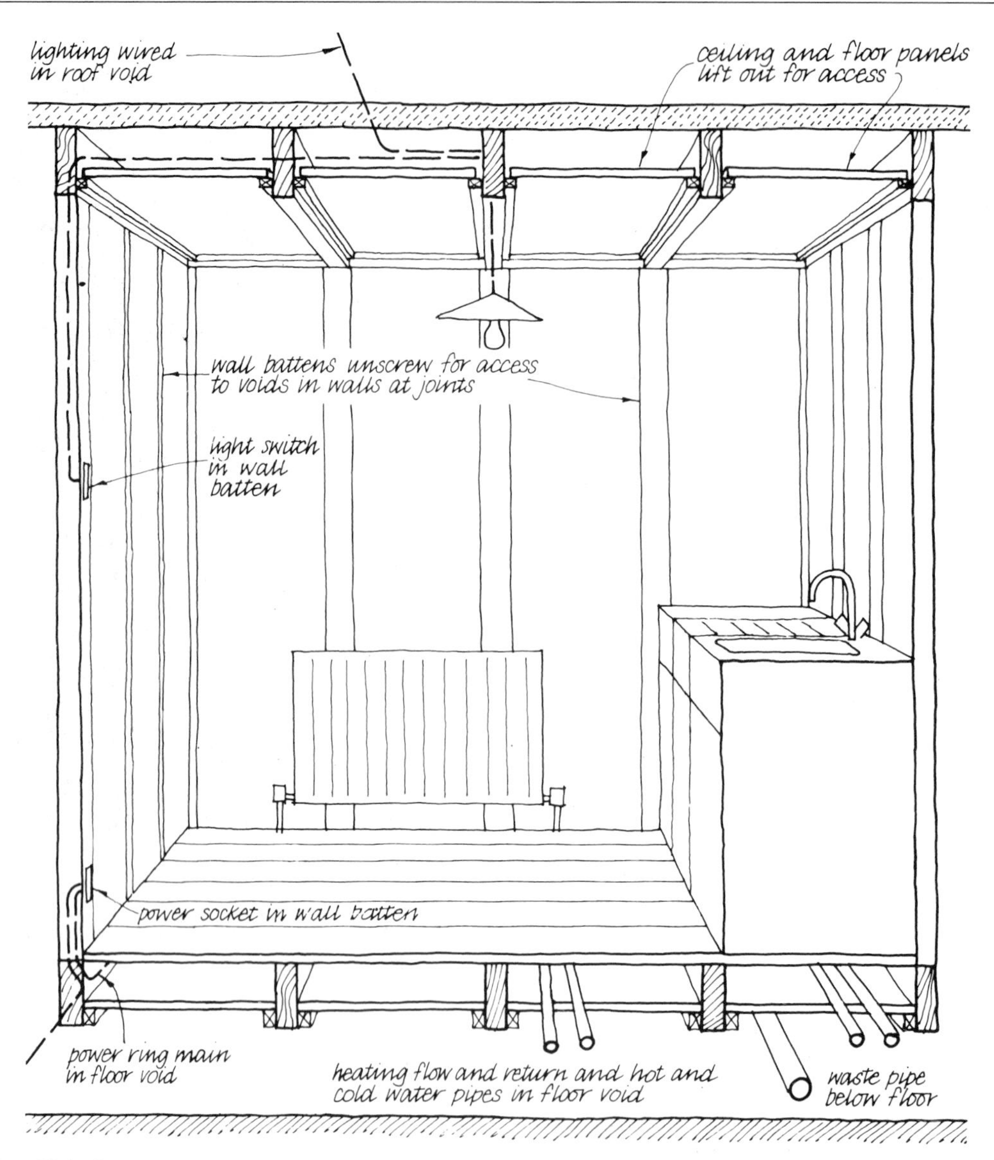

Services run in voids in the structure

'breathing wall'.

This allows moisture and air to diffuse through the structure. The interior surface of the wall can take up (and release back) a certain amount of moisture while not allowing too much humid vapour to pass right through.

The outside layers of the wall are made of porous materials that allow the vapour in the interior of the wall to escape. It follows that the external face of the wall has to be protected from the driving rain—the actual weathershield layer is spaced off the wall with a drainage and ventilation space behind it. An advantage is that extremes of humidity inside the building (too damp or too dry) are tempered by allowing a moisture exchange between indoors and out.

Conservatories and Porches

These items, simple to construct in timber frame, can transform a building.

The unheated buffer zone of the entrance porch (desirably on the north side) not only gives you space

A typical partition

for muddy boots but dramatically reduces heat loss in the winter.

On the east, south and west sides a glazed conservatory acts as a solar collector as well as a buffer zone and gives low cost living space in the warmer seasons.

Borer and Harris remind us of a potential disadvantage—that the conservatory becomes so popular it is used as a normal room through the winter too and then gets wastefully heated, having no insulation. It can also obscure the view from house rooms and impede their ventilation and natural light.

Verandahs

These too can extend useful living space in suitable weather and contribute enormously to the pleasure given by the building. To stand up to the weather the timber deck and supporting frame must be of good quality material, preferably a species with natural resistance to decay such as cedar, larch or oak.

Outside decks need to be inspected and will need periodic replacement.

Interior

Internal walls and services

Getting things in their right order is difficult. It is already too late to be having first thoughts about services.

At the layout plan stage you will have had to decide on locations for stove, or boiler and radiator positions, sinks, baths, basins, WCs and bidets; for lights, power points and switchgear positions.

Now you must allocate routes for the pipework and drain runs and the electric cabling.

It will all take up a lot more room than you originally thought and it is good to allocate generous duct space on the layout plan.

The timber frame and infill panel construction

A Segal-method staircase

makes this easy. The horizontal ducts, framed with nogging pieces, are covered with screwed boards to match the floor. Vertically they can be integrated with the wall fixing cover battens. The naturally occurring gaps behind the cover battens can be used as a route for electric cables to switches and socket outlets (but the greatest care has to be taken to keep cables clear of the batten-fixing screws). It is a great advantage to provide skirting ducts and to plan conduit runs where electricity cables are routed through thermal insulation.

Internal walls

These can be made of rigid panels clipped in to the frame or be conventional stud framing with plasterboard nailed on each side, in which case it is possible to have the smooth plaster faces that some people prefer (though the aesthetic result is uninteresting in that it gives no indication that one is in a timber framed building).

Construction of the internal partition walls can be just as it was in the first Segal buildings. The rigid panel has a core of 50 mm (2″) thick woodwool slab, 600 mm (24″) wide and sandwiched between two sheets of plasterboard. Partitions can be placed anywhere on the grid. The vertical timber coverstrips that retain them engage on a soleplate at the bottom and a head plate at the top on to which have been fixed 50 mm (2″) wide blocks on the tartan crossings. The regular rhythm of the module can add to the perception of proportion in the room (and it is convenient to have so many secure places to fix things to) but you have to like the look of a lot of wood .

If you have decided on smooth plaster walls it is perfectly simple to use conventional timber stud framing—it is just that they will not be so easily relocatable.

Ceilings

The underside of an 'upside down' roof needs no ceiling. Elsewhere ceilings can be of painted plasterboard laid in between the joists on battens, or tongued and grooved v-jointed timber boarding.

When you employ the centreline grid 'close fit' scheme, ceilings are completely sheeted with plasterboard in the ordinary way.

In a two storey house, the first floor has to be made fire resisting. Plasterboard 12 mm thick withstands fire for the mandatory thirty minutes. The inherent resistance of timber exposed to fire is given by insulating the layer of charcoal formed as it burns, which is at the predictable rate of 0.635 mm per minute. Either you can oversize exposed timber so that after half an hour there is still enough undamaged material to carry the load or you can protect it with a 'sacrificial' timber lining. The battens supporting the plasterboard panels can be arranged to provide this.

Stairs

In a Segal-method house the stairs are designed to be simple to construct. This is achieved by either suspending the treads on hangers from above; supporting the stairs on posts from below; or supporting them on bearers fixed to the wall battens. The essential feature is that the whole arrangement is assembled with screwed right-angled joints, avoiding the difficult raked stringers and complicated joints of ordinary stairs.

Stairs of many configurations can be formed in this way: straight flight, dogleg or winding.

Chapter 24

Segal-method documentation

Having decided on your structural design you have to commit the essentials of it to paper, as we outlined in Chapter 18.

Design documentation

You need to explain your scheme to the approving authorities, and you need working documents to give you information for ordering materials and for measuring all the various members on site and noting how they go together.

We listed the general requirements on page 158. For a Segal-method building the documentation has a special character.

Drawings

General arrangement drawings

You need to prepare drawings—plan sections and elevations to scale, and supporting sketch views. The fundamental drawing is the two-dimensional floor plan, generally understood to be the basic graphic representation of a building. It is a cut-away birds-eye view showing the arrangement of rooms, the positions of the walls, windows and doors, stairs, porches, verandahs and so on, and shows the modular grid. The sections are cut-away views in the upright plane and you need several of them to show all the vertical measurements. Sections are particularly important for showing stairs. In conjunction with the sections, elevations show the disposition of elements forming the walls, panels, windows, doors etc.

An appropriate scale for these general arrangement drawings is 1:50. Enlarged drawings, the working details showing the precise way things fit together—junctions at the corners of walls, roof and floors and how they relate to the frame—are drawn at 1:10 or even 1:5, and have grid references to locate them in the building.

Calculations

The frame of a Segal-method post and beam building is a piece of timber engineering. There is a point of difficulty here. Many people, after leaving school, say they 'can't do mathematics' and a certain amount of it is needed now. Nothing advanced, but there are a number of formulae to work through, and some concepts of static mechanics to be borne in mind.

The difficulty is that if the designer of the building opts completely out of the process and relies on the special skill of an engineer brought in at a late stage, the overall design may be flawed. Ideally the architecture and the structure are fully integrated. The engineering cannot be 'tacked on'; the early decisions on the disposition of framing members are crucial to the later constructional detailing. If you cannot summon up enough 'O level' physics and maths to cope, the next best thing is to go to an architect who does her or his own calculations (and beware, not all do—with conventional building the sizes of many structural members can be taken direct from tables published in the Building Regulations and it is customary for architects to go to an engineer for anything beyond this).

The appointed architect would work with you on the design of the whole project—we discussed the pros and cons of this on pages 136 and 137.

Once the structural layout has been determined it needs to be checked for structural adequacy by calculation and possibly modified. You will probably have done some very quick back-of-envelope sums to check roughly the critical elements of the structure already, so not too much modification should be necessary. You will also be designing the joints between the columns and beams and specifying grades of

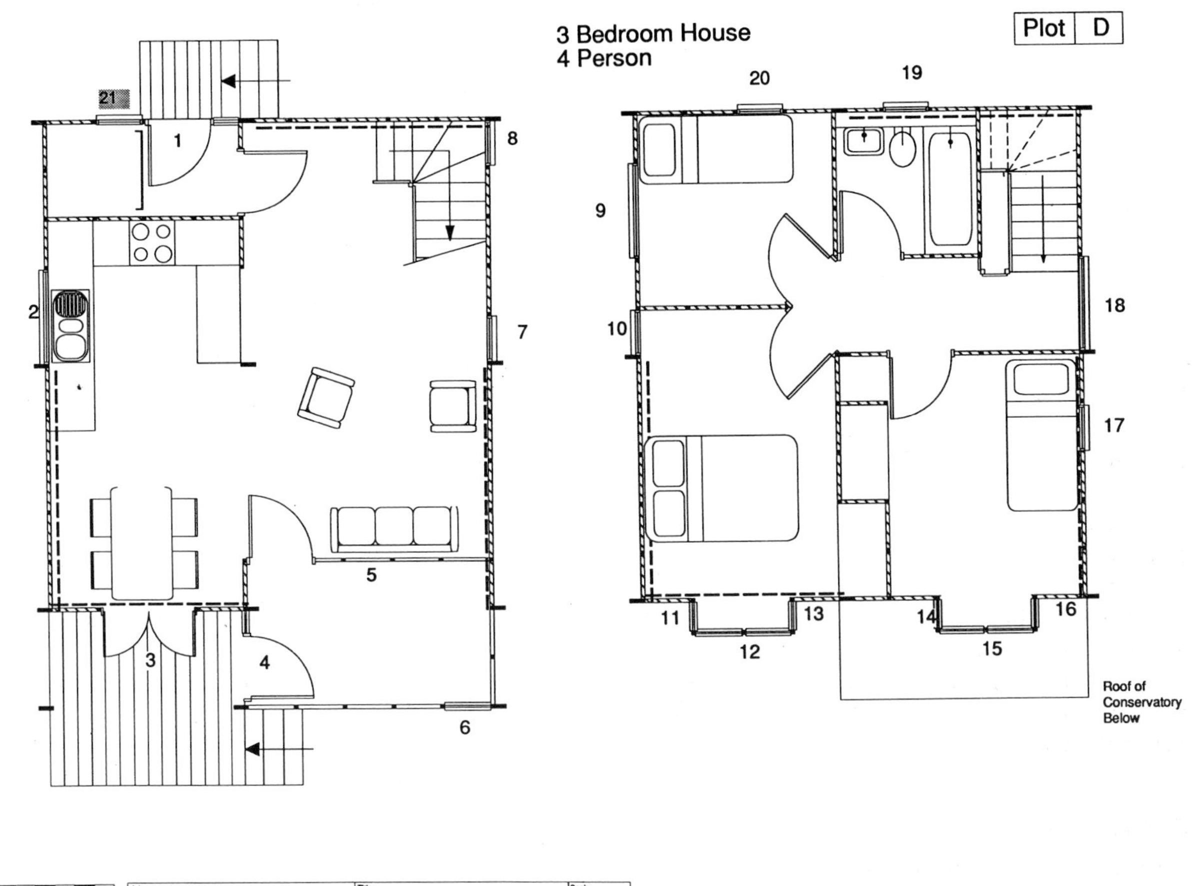

Layout plans

timber to be used for the various elements of the structure and determining the sizes of the various members.

The calculations are relatively straightforward. The basic information is to be found in a good book on the subject (we would recommend *Timber Designers' Manual*, by Baird and Ozelton[1]). Experience counts for a great deal in getting a good balance between the elements, knowing when to specify good-quality but expensive timber and when you can do with less, when to make an approximation and when something is very critical. This is the knowledge that takes experience to acquire and which cannot be learned from books.

The process is one of tabulating the various loads on the building, the self-weight of the building itself, the roof, walls and so on, and the imposed loads from the people, furniture and contents of the building and the weight of snow which will lie on the roof in winter. Work out where these loads are acting and how much of each load is carried by which element of the structure, and how it is acting, whether in bending or compression, shear or tension. Determine which are the critical elements of the structure—the most heavily loaded beam and joist, for instance. It may be that there are a number of potentially critical elements; a lightly loaded beam may be spanning further than a more heavily loaded one and it will be necessary to check that both are adequate. Having identified the critical elements, check each of them

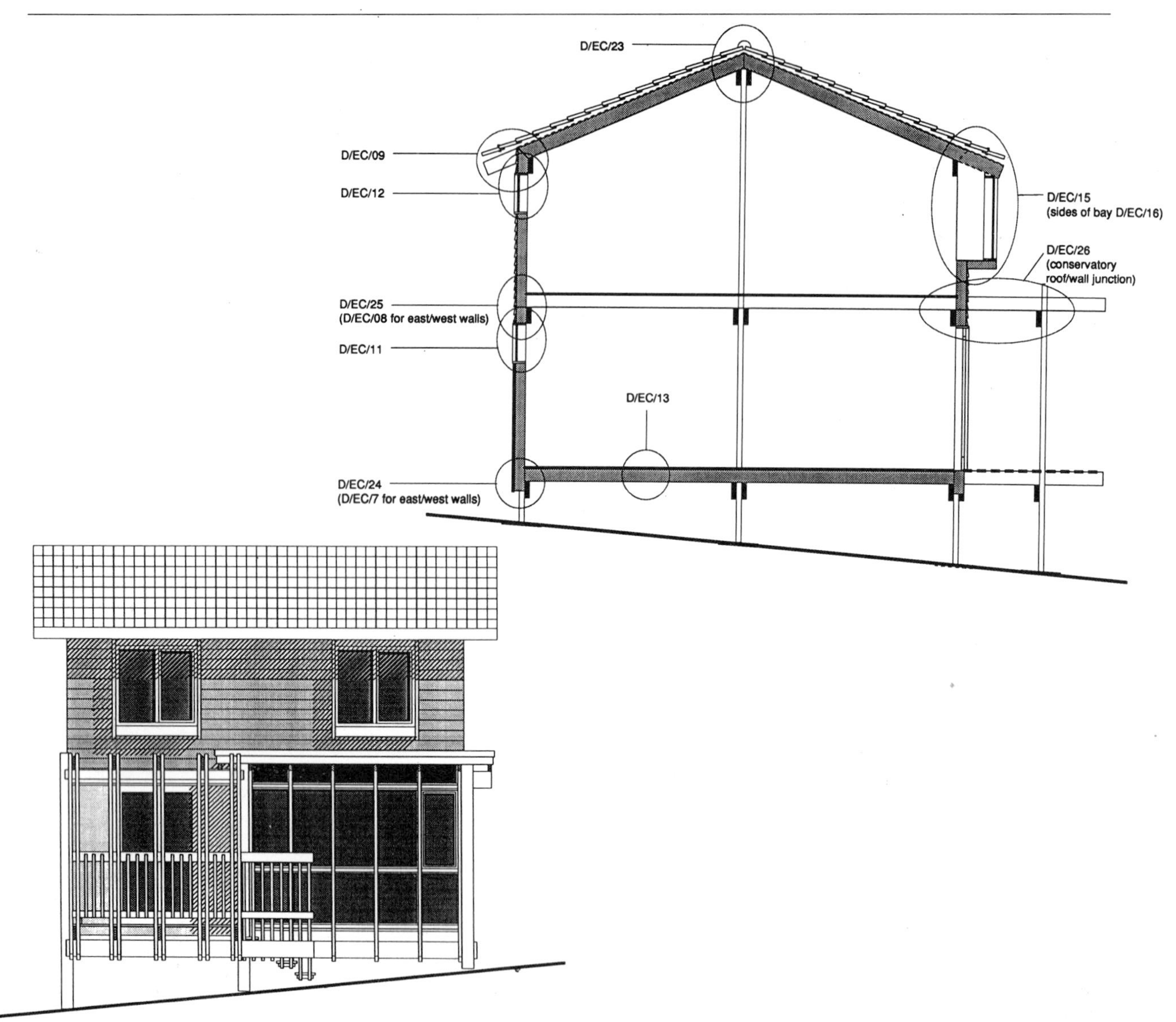

Typical layout, elevation and section drawings

in turn: beams at the different floor levels and with different spans, roof joists, floor joists, columns. Those beams and joists subject to bending forces will have to be checked for deflection as well as for bending strength. Columns will have to be checked for buckling. Select the most appropriate grade of timber and size of section for each element. It may very well be that the sizes of the timbers are determined by the design of the joints, which is often the most important factor in a structure of this sort. The number and size of the bolts required is determined by the load that they can safely carry without splitting the timber (the strength of the steel bolt itself is always in excess of the load that it can impose on the timber). This in turn is affected by the type of timber being used, its thickness, the duration of the load, and the diameter of the bolt. The other critical factor is the minimum spacing between bolts, and the bolt and the edge of the timber, the end of the timber and what is known as the loaded edge—that

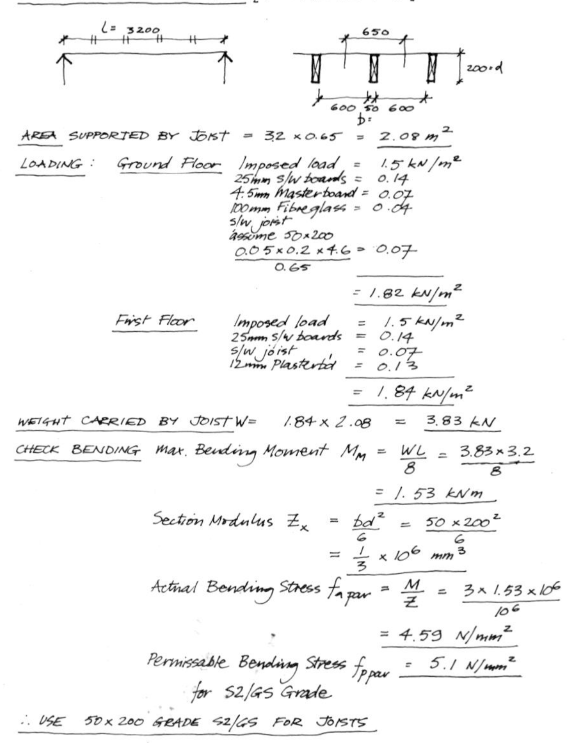

FLOOR JOIST OVER WIDEST SPAN [Ground & First Floor]

l = 3200

650

200 = d

600 50 600

b =

AREA SUPPORTED BY JOIST = 3.2 × 0.65 = 2.08 m^2

LOADING: Ground Floor
Imposed load = 1.5 kN/m^2
25mm s/w boards = 0.14
4.5mm Masterboard = 0.07
100mm Fibreglass = 0.04
s/w joist
assume 50×200
$\frac{0.05 \times 0.2 \times 4.6}{0.65}$ = 0.07
= 1.82 kN/m^2

First Floor
Imposed load = 1.5 kN/m^2
25mm s/w boards = 0.14
s/w joist = 0.07
12mm Plasterbd = 0.13
= 1.84 kN/m^2

WEIGHT CARRIED BY JOIST W = 1.84 × 2.08 = 3.83 kN

CHECK BENDING Max. Bending Moment $M_M = \frac{WL}{8} = \frac{3.83 \times 3.2}{8}$
= 1.53 kNm

Section Modulus $Z_x = \frac{bd^2}{6} = \frac{50 \times 200^2}{6}$
$= \frac{1}{3} \times 10^6$ mm^3

Actual Bending Stress $f_{a\,par} = \frac{M}{Z} = \frac{3 \times 1.53 \times 10^6}{10^6}$
= 4.59 N/mm^2

Permissable Bending Stress $f_{p\,par}$ = 5.1 N/mm^2
for S2/GS Grade

∴ USE 50 × 200 GRADE S2/GS FOR JOISTS

Structural calculations: a typical sheet

SCHEDULE of MATERIALS : 17 LONGTON AVENUE SE26 — 1.
BILL GOSBEE & GORDON LOCKIE

MATERIAL	DESCRIPTION + LOCATION	GRADE	SECTION/ UNIT SIZE	LENGTH + QUANTITY/AREA	FINISH	PRICE
	TERRACE + PERGOLA (Quantities for one house unless stated otherwise.)					
WOODWOOL	for both houses!	Type B	50mm thick × 600mm wide	8/2250 long (cut from 2400 long)		
GLASAL	for both houses		3.2mm thick	8/2325 × 600		
FRAMING TIMBER	Posts supporting deck. (End Centre End)	S2/50	50 × 150	1/1500 1/1200	Sawn PAR	
	Post to wall between decks	S2/50	50 × 50	7/3000 2/3000	Sawn Planed one face	
	Beams	S2/50	50 × 200	4/2100	PAR	
	Joists	S2/50	50 × 200	6/2700 2/3300	PAR PAR	
WALL TIMBER	Heads & cills to wall for both houses.	S/W	50 × 50	4/2700	Sawn	
	Battens to wall for both houses	Joinery S/W	25 × 100	20/2400	PAR	
	Blocks under cills below wall	S/W	50 × 50	1/2100	Sawn	
	Capping to top of wall	Joinery S/W	25 × 125	2/2700	PAR	
STEPS	Hangars to steps	S1/75	32 × 64 finished!	1/1500 1/3600	PAR	
	½ riser to steps	Iroko	21 × 71 finished!	1/2100	PAR	
	treads	Iroko	38 × 250	1/2400	PAR	
DECK	Boarding	Keruing	25 × 150	8/4200	PAR	

A typical page from a schedule of materials

edge of a beam towards which the load is acting. This is where you need a substantial amount of timber to resist splitting, which is generally on the top edge of a beam because the bolts are resisting the load which is pushing down on them by pushing up on the beam to an equal and opposite extent. This can be confusing because it may appear at first sight that you would need more timber at the bottom edge of the beam to support the weight rather than at the top. These spacings are also determined by the diameter of the bolt. You may need at this point to reconsider the sizes of the timbers to accommodate the bolt spacings required.

To complete the structural analysis you will have to determine the wind loads acting on the building, which are determined by where it is in the country, how exposed it is in its surroundings and the shape and height of the building. This will enable you to design the bracing and check the building for wind suction on the roof. You may be required to check against overturning and to check parts of the building such as the wall panels for their ability to resist wind forces. Finally, you will have to determine the pressure that the foundations will exert on the soil. You may also be required to check the behaviour of the building in a fire, to check that as the structure burns, it will remain standing for a sufficient time to allow people to escape.

A set of such calculations might run typically to about twenty pages. They will be checked by the building inspector.

Framing drawings

Once the structure has been checked, the diagrams showing the arrangement of the structural frame can be prepared. A typical set would contain:

▷ a foundation plan;
▷ plans of each floor level;
▷ roof plan;
▷ cross-section;
▷ long section;

▷ diagrams of each of the main frames;
▷ details of each of the different types of joints required to construct the main frames.

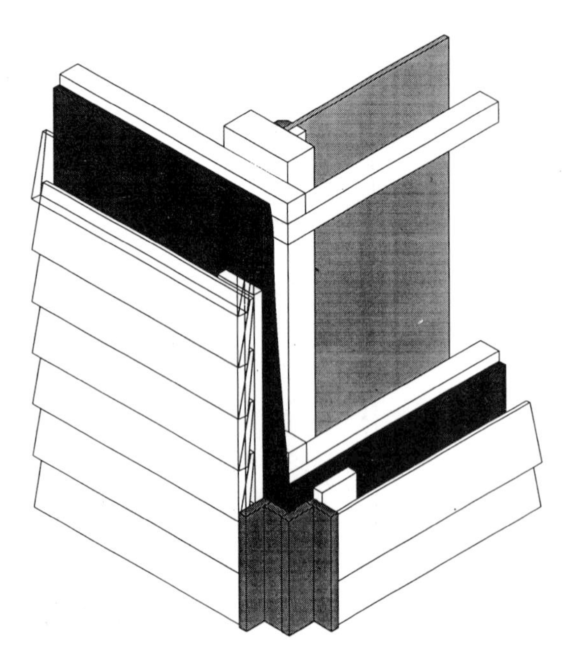

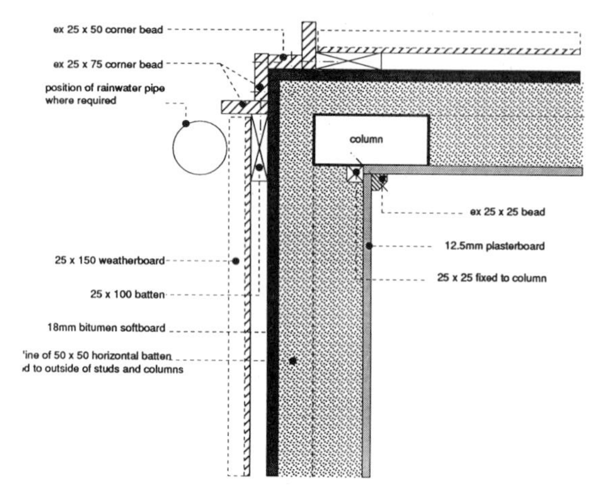

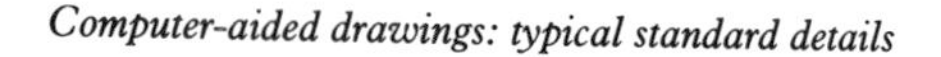
Computer-aided drawings: typical standard details

This would amount to about eight A4 sheets in all. The drawings contain the dimensional information necessary to build the structural frame. Once that is done they are no longer needed.

Working details

This is a set of A4 diagrams that is similar for each building. These plan and section details show how each element of the infill to the structure is joined to the other elements; how a window fits into a wall and how the roof joins the walls, for example. Using these standard details of junctions it is possible to form the envelope of buildings with all different layouts. They show how the parts are joined together, what kind of fixings are required and how the parts relate to the modular grid.

Schedule of materials

Once the drawings are completed, the schedule of materials is prepared. This is a detailed list of all the materials required for a particular building. Each component of the building is described with its size, length, quality and finish, together with its location in the building. Diagrammatic annotations are given where necessary to help to identify the parts.

This schedule is a combination of a normal bill of quantities such as a quantity surveyor would produce and a specification of the kind that an architect usually prepares. The difference is that in this kind of schedule all the pieces are described accurately. For example, a bill of quantities will tell you how many square metres of ceiling there are but not how many plasterboards 2400 × 600 mm (8′0″ × 2′0″) you will need.

The schedule is a key document because it not only locates all the parts that go into the building but also allows you to obtain a number of competitive quotations for the supply of the materials at the outset, in this way building up an accurate cost estimate that should not vary from the cost on completion. The schedule is the basis from which the materials are ordered and forms a check list on site against which to establish that all the material has been correctly delivered.

A typical schedule for a house would comprise about eight pages.

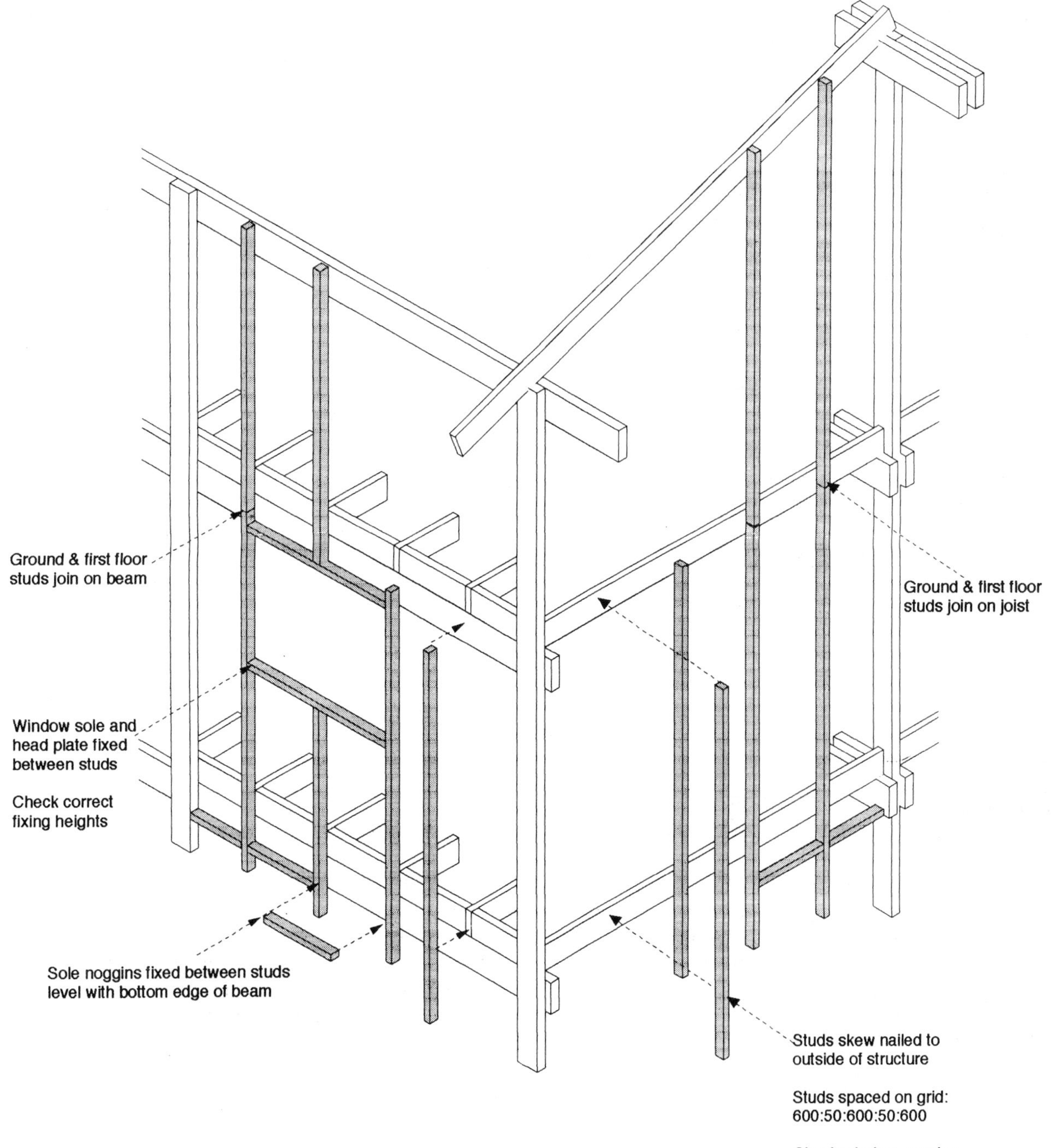

Set of computer-aided drawings showing typical assembly details

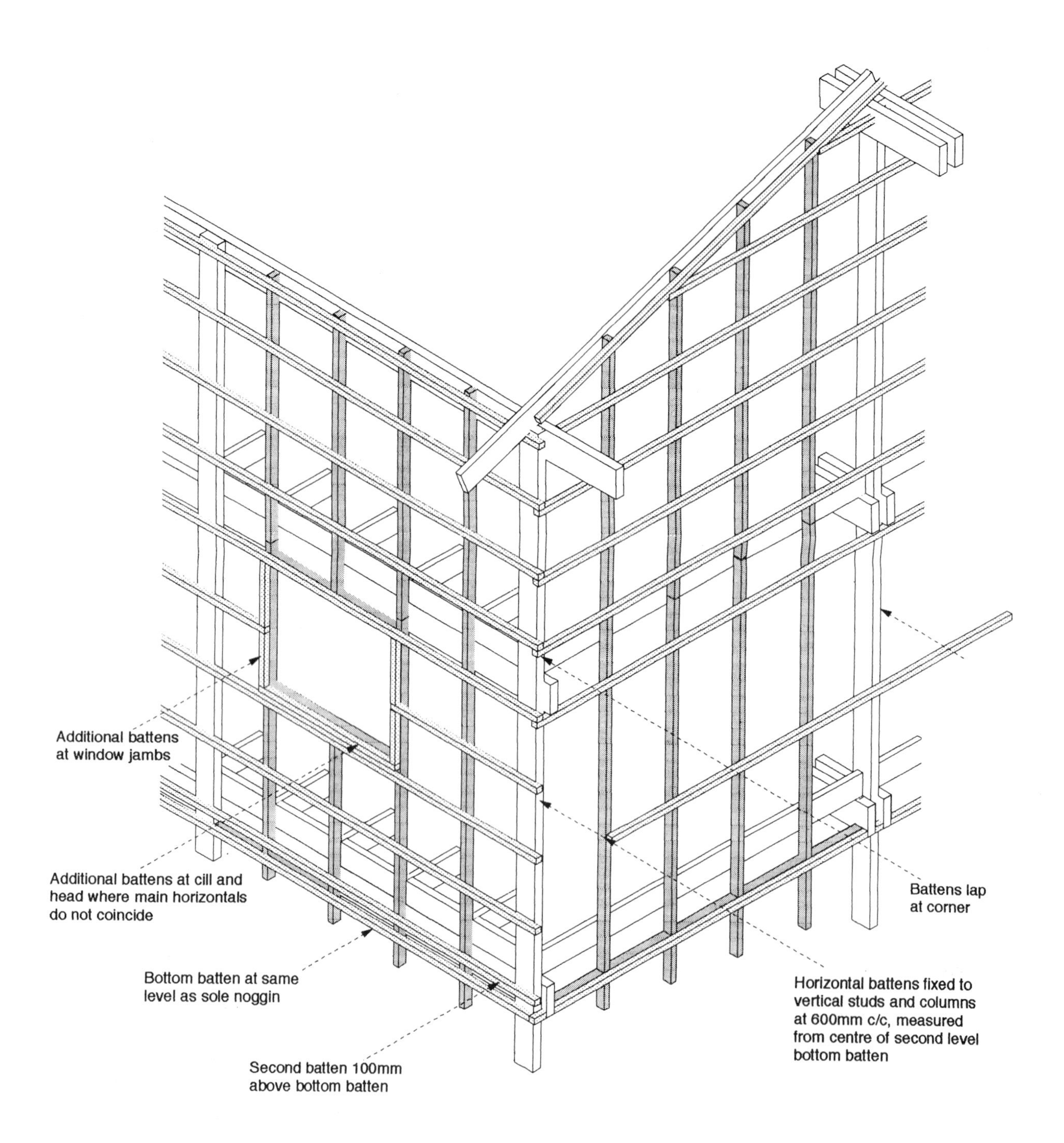
Additional battens
at window jambs
Additional battens at cill and
head where main horizontals
do not coincide
Bottom batten at same
level as sole noggin
Second batten 100mm
above bottom batten
Battens lap
at corner
Horizontal battens fixed to
vertical studs and columns
at 600mm c/c, measured
from centre of second level
bottom batten

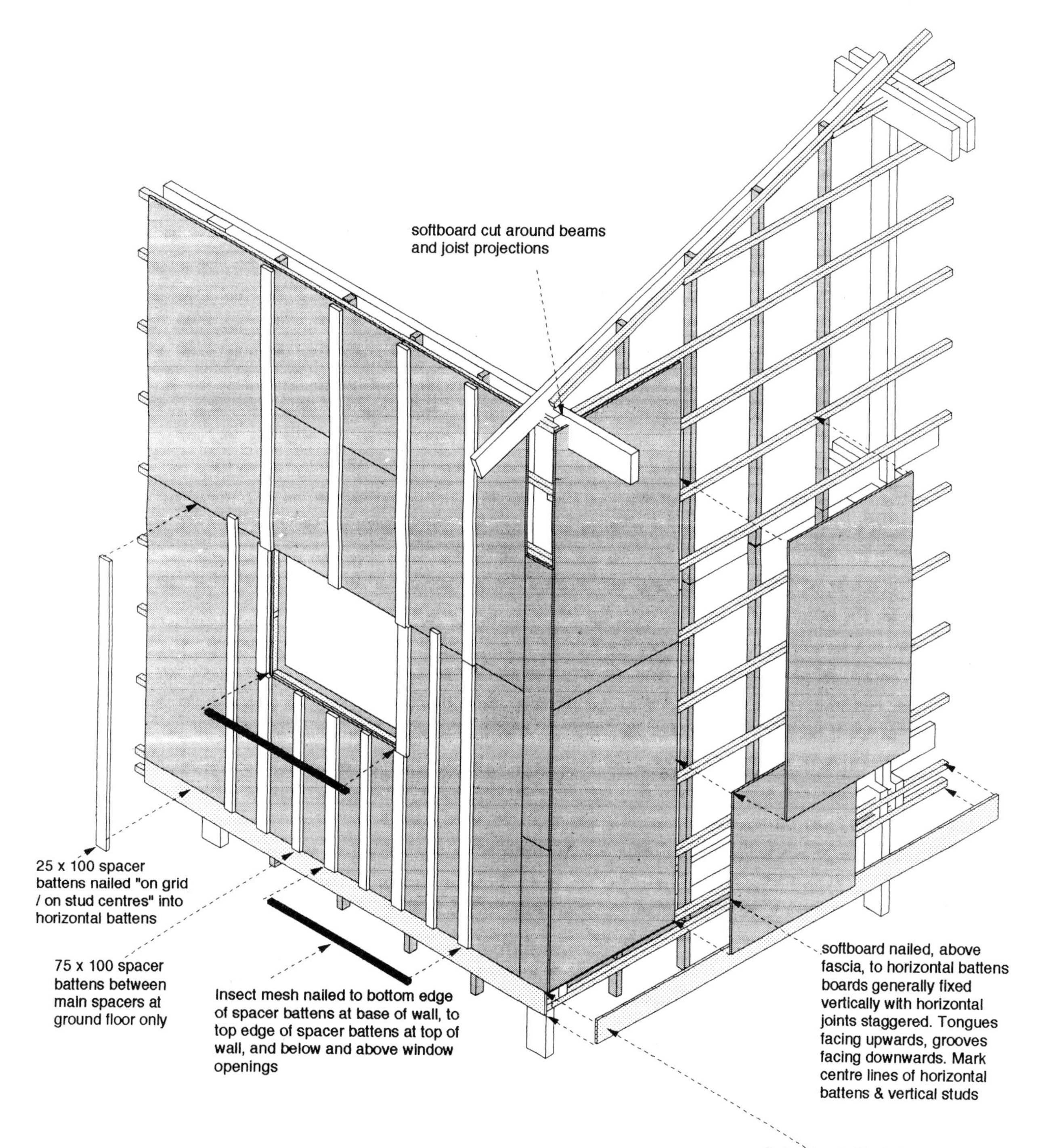
softboard cut around beams
and joist projections
25 x 100 spacer
battens nailed "on grid
/ on stud centres" into
horizontal battens
75 x 100 spacer
battens between
main spacers at
ground floor only
Insect mesh nailed to bottom edge
of spacer battens at base of wall, to
top edge of spacer battens at top of
wall, and below and above window
openings
softboard nailed, above
fascia, to horizontal battens
boards generally fixed
vertically with horizontal
joints staggered. Tongues
facing upwards, grooves
facing downwards. Mark
centre lines of horizontal
battens & vertical studs
timber fascia fixed at base of wall

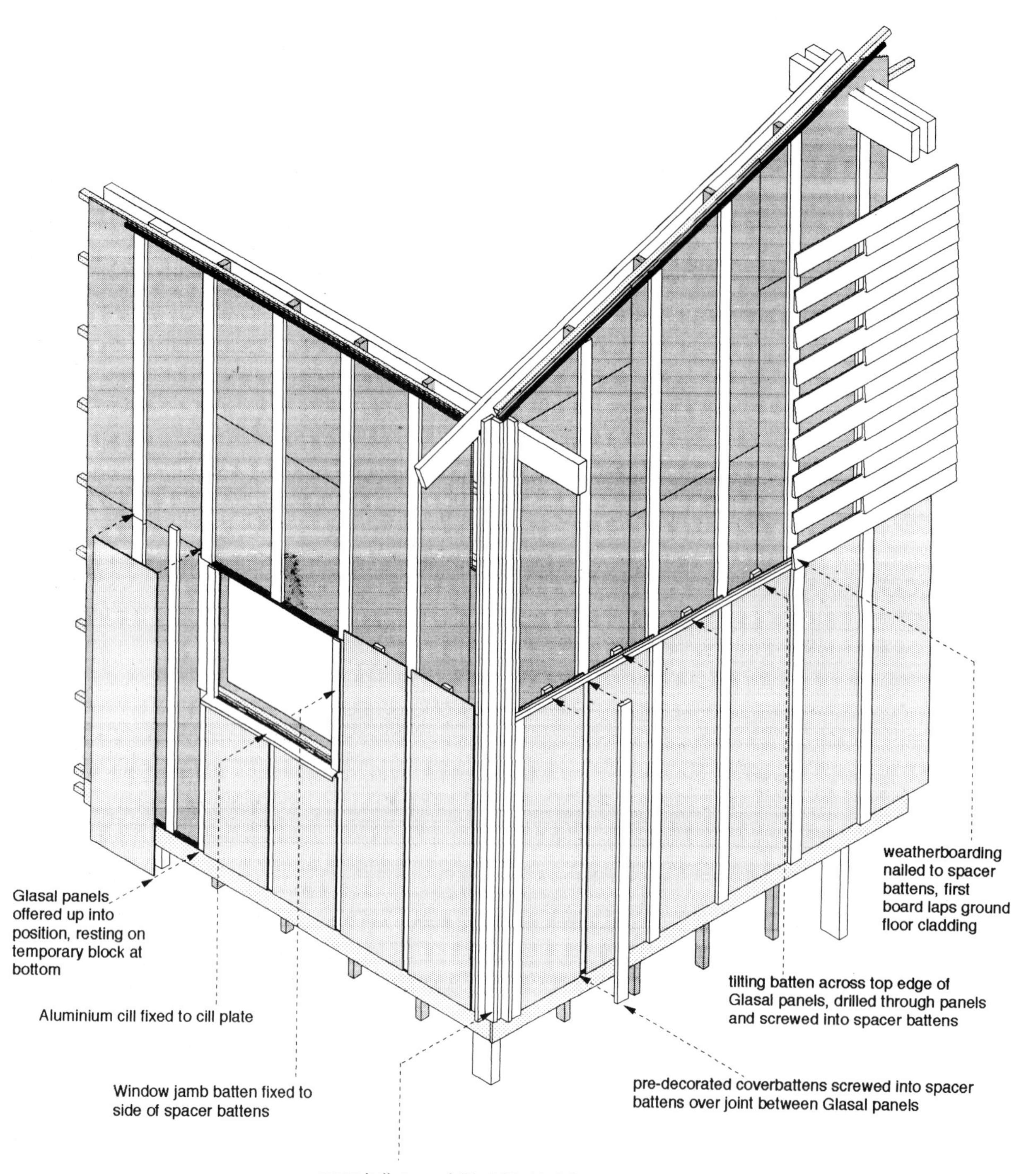
Glasal panels offered up into position, resting on temporary block at bottom
Aluminium cill fixed to cill plate
Window jamb batten fixed to side of spacer battens
corner battens run full height of building
weatherboarding nailed to spacer battens, first board laps ground floor cladding
tilting batten across top edge of Glasal panels, drilled through panels and screwed into spacer battens
pre-decorated coverbattens screwed into spacer battens over joint between Glasal panels

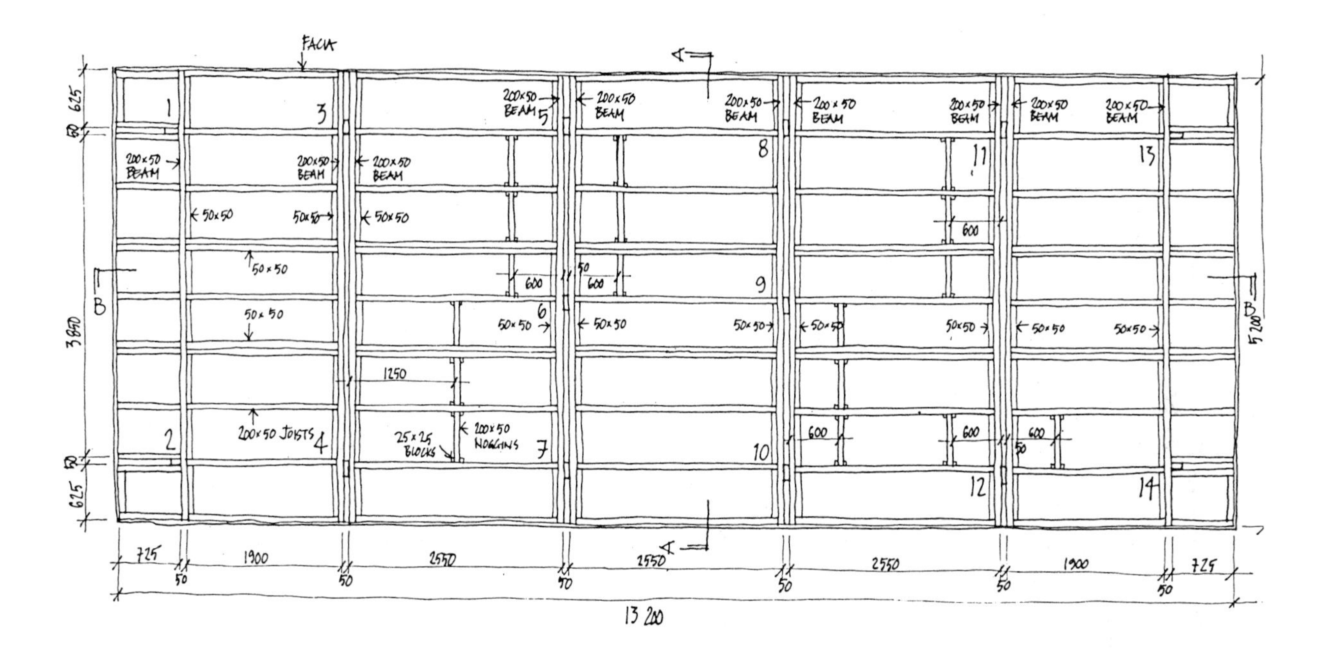

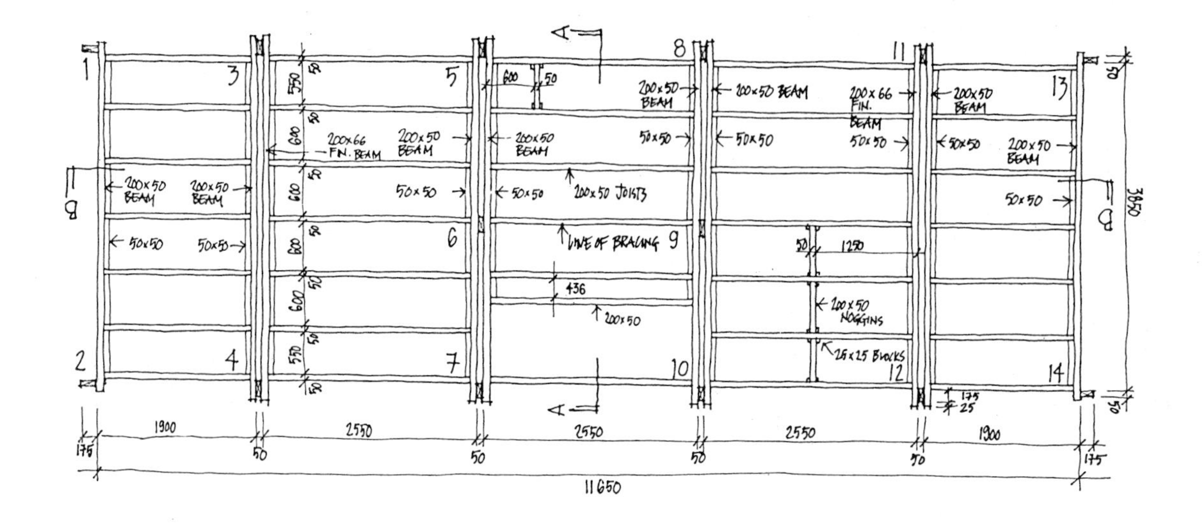

Typical structural frame drawings: plans at floor and roof level

Building instructions

The final element of the documentation is the instructions. They are straightforward, step-by-step instructions with diagrams as necessary. The example shows the start of making the main cross frames. These instructions need to be augmented with one or two good DIY books[2] to explain how to do installations that are common to all forms of building—the wiring, for example.

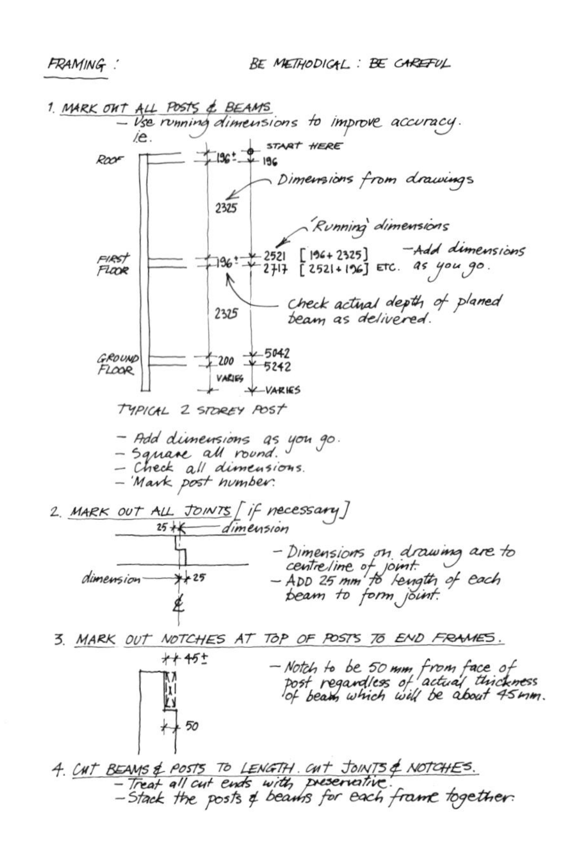

FRAMING: BE METHODICAL : BE CAREFUL

1. MARK OUT ALL POSTS & BEAMS
- Use running dimensions to improve accuracy. i.e.

ROOF
START HERE
196±
196
Dimensions from drawings
2325
'Running' dimensions
FIRST FLOOR
196±
2521 [196+2325]
2717 [2521+196] ETC.
– Add dimensions as you go.
Check actual depth of planed beam as delivered.
2325
GROUND FLOOR
200
5042
5242
VARIES
VARIES

TYPICAL 2 STOREY POST

- Add dimensions as you go.
- Square all round.
- Check all dimensions.
- Mark post number.

2. MARK OUT ALL JOINTS [if necessary]

25
dimension
dimension
25

- Dimensions on drawing are to centreline of joint.
- ADD 25 mm to length of each beam to form joint.

3. MARK OUT NOTCHES AT TOP OF POSTS TO END FRAMES.

45±
50

- Notch to be 50 mm from face of post regardless of actual thickness of beam which will be about 45 mm.

4. CUT BEAMS & POSTS TO LENGTH. CUT JOINTS & NOTCHES.
- Treat all cut ends with preservative.
- Stack the posts & beams for each frame together.

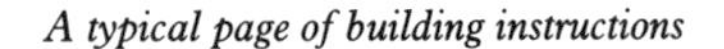

A typical page of building instructions

Chapter 25

Work on site

Having established the design of your house in fair detail and got it down on paper, you can at last get on with the building.

The building operation itself needs to be planned. The sequence of events which we shall go through next is not always the same as the order in which the design evolved. With self-build, designing does not stop when building starts. You cannot make radical changes to the concept once you have committed yourself but it is a joy to be able to refine and revise as you go along, in complete contrast to the 'contractual' job which can be fatally affected by any deviation from the agreed documents. We repeat our advice (p.182) that a successful building project of any kind relies on being planned well in advance of things happening, so keep thinking ahead.

Begin by determining your working methods.

Working methods

If you are constructing a Segal building as a member of a group, you will need to have decided on the balance between group working and individual working. Working as a group has the advantage that the weaker members will learn from the more experienced and morale can be kept up. What can be achieved is often more than the sum of the parts. Some operations, like raising the frames, require collective effort and others, like laying the main drainage system, are a collective responsibility anyway. Individual working, on the other hand, allows you to proceed at your own pace and in your own way without interference. The additional flexibility that this allows is of the greatest value when you come to enclosing the building and installing the finishes and equipment. You may decide to set up work teams, building a group of houses together. You may also decide to work together up to the stage of completing the structural frames or to the stage when the building is weatherproof with the roof on and the walls and windows in place. The finishing stages, working inside the house, would then be carried out on an individual basis, allowing the benefit of individual control of designs and standards.

Tools

We offer suggestions on the tools that you may need as follows. This list will need to be adapted to your own particular way of working and the number of each required will depend on the number of people building.

Group tools:

- ▷ 240-110 v transformer;
- ▷ 110v heavy-duty drill with long auger bits 12, 16 and 20mm diameter;
- ▷ 110v heavy-duty, portable circular saw;
- ▷ 110v electric concrete mixer;
- ▷ 30m tape measure with refill;
- ▷ wheelbarrow;
- ▷ shovel, spade, pickaxe, fork and broom;
- ▷ plumb bob and line;
- ▷ hacksaw;
- ▷ set of spanners and adjustable spanner;
- ▷ carborundum stone for sharpening tools;
- ▷ G clamps;
- ▷ rope.

Individual tools:

- ▷aluminium double extension ladder (could be shared between two);
- ▷crosscut and tenon saws;
- ▷claw and pin hammers;
- ▷set of chisels and mallet;

▷spirit level;
▷carpenter's square;
▷3-m tape with refill;
▷plane;
▷mitre box;
▷bradawl;
▷nail punch;
▷screwdrivers for slotted and Phillip's heads (Yankee type are good);
▷Stanley knife with blades;
▷set of high-speed twist drills and countersink;
▷pliers;
▷wire stripper and electrical test screwdriver;
▷paint brushes, roller with spare sleeves and tray.

Other tools that are good if you can afford them:

▷ orbital sander with sand paper (this gives a good finish to the timber);
▷ battery drill/screwdriver (very convenient tool);
▷ jigsaw (this could be shared among the group).

Site preparation

The first steps are clearing the site, carting away any rubbish, erecting the perimeter fencing and a temporary hoarding if necessary; getting any demolition work carried out, perhaps tree surgery, and building an enclosure for the temporary electricity supply.

Drainlaying

It is a good idea to get the main drain runs which may have to pass under your access road in at an early stage. You may also decide to lay in the water pipes at this stage to save digging later on. You will be using Alkathene plastic water pipe. It comes in a number of different grades of slightly different diameter and wall thickness, so make sure that you get the correct metal inserts to put into the end of the pipe when you make a joint and you have the correct adaptor to make the connection to the Water Board stop valve. Drainage can be one of the most difficult operations and you may decide to subcontract it out. If you decide to do it yourself, it will be a great help to hire a mechanical excavator. You can avoid brick inspection chambers by using plastic ones: more expensive in material cost, but quicker to make. Make sure that any excavations are properly supported; trench collapses are one of the major sources of accidents in the building industry. Achieving the correct fall to the drains is most important to avoid blockages in use. Make sure that the drains are on the correct bedding, that they have the correct surround and that they are protected with paving slabs where they have less than 600 mm (2′0″) of cover or with a surround of 150 mm (6″) of concrete where they have less than 900 mm (3′0″) of cover under a road or car space. You will have to test the drains and get the test witnessed by the building inspector before backfilling.

Roads

You may need to build an access road to serve a number of houses and it is a good idea to get the base in early on to allow deliveries on to the site without difficulty. This is an operation that you can do yourself by hiring the plant necessary. An excavator with its driver, 'muck-away' lorries to remove the spoil and a roller to compact the base are needed. The most neglected part of road making is good land drainage. No amount of fill rolled into wet ground will provide a solid base. Do not skimp on the thickness of crushed rock or other fill material to give you a base that will not sink as soon as the first removal lorry appears. It is almost impossible to make a road stronger once it is laid. The local highway engineer will advise you of what is required to suit the particular ground conditions. When you come to bedding kerb stones and road gullies, put enough concrete under and around them. Make sure also that you have provided enough ducts under the road for services and that you have marked where they are so that you can find them again later. (Much of the art of building is to do with having in your mind the whole process, and anticipating what comes later on.) Construction traffic over the next year or so will allow the road base to consolidate before you get in a subcontractor to lay the finish to the road at the end. Paving bricks give a very pleasant finish. They are relatively expensive to buy but you can lay them yourself and hire a vibrating plate machine to consolidate them.

Foundations

The essential feature of the foundation arrangement for a Segal building is that the building is built

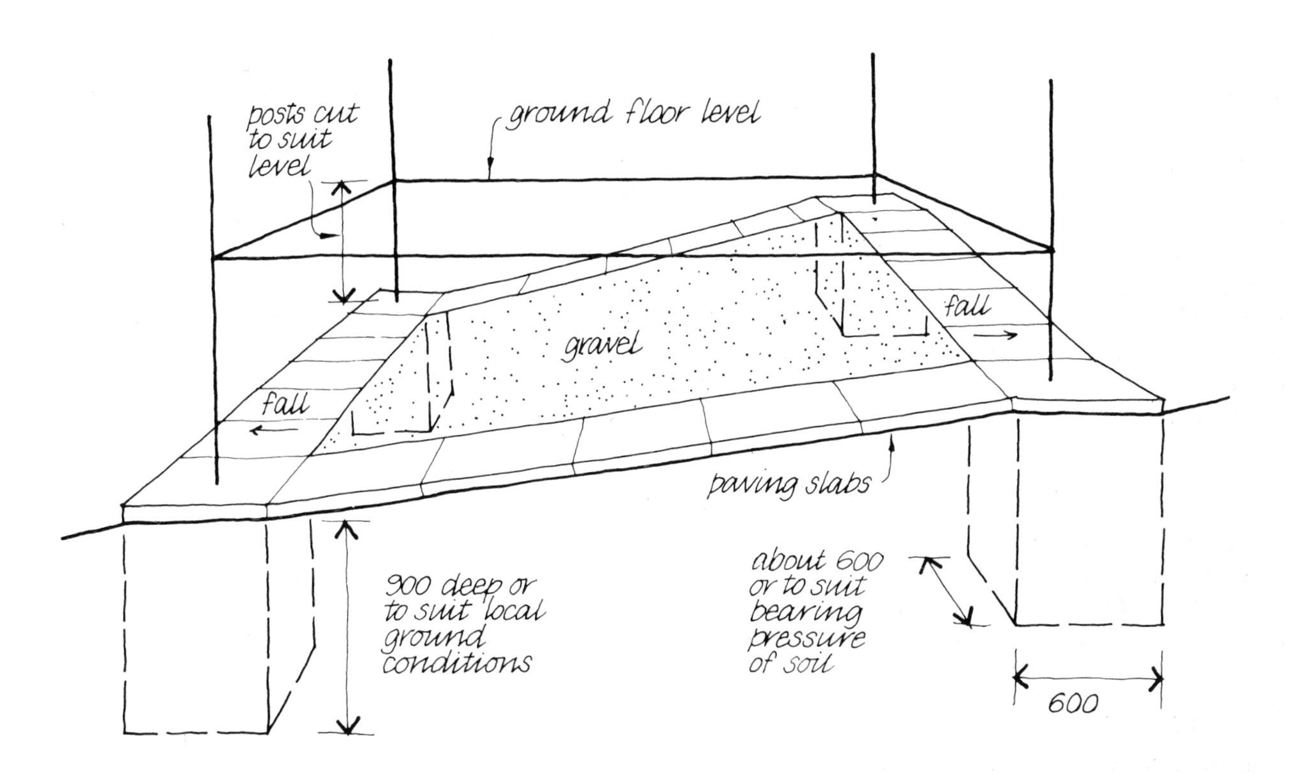

Diagram of foundation arrangement in uneven ground to support frame-posts of varying length

above the ground rather than *in* the ground in the usual way. Each post of the building is supported on a separate concrete base about 600mm (2′0″) square and about 900mm (3′0″) deep, to suit the local ground conditions. A bungalow might have eighteen and a two-storey house about twelve bases. The bases are constructed at the prevailing ground level without levelling the site. This removes the necessity for getting the foundations in at a particular level—always difficult when you are wading about in the mud, as you will be at this stage. It eliminates much earthmoving and means that you can build on a steeply sloping site without extra cost and close to trees without damaging them. The topsoil is removed below the building and replaced with gravel. There is no need for an oversite slab in the usual way. The reduced extent of the foundations reduces their cost and makes this a job that is relatively easy for self-build labour. The holes can be dug by hand and the concrete placed by wheelbarrow. This eliminates the need for machine excavation and ready-mixed concrete and thus enables you to build without difficulty on sites with restricted access. The posts stand on the foundations but are not anchored down to them (the weight of the building is calculated to ensure that it is secure). In this way the building is to some extent independent of the foundation, the frame can be erected and the critical dimensions between the frames adjusted at a later stage, independently of the foundations. In other words, the foundations can be put in without the same level of accuracy that is normally required. This makes the job easier for unskilled people, who often have considerable difficulty achieving accuracy at this stage when there is nothing fixed to go by.

If you are building on particularly poor ground, short bored piles can be a relatively economical solution. They are in effect giant concrete pegs in the ground under each post of the building. You can get a piling company to prepare a design for the piles after a site investigation has been carried out to determine exactly what ground conditions there are by digging pits or drilling holes in the ground and taking samples of the soil for testing in the laboratory. The piling company will auger holes in the ground with a machine, place a reinforcement cage in each one and fill it with concrete. This operation is very quick and relatively economic, provided that you have a number of houses to do in one go, as a large part of the cost is in setting up the machine on site. You will have then to construct a pile cap on top of each one. This is a lump of concrete with a paving slab on top, on which the post stands. With Segal-method framing you avoid the necessity of constructing reinforced concrete ground beams which are necessary to support load-bearing walls, and this saves a great deal of expense.

With work below ground now completed you have a convenient base to make the timber frames on.

Frame making

This is straightforward work for one, or even easier, for two people,—sorting material, marking out, cutting to length (and scarfing together where necessary), then laying the beams and posts across each other, squaring them up, and drilling and bolting them together. This is all done with the frame horizontal at a convenient height from the ground. Make the last frame you want first, as you will be stacking the later frames on it—so you will finish by making the first frame in the erection sequence (someone has dubbed this the reverse domino effect). Take careful thought to make the frames the right way up for lifting straight into position—once made they are very hard to turn over. Even frames that will need no bracing when they stand in the finished building need to have a temporary diagonal member screwed on, to hold the frame square during handling and erection (this timber invariably gets used later).

Whilst the frames are at a good working height is the right time to mark out the locations of joists and rafters, and to attach any bearers and spacing blocks that will eventually be required. It is so much easier on the ground. Apply any preservative stain on the parts of the frame that will be exposed to the weather at the same time. Lastly the posts need to be cut to the appropriate lengths for their individual foundation pads. What is this measurement?

Levelling

To find out, you must measure the relative levels of the pads. This is readily done once you have borrowed or hired the right instrument and its tripod —a 'dumpy' or 'quick-set' level, so called because of its compactness and simple hemispherical clamping mount.

You do not need a staff, which is marked in a peculiar way and is all too easily misread. You just set aside a piece of straight wooden lath for each foundation pad/column and mark it, the pad and the frame post, with the grid reference from the plan.

Set up the level in a position from which all the pads are visible. Stand each lath on its pad in turn, and focus the level on it—adjust the instrument with its built-in spirit bubble and sight through it—you will see a horizontal hair-line striking across the lath. the lath 'holder-upper' now slides a carpenters square slowly up the lath until it comes into the surveyor's view and is aligned with the hair-line—the surveyor gives the signal and the holder draws a pencil line on the lath. When the square is removed the line can be checked for accuracy—the hair-line should obliterate it.

When all the pads have been done (checking the bubble at each new setting) the laths are laid together with the pencil marks in line. The lath ends will be seen to protrude by various amounts and these variations represent the levels of the pads. Choose the shortest one—this will of course belong to the highest pad, which is the one you must relate your new floor level to.

It is convenient to identify as the datum mark, to which all vertical buildings will be checked back, the top surface of the ground floor joists. On the post at the highest foundation, mark this datum line and measure down the post to allow for the depth of the floor joist plus the depth of the floor beam (plus the minimum clearance to the ground, say 150mm). That decides where you must cut this post. Lay the foot of the lath against this mark and transfer the datum line from the post to it. Take it back to the other laths, realign the 'hair-line' marks and scribe the new site datum line across them all (you realize that from this time onward the hair-line level is redundant and you can safely dismantle the level from its tripod and return it to its owner).

Assembling a frame flat on the ground

Take each lath to its post, align the datum marks and the spare length of the post that sticks out beyond the lath is the part you saw off.

There has been no calculation required, no arithmetical mistakes can be made, the operation can be visually checked. The posts always come out the right length for the frame beams to lie truly horizontal.

Frame raising

Prepare for erecting the frames. You will need:

▷ timber for temporary bracing;
▷ stout pegs in the ground to fix temporary braces for the first frame (or you may be able to fix it to a convenient tree);
▷ timbers marked ready with the correct modular distances between the frames that can be used to space the frames apart as they are erected;
▷ some long lengths of timber to fix to the top of the frames to push them up;
▷ plenty of hammers and nails;
▷ spirit levels;
▷ ladders;
▷ and about nine or ten strong helpers for an hour or two (more than ten and people just get in the way).

On the erection day make sure that everyone is clear what they have to do and that there is only one person giving the orders. Carry the first frame into position and loosely bolt the long 'pushers', fixed with lock nuts, in such a way that they can rotate as the frame goes up. Detail one person to put their foot on the bottom of each post as it goes up to prevent them sliding. Lift *together* and up it goes. A moment of great excitement. But make sure that everyone keeps hanging on until the temporary braces are securely in position and that they don't stand back to admire their handiwork too soon. Remember to brace back the feet of the posts as well as the tops. Check that the frame is upright and in the correct position on the foundations.

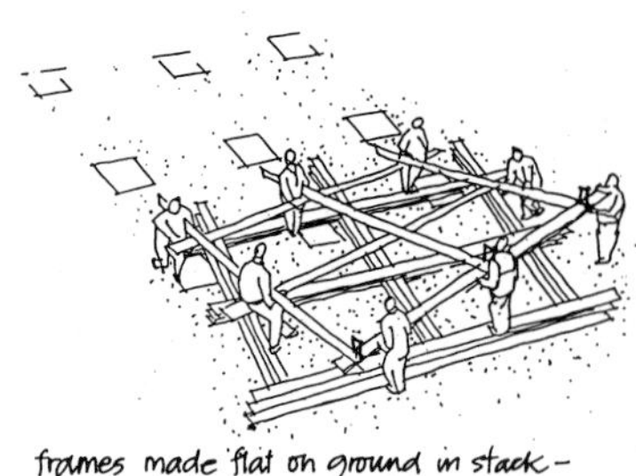

frames made flat on ground in stack –
last frame made is first frame wanted –

first frame raised on foundation pads –

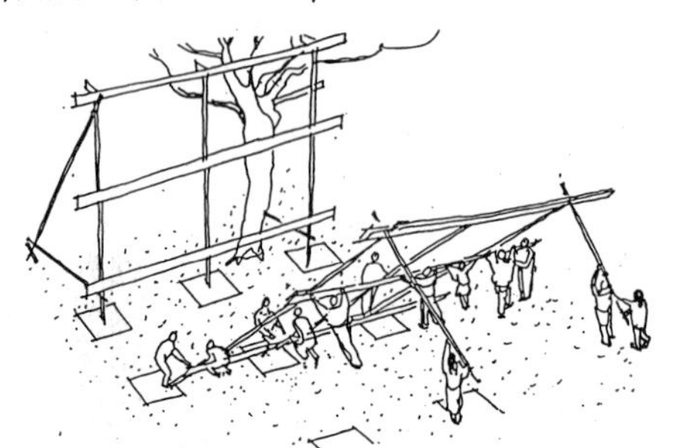

first frame plumbed upright – strutted to ground
and any convenient tree – second frame raised –

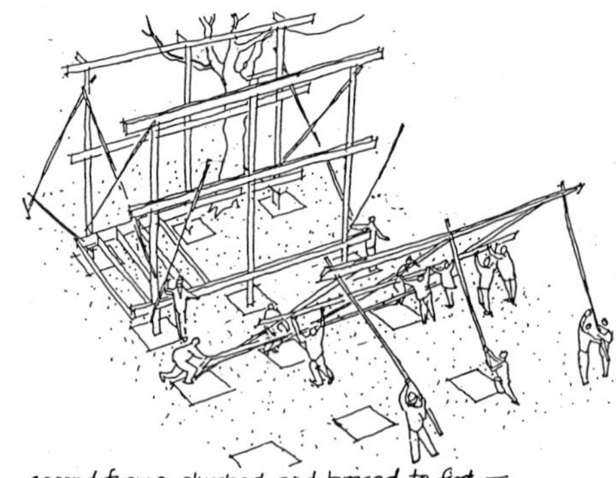

second frame plumbed and braced to first –
third frame raised –

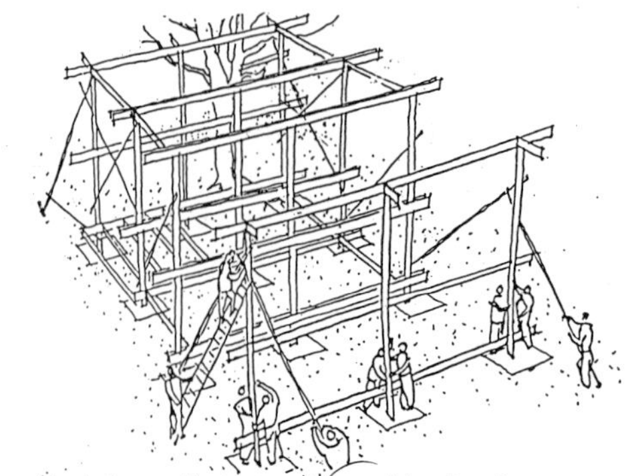

third frame plumbed and braced to other two –
last frame raised – after bracing, next comes the roof...

Erection day

Repeat the procedure, bracing the second frame off the first to form a rigid structure and using the spacer timbers to maintain the correct spacing between the frames.

When all the frames are up, check that they are all in line by running a string line along the edge of the whole structure, check that they are square by measuring the diagonals, check them to ensure that they are upright (with a long spirit level held vertically) and that the spacing is correct. You can take the temporary bracing members off very carefully one at a time to make any adjustments that may be necessary. On no account should more than one brace be removed at a time or the frame will collapse, causing serious injury.

That done you are now ready to fix the joists. These may be fixed either:

▷ on top of the beams (usually at ground-floor or possibly roof levels);
▷ between the beams (usually at first-floor or possibly roof levels).

Where the joists are fixed between the beams they may be supported by:

▷ a 50 × 50 mm (2″ × 2″) bearer fixed to the side of the beams; or by
▷ joist hangers.

If you are notching out to fix joists on to a bearer, make sure that the top edge is slightly above the top of the beam when it is fixed or you will get a slight hump in the floor over the beam. A useful tip for marking out the notches on the joists is to place the joist in position on top of the bearer and mark it at the level of the top of the beam. In this way when the joist is turned over, the depth of the notch will be correct every time. The joists are fixed by skew nailing from each side with another nail driven in at an angle at the top to keep the joist upright. You can use timber that is to be used later on in the building process nailed on to the posts to construct temporary staging at a convenient working height for fixing the joists.

When the joists are in position, you can fix the permanent bracing. The braces should be a good fit. Measure the height from floor to ceiling and the width between columns. Calculate the length of the diagonal using Pythagoras' theorem and measure this on the diagonal of the piece of wood. Cut the brace to length and offer up into position. Mark the angle cuts at the ends. Drill the bolt holes, which should be in a position that ensures the maximum amount of timber beyond the bolt towards the end of the piece of timber to give the maximum strength. Cut the notches to give a seating for the nuts and washers so that they are as small as possible for the same reason. Offer up the timber and, using the holes already drilled in the brace as a guide, drill the columns. The point where the braces cross is halved together and fixed with a nailplate on either side. If the building is more than a couple of feet above the ground at the position of the bracing, then braces should be provided below the ground floor level to ensure that there is no sway on the structure.

Fix any noggings that are necessary to support the top of the partitions before fixing the roof or floors. These are nailed into place between the joists on the modular positions.

Roofing

The braced up framework of the house only takes you a little way into your construction programme in terms of timescale but the achievement is a huge morale booster and marks a significant stage of progress.

You now know that you will not turn back—nothing can beat you. You can put a roof on the frame straightaway and give yourself covered workspace for the rest of the job.

While the rafters are going up, and before you underfelt and batten, stretch the scrim between them 50 mm (2″) down from the top edge to retain the Warmcell—it is difficult to put it in from below afterwards.

The underfelt held down by battens makes a quick and secure temporary roof, tiling can be done when convenient. The steep roof is a dangerous place to work—you must have scaffolding.

First-fix plumbing

With the roof on and the houseframe still bare is the time to begin putting in plumbing services. This encompasses water supply, storage and distribution equipment, drainage, heating pipework, flues, gas service pipes and so on.

We already mentioned that at the design stage you will allocate duct space where service traffic is heavy —now is the time to start framing them up.

All these services need careful thought but it is all based on common sense and simple physics—no different in Segal-method building than any other, so we won't duplicate here what is well covered elsewhere.[3] It is just easier to accommodate services in timber-frame than in brick and block construction.

First-fix wiring

Wiring for power points may have to be installed in the ground floor construction before the Warmcell insulation is laid in and the deck fixed down. A skirting duct can be a help but there are still door openings to get around. A good measure is to run the cables in conduit wherever they have to dive into the floor void. It is much easier to rewire when that inevitable day comes. Ideally, *all* services should be accessible for future maintenance or renewal.

With the electrical installation as with plumbing there is nothing special about Segal method. You can easily do it yourself, but not from this book! You will have to go to one of the specialist handbooks.[4]

Flooring

You need a floor deck to work from as soon as possible. Either you can put down the permanent floor and keep it clean and dry (you will have to sheet the

The group mobilized

walls temporarily) or you can put down your flooring material loosely, possibly upside down to keep it clean, and fix it permanently near the end of the job.

This is double handling, but has the advantage that you can delay putting the Warmcell insulation in until the building is fully enclosed, and the walls are ready for the Warmcell too. Also the flooring material will by then be better seasoned and will cramp up better. It is only safe to secret-nail floorboards after they have reached their lowest moisture content or, however tightly you cramp them, they are bound to shrink and open up hideous cracks.

The only alternative is to fix with screws in the centre of the boards, allow them to shrink, and cramp and refix in a years time (another of the joys of the cordless screwdrivers and the posidrive screw form—this has only now become a practical proposition. You cannot un-nail a floor and refix without damaging it, but you can unscrew it).

External walling, windows and doors

The floors give a good working platform from which to construct the walls. Because you are infilling a framed opening, the panels are non-structural and the sequence of work is open to choice. If you are using the loose-fit relocatable principle you will start by making up the spaced studs and screwing them into place on the grid lines. You can then fasten the inner wall-face of plasterboard, with or without a woodwool slab behind it, with battens screwed to the studs. The external sheathing of bitumenised fibreboard is screwed to the outer face of the studs.

Once the space between is sealed at the edges by the insertion of door and window linings the loose-fill insulation can be pumped in at any convenient time.

Watertight flashings are fitted around the frames before the final external rain-shield of capped boarding or panelling is screwed on. This work has to be done carefully, always ensuring that any water driven in by the wind can find its way out again.

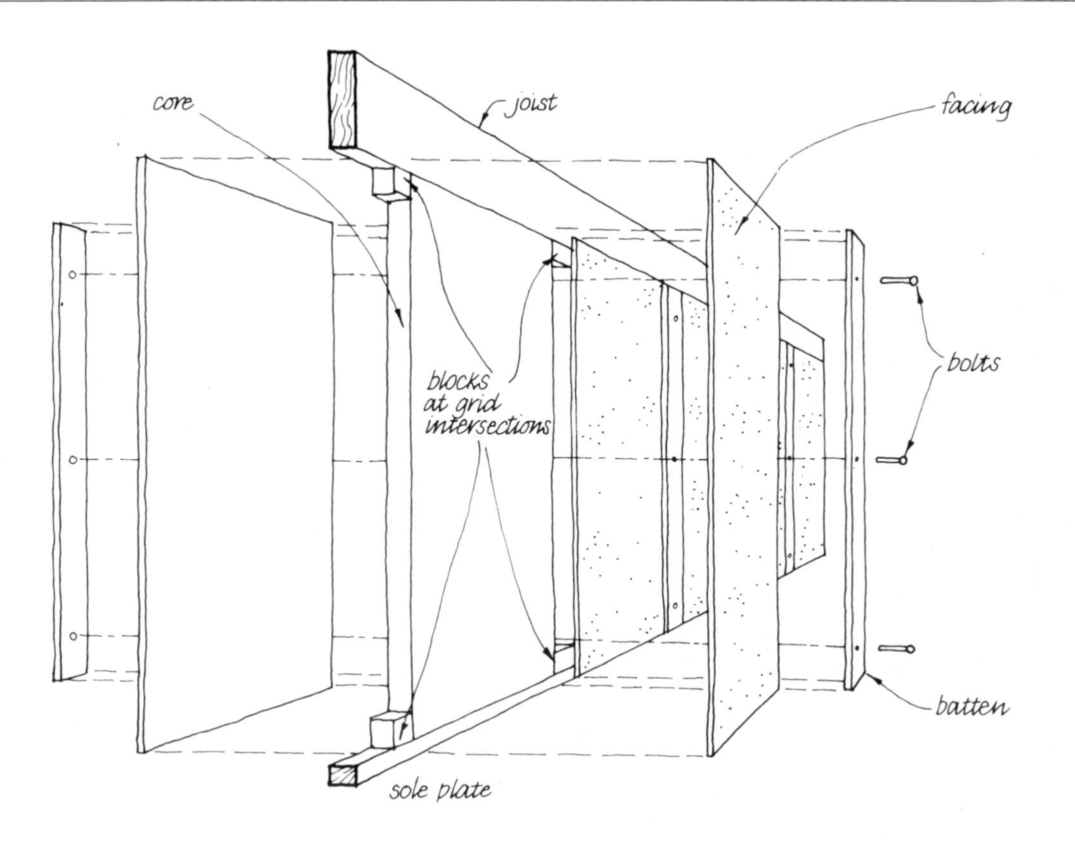

Internal wall panels clamped into frame

The doors you can buy in ready made or have fun making your own, spending time rather than money.

The windows are more difficult to make yourself. It may be better to buy in the elaborate joinery bit, the storm-proof weatherstripped opening light (it may or may not come factory glazed), and make up the fixed lights yourself, putting double-glazed panes into your own framework with 'Driglaze' rubber glazing tape and screwed timber beads. You must use safety glass in all doors and in window panes at low level.

Internal walling and doors

Once you have the windows and external doors in position you can then work fully protected from the weather. The next operation in sequence is to fix the internal subdivisions, although you may find it more convenient at this stage to fix only those partitions that have to be in position to form the bathroom and kitchen and support the radiators, and leave the other divisions until such time as the plumbing is completed and the ceilings are all in position. In this way you can put off the moment at which you have to lose the benefit of the extensive and largely uninterrupted working space that you will have been enjoying up until now. In particular, this makes manipulating the ceiling panels so much easier.

Door linings and linings for any internal windows that you may decide to incorporate are made up on site in the same way as the external window linings.

Second-fix plumbing

Having fixed those partitions that have radiators and other plumbing fixtures attached to them, you can complete the plumbing installation. Particular attention has to be paid to the details of the soil drainage system, which has critical pipe sizes, falls and horizontal distances from each trapped appliance such as basin, bath, sink and WC. Refer to a good text book,[5] peruse the building regulations and consult

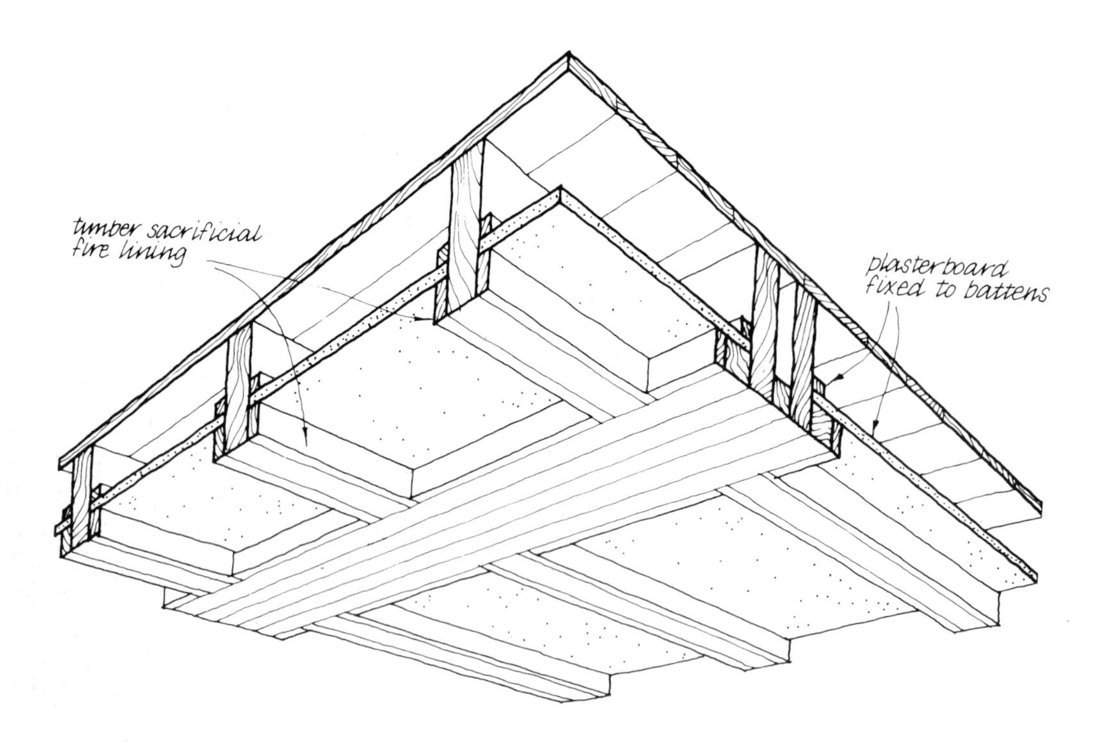

Ceiling construction

your building inspector to verify that your proposals are satisfactory. The waste pipes can be run below the building, which is very often a very simple procedure, but do make sure that the *traps* are not outside, as they will freeze up in winter and make the drains flood over the floor.

Overflows to the lavatory cisterns and to the water storage or heating header tanks if you have them can most easily be routed vertically to discharge in an easily observed position below the building. Fit a separate overflow for each of the tanks and cisterns. You can also obtain a fitting that will allow the WC cistern overflow to discharge into the bath.

The gas pipework can be most conveniently routed below the building, where it is well ventilated (remember that gas pipes must not pass through wholly unventilated voids). The gas meter can be located in a ventilated plywood enclosure in a convenient position for reading below the building or alongside the front door beside the enclosure that houses the electricity meter. In this case you must ensure that the two meters are in completely separate compartments and that no cables pass through the gas compartment. The hot and cold-water pipework is best located above the floor behind the bath or sink unit or in a purpose-built duct to give a neat appearance in the bathroom. Provide the duct with enough rigidity to support the washbasin and access panels to get access to the plumbing. The water main must be protected from frost where it enters the building. This can be simply achieved by splitting a 100 mm (4″) plastic soil pipe along its length, slipping it over the incoming main and stuffing it with insulation quilt. Install a stop valve in the main with a drain-off cock which is accessible from outside the building.

Before fitting appliances such as the bath, shower, washbasins and kitchen units, fit the washdown linings into the wall construction and fix filler pieces between the wall battens to provide an upstand—at the back of the sink and the bath, for example. Use a good-quality silicon mastic behind the shower, sink, bath and washbasins rather than the cheaper mastics

A Segal-method house at the Glasgow Garden Festival with conservatory, balconies and external stairs

that are on the market, which do not last.

The radiators can be fixed at this stage. The fixing brackets can be plugged and screwed into the wood-wool core of the partitions in any convenient position. The holes for the pipes feeding each radiator can be drilled at this stage in exactly the right position to ensure that these pipes are vertical. When plumbing in the boiler you must use a 380 mm (15″) length of copper pipe before changing to plastic to protect the plastic pipe from excessive temperatures. Water test all the installations. From this point on you will be working with the luxury of heating.

Second-fix electrical

You can now fix the electrical accessories. The cover battens provide convenient locations for socket outlets and wall (or architrave) switches, with space for cables behind. But great care must be taken to clip cables tight to one side of the gap so that the centrally located screw fixing the cover strip is well clear of the cable. Make sure that all the boxes are securely fixed as they have to resist someone pulling out a stiff plug.

Ceilings, finishings, stairs

Once the plumbing has been tested it is safe to fix the ceilings, blow the Warmcell into the ceiling, wall and floor cavities, hang the doors, lay floor finishes, fit duct covers, build in cupboards and fittings and build the staircase.

Stairs are notoriously difficult to visualize from the drawings. The notion of suspending the treads from hangers, closely spaced so that they serve also as balusters (less than 100 mm apart for safety) makes it reasonably simple to work out in three dimensions. It is rather like building a series of related hanging shelves. The vertical spacing of the treads must be even, remembering that the tread itself has a thickness when marking out the level at which the bearers are to be fixed. Each tread has to overlap the one in plan below by about 25 mm and you need a half riser at the back of each tread to reduce the gap to 100 mm—this also stiffens the tread. Make templates for cutting the wedge shaped winding treads. These can be in two parts from the same piece of wood with a single-angle cut. Outside steps to the external doors can be formed by cantilevering the floor beams and hanging the steps from them. The internal staircase is the one place in the house it is desirable to glue as well as screw—otherwise the staircase can get squeaky as the wood shrinks.

Staircase suspended from balusters

Finishes

At this stage, Segal-method self-build is not very different from any other building type. You may think the house practically finished but be dismayed at the time it takes to fix all the trim, apply decorative finishes and put in all the embellishments that go with your personal possessions. We do not need to exhort you to take on these DIY tasks; you will by now be handy with tools and brimful of confidence. Exhaustion is the main hazard! But once the Building Inspector (or architect) has certified that your building is completed in accordance with the submitted plans, and is habitable, you can if you wish move in and then take all the time in the world to achieve whatever degree of finish you aspire to.

What does make Segal-method lighter work now is that you will have been able to do so much pre-decoration during the construction stage. True, there is usually a lot of exposed wood to treat, both indoors and out, and some of this will have to be done (or at least touched up) at the end of the job. The environment-friendly timber treatments we recommend (as produced by Auro Paints and O.S. Color for instance), made of natural oils and waxes, are a joy to handle and give a lovely finish with just one coat.

At this stage you will be getting good value from your Collins *Complete Do-It-Yourself Manual*, but eventually that will take its place on the bookshelf and normal life can be resumed.

Outdoor features

The sequence ends with those secondary but so essential items as verandahs, pergolas, balconies, conservatories and porches. The structural work on them must be included in the programme at the appropriate time but they tend to get finished off towards the end. Provided that they have been allowed for in the design it can sometimes be possible to add them afterwards.

The fact that you can go on messing about with the building after you are living in it—adding bits and embellishing others makes building rather like gardening, a never completed process. Why should it be?

External works

Some items can with advantage come early in the sequence—we have already mentioned the access road. If your scheme incorporates a workshop or even a garden shed it can be very useful during the building programme if it is done at the beginning. It is also a good way to rehearse post and beam carpentry techniques on a relatively small scale before committing oneself to the house.

The exterior spaces around the building and created by it are as important as the form of the building itself, as Alexander stresses in *The Timeless Way of Building*. There are good books dealing with this aspect of design and construction.[6]

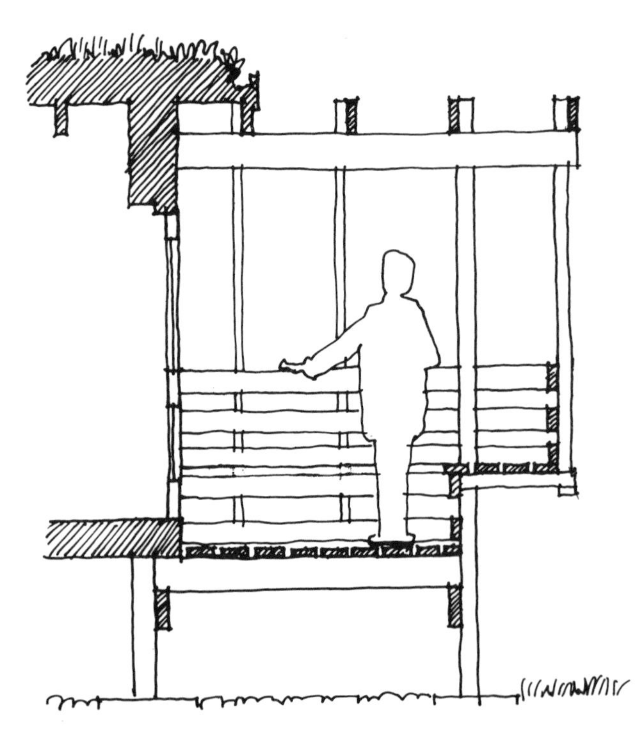

Balcony and pergola

Part Five

Why don't we all self-build?

We have recounted the rewarding experience of building our own houses and we have told the story of others who have built successfully. It is going on up and down the country, but people generally do not yet seem to feel that it is normal to engage in home building for themselves. Why is this? We examine the history behind the way housing is provided nowadays, note some of the shortcomings of the current state of affairs and put forward ideas for a new approach to housing.

Chapter 26

A flawed system

The ideal of a free and prosperous society where 'a man's house is his castle' has never quite been realized in terms of secure, sound homes for everybody in Britain. After various attempts throughout history to raise standards to a decent level we are left with a housing shortage, lack of variety and often poor quality. The modern tendency to leave things to the experts has produced another housing crisis. We show how people are involved in the housing process in other parts of the world.

The shortage of good housing is a source of misery practically the whole world over, and for all our good fortune in Britain, being one of the wealthiest nations in the affluent Western world, we have never succeeded in achieving an abundance of good housing here. Now we are in the closing decade of the twentieth century, we are as far from it as ever, with acute homelessness affecting the poor and with much of our housing stock in appalling condition.

Good feelings about the place we live in are fundamental to a contented and stable social order, and the longing for these qualities goes hand in hand with the rest of the struggle for social justice. The search for an equitable degree of comfort, privacy and security for all members of the community has gone on so long that it could be thought of as part of our folk history. As Dennis Hardy says, 'When aspirations fall short of reality, people's thoughts are stimulated towards conceiving Utopian ideals.'[1]

Examples of attempts to realize these dreams occur throughout our history, and many of the old buildings that gave expression to them still survive.

A history of discontent

In medieval times there was some splendid building done, most of it in praise of God, for military purposes or for the comfort of the nobility. The common people were poorly housed and few of their dwellings survive. Social injustice prevailed then as now but the Peasants' Revolt of the fourteenth century shook the feudal system that ordered things so inequitably. Although the peasants did not achieve all their revolutionary aims, afterwards serfdom was doomed and paid labour gradually took the place of the propertyless bondsman, who became a copyholder, enjoying practically the same rights as the freeholder. It became easier for peasants to obtain an area of land sufficient to build a house for themselves.

In late medieval times, many peasants seeking to establish a family home and cultivated garden were able to stake out a plot in the 'waste'—land that was neither enclosed private territory nor privately owned open land with common rights over it—and build on it.

One hears, again and again, local variants of a tradition that a right of occupancy was gained if smoke could be seen to issue from a chimney between dusk and dawn. Certainly in the part of Herefordshire

Our folk dream—the cottage at the edge of the wood

The Diggers were driven off St George's Hill

where Brian lives there are many examples of cottages apparently randomly sited away from roads and farms, and on the edges of woods and commons. Often, too, they are built of sturdy oak frames. A modern self-builder will recognize the advantage in speed of building with timber frame and infill, and the possibility of improvising a quick solution to enable early occupation with the promise to oneself that a more careful elaboration will be carried out later!

Another notable resurgence of Utopianism occurred in the seventeenth century. The Commonwealth was failing to restore land to the people after having dealt the death blow to the Divine Right of Kings. The Church continued to preach patience to the poor, saying that riches would come in the life hereafter. Gerald Winstanley asked, 'Why do we not have our heaven here?', and, as his questions to the Protector went unanswered (Cromwell advocated preservation of 'the ranks and orders of men whereby England hath been known for hundreds of years: a nobleman, a gentleman, a yeoman'), he declared, 'If ye will not provide for the poor, the poor will provide for themselves.' He put words into action at St George's Hill, Surrey, in 1649, when he and the Diggers occupied, tilled and built houses on the common land there. 'No man is free unless he has land and work to feed himself and his family,' declared Winstanley.[2] Their self-built timber houses were repeatedly sacked by the soldiers of the New Model Army, brought in to protect the interests of the landowners and their hangers-on, and after a hectic struggle their enterprise failed—but the Dream lived on.

It revived during the Industrial Revolution, when workers were becoming wage slaves, alienated from the land, enduring wretched and unhealthy housing in the great towns. In 1840 Feargus O'Connor produced brilliant schemes of land settlement for the Chartists—rural communities with generous land allotments and high-quality houses. Five of these colonies were rapidly built and occupied by working people escaping from the horrors of factory and town. The idea of the land tenure system devised by O'Connor was unprecedented and threatening to the landowning class. It was challenged and overthrown in the courts, the general fund and lottery for places were wound up, and by 1847 the plots were resold into orthodox private ownership. Traces of their layout can clearly be seen in the landscape today (good examples being at Staunton and Lowbands in Gloucestershire), although the social experiment was short-lived and the little cottages survive only as individual private dwellings.[3]

The enclosure of the countryside prevented people from building on waste and common land just at the time the Industrial Revolution created an enormous demand for homes. Private investors built houses for rent and a few employers built dwellings for their workers. The vast majority rented from a private landlord. A minority of skilled artisans were, in the late eighteenth and early nineteenth centuries, able to free themselves from landlords and rent by setting up the first building societies as mutual savings clubs and building for themselves.

The drive to realize Utopian dreams persists into present times. At the turn of the century, Whiteway was founded as a Tolstoyan land colony. The soil was

Still surviving in Gloucestershire; Chartist cottages from the 1840s built in a crescent with their two-acre plots fanning out behind them

poor and the site treeless and unsheltered, and the colonists soon abandoned their attempt to be self-sufficient. So they modified their total commitment to communal work and ownership of property and practised instead individual use-possession of land communally owned (at the outset the title deeds had been ceremonially burnt). Until the present day, continuous self-building and improvement of the infrastructure, communal buildings and individual dwellings has gone on. The bleak landscape has been transformed into thick woodland sheltering beautiful and fertile gardens.

An idyllic and lasting environment has been created at very low cost to society at large.

Less consciously planned but in their way equally inspired are the settlements of the post-First World War plotlanders:

> In the first half of the twentieth century a unique landscape emerged along the coast, on the riverside, and in the countryside, more reminiscent of the frontier than of a traditionally well-ordered English landscape. It was a makeshift world of shacks and shanties, scattered unevenly in plots of ranging size and shape with unmade roads and little in the way of services.
>
> To the local authorities (who dubbed this type of landscape the 'plotlands') it was something of a nightmare, an anarchic rural slum, always one step ahead of evolving, but still inadequate, environmental controls. Places like Jaywick Sands, Canvey Island and Peacehaven became bywords for the desecration of the countryside.

Whiteway: owner-built houses in productive gardens surrounded by cherished wilderness

Arcadia in converted railway carriages

> But to the Plotlanders themselves, Arcadia was born. In a converted bus or railway carriage, perhaps, and at a cost of only a few pounds, ordinary city dwellers discovered not only fresh air and tranquillity but, most prized of all, a sense of freedom.

So write Colin Ward and Dennis Hardy in the preface to *Arcadia for All*.[4] The book traces the history of this extraordinary example of self-help housing enterprise, and describes how:

> everywhere the plotland hut has grown to be the retirement home of its owners . . .there is scarcely a plotland area . . .where we have not seen some ultimate legatee of the old cabin building a new house around it, thankful for the land and the existing accommodation around which to do so . . .[5]

These are contrasted with:

> the expensive municipal housing, built to Parker-Morris standards, often despised by its inhabitants and deteriorating at a terrifying rate, so that in a growing number of instances it is obsolete and uninhabitable many decades before the money borrowed to pay for it has been repaid.[6]

The authors conclude:

> It is arguable that a society whose industrial base is slipping away, which cannot provide employment for its population and where house building, public or private, has reached its lowest ebb for

decades, might well seek to encourage, rather than deter, those who choose to turn their own labour into capital, in housing themselves.[7]

But deterred we have been, by state interference and market pressure. The result is that most people now live in either publicly or privately produced mass housing as opposed to the individually produced houses of the past.

The modern housing catastrophe

Just one of the disadvantages of any attempt to provide mass housing is that if mistakes are made they are made on a massive scale. The huge efforts made by both Labour and Conservative regimes in the 1950s and 1960s were characterized by the need to produce numbers of 'units'.

To boost numbers, central government and the giant building contractors persuaded each other that industrialized system-building was the panacea. The idea was that mass-produced building components for walls, floors and roofs could be made in concrete from standardized moulds. There could be 'economies of scale' which would pay for the heavy industrial plant involved. Traditional tried and proved craft practices were dispensed with in favour of mechanisation and the 'technical fix'. The architectural and planning professions suppressed their misgivings about the wisdom of the experiment and the public stood by, trusting in the specialists to get it right for them.

The result was a catastrophe. As Colin Ward said to a housing conference in 1981:

> In the days when we thought we were rich, it was thought that any technocratic innovations could be provided, at hideous cost, to house all those who hadn't any other choice, and who would be the grateful recipients of all that expensive expertise. Now these policies have exploded in our faces.[8]

The system-built new stock, far from 'solving the housing problem', suffered such appalling technical defects that by 1984 the Association of Metropolitan Authorities submitted to the Inquiry into British Housing the statement:

> The Association estimates that about £25 billion is needed to repair all the substandard housing stock in England; another £10 billion is needed to put right design defects, and £15 billion to meet the shortage of housing in Britain. That is a total bill for both private and public sectors of £50 billion —the equivalent of about £1,000 for every man, woman and child in the country.[9]

Though private affluence is said to be increasing, public spending on housing has been cut and cut again. The necessary £50 billion was not spent and the housing stock continued, and continues, to deteriorate and there are more people homeless then ever.

Although mass-produced housing does fulfil many needs, the individual requirements of the people who live in it are low on the list of priorities. Such housing is initiated by politician or developer, financed by bankers for profit, designed by the architect within cost limits to the brief of the marketing agent or housing officer, and built by the contractor for maximum profit.

Housing produced in this way may at best be sound and decent in that it conforms to a certain standard, but that standard will be a rigid and fixed one that takes no account of people's different attitudes to where they live.

The contribution self-build can make

Not all houses need to be of the same standard, too expensive for some, too mean for others. Different space standards, materials, finish and equipment can be employed by self-builders to achieve their personal requirements. For some home is just a 'base' where they spend little time, and as such it needs to be small and easy to maintain—a garden would be a considerable burden, for instance. Other people, of course, may need lots of space both indoors and out for children to run about in; some people want a place to work or pursue a hobby. Mass-produced housing is not able to supply the variety of standard, size and type necessary to satisfy people's individual requirements.

In self-build, costs and standards can be balanced as the occupant decides. Nor is the balance fixed for all time. Housing is not a once-and-for-all 'thing'; it is a process. Houses need maintenance and improvement during their lives. These are particularly difficult tasks for the mass landlord. The self-builder has both the motive and the skill to attend promptly to maintenance. Improvement is likely to be a continuing process.

The limited choice of high cost private ownership on the one hand and lower (but rapidly rising) cost public renting, with very limited availability, on the other does not cater sufficiently for the different and changing needs and financial circumstances that people experience: the need young people have to be able to move on as opposed to the need for security of an established family; the change from the relative wealth of a single, young, working person who on getting married suffers the special financial hardships of the young family, back to relative security as the family becomes established and, finally, the possible hardships of old age.

There is a need not only for flexible tenure arrangements and a range of cost but also for dwellings that are constructed in an adaptable manner and which can be added on to easily so that the house is able to develop progressively as people's needs and financial circumstances change. Houses do not have to stay the same size. The patterns of family life shift and different accommodation is needed at different times. It is not always desirable to move house to achieve this. In 1961 the Parker-Morris Report stated:

> With the greatly increased rate of social and economic change, the adaptable house is becoming a national necessity . . .We see the investigation of the practical possibilities of doing it easily and at reasonable cost as one of the most important lines of future research into the development of design and structure. The sooner it is started the better.[10]

True now as when it was said, and little done about it. We have described in earlier chapters how easily self-build designs can incorporate this notion of adaptability—from the loose-fit, generous-area concept of Robin Heath's Lightmoor shell houses to the easily dismantled and reassembled components of the Segal method of building. Extensions and divisions

Self-build at Netherspring: a recent instance of people using their talents to create and manage where and how they live

can be created with no more difficulty than was involved in bringing the originally conceived house into existence.

Large public housing estates are often described as 'bleak', 'inhuman' and 'monotonous'. We also live with private developments of 2,000 houses with no more than a semblance of diversity introduced by changing the colour of the bricks here and there. Where once a street was the product of many minds, now a single mind works on a whole neighbourhood. Homes for a democratic society should arise from a complexity of design interests rather than be the result of submerging them into identical forms.

People are largely alienated from the processes that shape their surroundings. We do not have enough opportunities to make the things among which we live, to use our energy and creativity in this way. The tenants of public housing have not had control over the management or maintenance of their homes, which has led to indifference at best and damaging vandalism at worst.

In short, we need more individual housing in a range of size, standard, type, cost and tenure; an architecture of diversity within an ordered framework which encourages people to be involved and use their talents to create and manage where and how they live.

Present housing policies, however, depend not on people being involved in creating housing, but rather on their buying housing created by private developers. Scarcity and rising values encourage speculative building and costs are high. Public housing is being sold and little new public rented housing is being built. The rents of public housing are being raised and obtaining a council or housing association tenancy is very difficult.

These policies make it more and more difficult for people on low incomes to obtain good housing in Britain. Nor do they address the question of a more appropriate form of housing. We have left the task of 'providing' housing to the experts and they have failed, as they were bound to do without an input from the dwellers.

Self-build worldwide

We have seen a little of how people in Britain have housed themselves in the past and we have described our belief that they should have the opportunity to do so again in the light of the shortcomings of the present housing system. The 12,000 people who built their own homes in Britain in 1987 did so in the face of the considerable constraints of the present housing system. In other parts of the world the situation is different. In the developing world sheer necessity has impelled countless homeless people to provide for themselves, and in the developed world there are countries with a different tradition of building to ours where a significant number of owner-builders contribute to the housing stock.

The developing world

The United Nations declared 1987 as the International Year of Shelter for the Homeless. It was, among other things, an exercise in mutual education between the rich, industrialized nations and the poor, developing ones. We learned that approximately one-third of the world's population house themselves with their own hands, sometimes in the absence of government and professional intervention and often in spite of it. We learned too of the failure of large-scale housing projects that offer inhuman living conditions and have fallen into ruinous disrepair. The expense and inefficiency of such an approach is plain for all to see in these fragile economies.

Over the years the United Nations, the World Bank and other aid agencies have changed the emphasis of their work in housing away from the grand project and towards policies that make resources of land, finance and advice available to people on a small-scale, local level so that people can improve and build their own dwellings. The role of governments is to provide a legislative framework that will give security of tenure, low-cost loans and the basic services of electricity, water, sewerage, roads, transport and schools. The residents control the building of their houses within this framework.

It is striking how much can be achieved. In Colombia there are over 500 organizations active in self-help housing, whose members are building about 90,000 houses. There is a national federation and a school of self-build. There are schemes for many thousands of houses to be self-built, using a wide variety of building techniques. Most are of concrete block construction but there is one using prefabricated concrete panels and another with single-storey steel-framed houses. The government in Malawi, concerned about the low quality of rural housing, organized training, building materials and credit. These resources were used by people who had not been involved until then —the women—who have gone on to build new houses throughout the country.

The developed world

What more could be achieved in our own wealthier circumstances if positive policies were to be developed and resources made available? A great deal. In the United States, the richest nation on earth, the idea of people building their own houses has not been lost, as it has in Britain. Land is more accessible, and twenty per cent of all single family dwellings are built by their owners. There is a thriving industry that provides the tools and materials they need. Small one-man-band subcontractors thrive, offering specialist services of all kinds. There are a number of magazines that give examples and advice and share self-build experiences. There are a host of schools and courses for aspiring self-builders across the country.

Another example comes from Sweden and is described here in a little more detail as we think that it has particular relevance to the situation in Britain. 12,000 families built their own houses under the auspices of Stockholm City Council during the period from 1927 to 1976. Completions are currently running at between 200 and 250 houses per annum. The 'Stockholm City One-Family House Department',

Swedish self-build

SMÅ, has a waiting list of 10,000 families and it has been responsible for building around thirty per cent of all the single-family houses in the city.

From the turn of the century, a number of garden cities were established around Stockholm. However, it soon became apparent that only the relatively well-off section of society could afford to live in them. For this reason, in 1924 the City Council resolved that the working class should be able to live in the garden cities and decided to establish an organization to enable them to build their own houses without the need for a large capital payment. The authorities provided the finance and assumed that the people would be able to build competently. The organization has been and still is entirely self-financing. One problem that they did face in the pre-war period was that of opposition from the trade unions, because of the threat, as they saw it, to their jobs in a period of high unemployment. There is an upper income limit imposed on those who wish to take part. The houses are held on a lease and costs are often lower than the rent on an apartment. Mortgages are made available at between five and a half and six and a half per cent interest. This is about half the commercial rate.

SMÅ has developed a well-worked-out procedure over the years. They design the schemes and construct the infrastructure which is sometimes installed as part of a larger council development alongside the self-build scheme. SMÅ issues a detailed programme covering the nine-month building period and a detailed manual on the construction. There is a site office where staff are available to answer queries on a day-to-day basis and during the construction period four general meetings are programmed between the SMÅ staff and the self-builders. The site office has an exhibition of the choices of finishes that are available. The staff inspect the work as it proceeds. Materials are bulk ordered by SMÅ, and this entails a high level of standardization and industrialization in the building process. The houses are all timber-framed and arrive on the site as factory-made components. The programme allows, for instance, for the ground-floor walls to be delivered on a particular day with a two-week period for them to be erected before the first-floor panel are delivered. (The building supply industry in Sweden is clearly more reliable than it is in this country, where you can wait weeks for materials to be delivered.) Formerly the houses tended to be detached but recently most of the designs have been for terraced houses. They are to generous space standards of between 80 and 140 m^2. The self-builders are organized into teams of between ten and fifteen. The service installations are all carried out by subcontractors. Only one self-builder in a thousand has failed to complete their house once started and less than one in a hundred has been more than two months late in completing—a much better record than the professional building industry in this country can boast.

The areas that have been developed in this way all have a lively residents' association and there is an elected council that manages the neighbourhood. Very few people have moved on from these neighbourhoods—1.9 per cent according to a study carried out in 1977. People were shown to have a very high level of satisfaction and cited the high quality of the houses and the neighbourhood, the self-fulfilment that they had enjoyed, the spirit of community and the wealth of knowledge that they had acquired as the chief benefits. The chief disadvantages were given as the time commitment involved, the lack of choice arising from the highly structured and industrialized process adopted and the lack of flexibility and adaptability of the stud-frame terraced houses that have been built.

We would like to think of an improved version of the Stockholm City Council policy as a model for local authorities in this country, providing access to the basic resource of land and finance to enable a wide range of people to build their own houses, designed to suit their individual desires and requirements. Our view is the same as that held by John Turner, advocate of dweller control as the first principle of housing: 'While local control over necessarily diverse personal and local goods and services—such as housing—is essential, local control depends on personal and local access to resources, which only government can guarantee.'[11]

What has to happen in Britain to make this possible?

Chapter 27

Self-build opportunities for all

We suggest that the opportunity to self-build should be part of housing policy in Britain. We outline the obstacles that there are to a more widespread adoption of self-build and suggest what needs to be done to overcome them. We suggest what town and country might be like if people building their own houses were to become major contributors to housing in Britain.

There is a tremendous potential for mobilizing the experience, skill and will to succeed of the whole community in housing itself. We can, and we must, make use of this great untapped resource. There is a housing crisis. There are people ready and willing to play a part in providing housing for themselves. There is a way of building available to people outside the specialist construction industry. The problems of access to land, planning control and finance are surely not insuperable.

Let's get on with it!

Self-build and housing policy

We believe that the direct provision of public or private mass housing has great limitations. We have argued the need for greater diversity in housing—for houses to be built in a range of size, standard, type, cost and tenure. We also advocate that people should be involved and encouraged to use their talents to create and manage their living environment. We believe that housing control and activity should be based at a local level. Many enlightened local authorities currently support a wide range of housing opportunities, including squatting, co-operatives, housing associations, home improvement, shared equity or outright purchase. In this spectrum, we emphasize the potential of self-build to provide the full degree of variety necessary. There are also other less obvious benefits of self-build: the self-confidence that can follow from completing a house, the community-building process of working with others, the satisfaction that can follow from controlling the design, construction and management of one's home and the pleasure that can come from creating something with one's own hands. We could look forward to satisfied citizens living in high-quality, appropriate homes at prices that they could afford. Public resources should be made available to people to enable them to get what *they* want. We therefore propose that self-build be incorporated as a significant strand of housing policy. Not everyone will want or be able to build for themselves, but the opportunity to do so should be an established part of housing policy in Britain.

The obstacles

We have shown that financial, organizational and constructional methods exist to enable people without building experience and earning low incomes to build their own homes, thus opening the way for almost anyone to obtain the benefits of a purpose-made house that were previously available only to the wealthy. Difficulties exist, however, that prevent self-build opportunities from being more widely available.

We have described some of the methods that are being developed to gain access to land for self-build at prices that people can afford, such as deferred payment arrangements and low-cost disposals by local authorities, but in the long term access to land remains one of the major obstacles and, in our view, this requires a radical shift in land policy.

We have outlined the different types of organization available for self-build schemes—co-operatives for rent and shared ownership, for example—but an inordinately long time, up to two or three years, is spent in getting approval for these arrangements from official bodies like the Housing Corporation and the Department of the Environment.

Finance for self-build and the subsidies that are possible for low-cost self-build—HAG, for example—are subject to the vagaries of the market and lengthy bureaucratic delays.

Many people, and professional planners and financiers in particular, are prejudiced against the simple, lightweight construction methods that make self-build so much easier and more accessible to people without building skills.

Before self-build can become an important element of housing policy, the mistaken idea that it is a slow and inefficient method of producing houses, that standards of construction are inadequate and that it might lead to disorderly and uncontrolled development, must be dispelled.

Overcoming the obstacles

A number of things need to be done for self-build to play a more significant role in housing policy.

There has to be a change in attitude among politicians and the officials who implement their policies. It is necessary to demonstrate to financiers and landowners what can be achieved, and to convince them that for people on low incomes to build houses for themselves is both a workable and worthwhile aim. The preconceptions of planners and architects need to be changed and new ways of thinking about how to plan and build developed. It is also necessary to make people as a whole aware of what the possibilities are and to articulate the latent demand that undoubtedly exists. Aspiring self-builders have to be persuaded that it is possible to obtain the resources they need to build for themselves by knocking on the town hall door and demanding an opportunity to build.

The politicians

Although the idea of self-help is part of Conservative philosophy, they have been unwilling to make the opportunities available to people who do not have resources of their own. Most ordinary people cannot hope to purchase land on the open market in competition with developers. Neither can a lot of people obtain a commercial bank loan. Many Labour politicians, on the other hand, consider that making these resources available to individuals for their own housing needs is providing public resources for private gain. To balance the allocation of resources is the stuff of politics, but it does seem that making resources available to enable people to build their own houses should be generally acceptable on the common-sense grounds of both long-term value for money and freedom of action, and therefore justifiable on both sides of the political divide.

The officials

We all, inside and outside public employment, like a quiet, easy life. As Colin Ward has said:

> We all find it easier to do business with big reputable firms like, say, Barratts, Wimpeys or Wates. We know we can't push them around and therefore accommodate them. From the 1950s up until the present day I have seen sites in cities up and down the country emptied of everything except the pub left behind in a no man's land, not because local authorities want to encourage drinking, but because you can't mess around with the big brewery-owning firms. I'm afraid I have to say that British local authorities always favour the big battalions.[1]

The many hurdles put in the path of the Lewisham self-builders by officials of all kinds—planners, building control officers, engineers, civil servants and even the Inland Revenue—arose largely out of their basic mistrust of ordinary people, rather than deliberate obstructiveness. (There were some absurd moments, mind you: for example, the building control officer who swore blind that he had seen people overcome by 'noxious vapours' accumulated between beams in the ceiling of a house; the discussion was about the minimum ceiling height permitted in a room, a control that has since been abolished.) The nation cannot afford the misuse of public funds spent on all these people checking and counter-checking. The effect is to prevent people from doing new, worthwhile things, not to prevent bad decisions. Local officials do not have the authority to take decisions; the approval for funds for eight houses in south London had to be referred to Whitehall, who are still passing it backwards and forwards years later! Our experience is that people want to do things for themselves properly and want to make sure that their house is safe and well built. Officials must put their trust in them to do so and not keep them on tenterhooks for years on end.

The financiers

Finance for self-build was relatively readily available during the years of the 80's boom in the housing market. The building societies invested in self-build as their traditional markets became more saturated and the Housing Corporation started taking its statutory role of promoting self-build more seriously. However, a cash crisis in the Housing Corporation and substantial losses by the building societies caused by a collapse in the housing market have meant that development loans have been frozen by these bodies for an unspecified period. They are also sceptical when it comes to putting money into schemes that fall outside the conventional self-build format. One building society recently withdrew their offer of funding, after having made an offer one year previously, on the advice of their valuers, who maintained that the scheme, which had been designed to be easy for people without building experience to build, could be built only by professional builders. This was demonstrably untrue and only goes to show the prejudices and lack of knowledge that some professionals have. The real tragedy is that sixteen households had already invested countless hours and a great deal of nervous energy in developing the scheme so that it was ready to go on site. After some delay the non-commercial sector in the form of Mercury Provident PLC have provided a development loan. The irony is that funding is not available for self-build because it is considered a high risk and yet the current crisis is a product of massive investment in the speculative housing market, where risks are taken on a grand scale. Self-builders, on the other hand, cannot afford to take risks. The people who control the financial institutions have to be persuaded that self-build is safe and worthy of sustained investment.

The landowners

Lack of access to land at a cost that people can afford is probably the greatest impediment to more widespread self-build opportunities.

At root the difficulty lies with the inequitable distribution of land in Britain and the operation of an unrestricted market. As Richard Norton-Taylor observes in his book *Whose Land is it Anyway?*, 'Just over one per cent of the adult population owns almost seventy per cent of the land. The concentration of land ownership in Britain, encouraged by the Enclosures and the Industrial Revolution, is unique in Europe.'[2] He chronicles the failure of the attempt to control the activities of private landowners and speculators through the Community Land Act of 1975, which was designed by the then Labour administration 'to restore to the community the increase in value of land arising from its efforts'.[3] One other idea to adjust this situation is described by John Seymour:

> So is there any practical and fair means of, firstly, securing the fair division of the land of a country, and, secondly, securing that the pattern of land ownership stays that way? Well yes there is, but so far it has never been tried. It has never been tried for the simple reason that it would work. The system was proposed by the American economist Henry George back at the turn of this century. It is simply to impose a graduated tax on land. In other words you pay more tax per acre according to how many acres you own. If you only own a few acres you pay nothing. The beauty of the graduated land tax is that it could be applied gradually. There need be no violent upheaval. The screw could gradually be turned. The person with ten thousand acres would soon find it expedient to put nine thousand of them on the market and thus reduce the tax per acre. The person with no land at all would soon find, that after a little effort to earn some money,

> he or she would be able to afford to buy some. If they only bought their fair share there would be no tax.[4]

Meanwhile, to generate pressure for reform we must agitate on our own narrow platform for the release of land for self-build now. Large landowners —industrialists, corporations, the churches and the landed aristocracy—must be persuaded of the philanthropic benefits of making land available for self-build.

One opportunity that exists concerns the quarter of a million acres of derelict land that lies unused in our cities. People squatting in empty property in the 1970s opened up a debate which has led to licensed squatter groups and then to short-life housing associations receiving government grants to bring vacant property into use. Permanent housing associations have in turn grown from many of these short-life associations.

One could imagine people in need of housing, equipped with easy-to-erect instant houses designed to go on the roof of a car, establishing themselves on vacant infill sites where services are already available in the street. There would be confrontations with the police no doubt, as there were in the 1970s, but the surrounding publicity would bring the whole question of land to the forefront of public attention.

One could further envisage short-life housing associations occupying vacant land under licence, building lightweight timber-framed houses that were designed to be capable of being demounted and putting them up in a different location if and when the site was required for a long-term development project. There is a proposal to finance such a scheme, using Mini-HAG funding, which simply requires a local authority to grant a ten-year licence for a site that they could not develop themselves. This idea is not very different from the relocatable houses built on vacant sites by the London Borough of Camden, among others, for homeless families.

This in turn points towards serviced sites being made available on leases by local authorities for the self-help erection of permanent houses. The local authority would provide the resources necessary for people to build what they want, need and could afford. These developments could be on infill sites in the towns and cities, extensions to existing settlements in the countryside or new developments incorporating smallholdings and workshops.

The planners and architects

As well as land being difficult to obtain, there are restrictions put on its use by planning control. Since 1947 no one in Britain has been allowed to develop or redevelop land without government permission. The first Town and Country Planning Act was widely welcomed as necessary and enlightened in intention. In that a significant factor leading to its introduction was a desire to suppress the plotlands development described in Chapter 26, we can now see a flaw in the intention. For the freedom of the plotlanders has produced success in the long run, whereas under the control of the planners have emerged countless building schemes that have desecrated our old towns and lovely countryside alike. In Richard Norton-Taylor's view:

> The post-war history of planning in Britain offers ample ammunition to those, whether farmers, property developers or politicians, who argue that what are needed are fewer controls, not more . . .There must be something seriously wrong with our priorities, with a system in which controls and endless correspondence with local authority planners are needed before the owner of a house in a town can make small alterations to a roof or porch, but which allows historic buildings, officially preserved, to be neglected and pulled down, prime farm land to be torn up, and landowners to scar the countryside.[5]

We can all think of recent developments which have been granted bureaucratic approval despite overbearing size and lamentable appearance.

We advocate giving substance to the 'inalienable right to a decent home and a healthy and attractive environment' which Richard Norton-Taylor and we believe in. There should be at least a presumption of planning approval for a family intending to provide its own house on its own land. On the other hand, all other development proposals would be subject to planning control and should be strictly vetted by local community organizations. This is similar to the situation that prevails in Switzerland, with its still-strong tradition of local democracy, where any proposed development can proceed unless objections are made when those people who would be affected are notified of the proposal.

Meanwhile, architects need to be convinced of the need for developing methods of construction that are easy for unskilled self-builders and that will be acceptable to planning officers in various circumstances. One approach would be to draw up an overall development plan which defines individual plots and their relationships with one another, with roads and services, with open space and so on. Plot development rules are drawn up which define the limits of height, density, overlooking and other matters that affect neighbours. A general planning permission is sought which allows plotholders freedom to plan their own dwellings within the rules. The method of building is determined and a general building permission sought that allows individuals freedom to construct their particular layouts within the general consent. Aspiring self-builders come together to form a group and plan their houses. Training courses based on the local adult evening institute are organized and a building, used initially as a site workshop and ultimately as a community centre, could be built as part of the learning process. A manual is prepared that allows individual plotholders to:

- ▷ Plan their dwellings and determine the specifications to suit their particular needs and financial circumstances.
- ▷ Select the most appropriate form of tenure from a range that would include renting and outright or shared ownership.
- ▷ Decide whether to carry out all or part of the building work themselves or employ local tradespeople or a builder.
- ▷ Assemble the components of the building method to build their own houses if they so decide.
- ▷ Adapt and extend their houses.
- ▷ Maintain the buildings.

The self-builders

Just as the politicians and those who control resources need to be convinced that a self-build policy is workable and worthwhile, so too members of the public need to be made aware of the possibilities. They must be convinced that building your own home is a realistic aim for anyone who wants to. They need to be made aware of the arguments that will persuade local officials to support their aims and be motivated to put the case for self-build.

A vision of self-build in Britain

In the countryside

Should planning approval for individual houses built by their owners become almost automatic, land currently sterilized by planning decisions would come on the market and land prices would be eased. The corollary is that people would build all over the place. This is not such a shocking possibility as we have been conditioned to believe. It is after all the way our architectural heritage came about in the first place. The great objection is said to be that we are a tight little island and there is simply not enough space, that the countryside would become congested and ruined. We think there is enough space, and that people building for themselves would fit into it better than the planners have been fitting them in. We have a beautiful countryside, but putting houses and gardens into it does not necessarily desecrate it. Who would be more likely to respect it in the long run—the owner-builder or the developer seeking to build massive estates in unspoilt countryside? And who would be more productive—the agri-business farmer or village-sized groups of families in their own houses and gardens with smallholdings and workshops to hand? Peter Kropotkin, in his visionary *Fields, Factories and Workshops Tomorrow* proved the latter. In his commentary on the 1974 edition of this early-twentieth-century book, Colin Ward says:

> The advocates of high-density housing have always cited the 'loss of valuable agricultural land' as a factor supporting their point of view. Sir Frederick Osborn, with equal persistence, has always argued that the produce of the ordinary domestic garden,

even though a small area of gardens is devoted to food production, more than equalled in value the produce of the land lost to commercial food production. Surveys conducted by the government and by university departments in the 1950s proved him right.[6]

Another vision of how workmanlike and attractive such a countryside would be comes from William Morris. In *News from Nowhere*, having gone to sleep in the 1880s he awakes in the twenty-first century in a transformed, beautified London reintegrated with the countryside. Morris found the garden-like countryside thickly populated; except in the deliberately kept forests,

> It is not easy to be out of sight of a house . . . the population is pretty much the same as it was at the end of the nineteenth century; we have spread it, that is all.[7]

Development in the countryside can be intelligently planned to be fitting and unobtrusive. Ian McHarg, in *Design with Nature*,[8] describes what should be done in the open countryside. McHarg issues the challenge, 'Can we not create, from a beautiful natural landscape, an environment inhabited by man in which natural beauty is retained, man housed in community?' His approach is to draw up physiographic principles for both the conservation and the development of a region. These are based on an assessment of its development needs, its historic character and its natural features —i.e. topography and subsurface geology, surface and groundwater patterns, flood plains, soils (with particular reference to their degree of permeability), steep slopes, forests and woodlands.

For instance, in a beautiful area of North America not unlike the English countryside—the Valleys of Baltimore—which is under pressure to accommodate much population growth, he has made plans which the local community has adopted in order to prevent despoilation. He proposes that the rich soils forming the flood plains of the valley floor should remain undeveloped apart from agriculture, and that where the valley walls are bare they too should be unbuilt on until they have been planted to forest cover.

The slopes already wooded, where they are not so steep as to be subject to erosion, can be developed at a density of about one house per three acres without losing their character, and the level forested plateau at the top can take housing at about one per acre. On the open plateau behind the wooded escarpment much more development at higher densities (with sewerage provided instead of septic tanks) could be concentrated in hamlets, villages and country towns.

He concludes that there is abundant land to accommodate the proposed regional growth without fouling the water supplies, losing agriculture or woodland, and still preserving the beauty of the landscape. He further holds that such a plan enhances land values over those resulting from haphazard growth.

And in the cities

Just as one could envisage self-built houses set in the countryside, so too we could look forward to more diverse, smaller-scale residential environments in the towns and cities. The cities remain places of opportunity for young people, poor people and the ambitious alike. They are the magnets of enterprises of all sorts, the centres of learning and intellectual endeavour. However, young and poor people in particular are forced to live in bad housing conditions. Access to good housing is beyond the reach of many. Young people coming to the city for a new life are often forced to sleep in the streets.

The commercial market cannot help them because it actually makes the situation of low income people worse, simply driving them out of the market. Large areas of our cities remain derelict. Apparently they are more valuable awaiting the right moment for some grand scheme to become viable than being put to use now. Their owners are waiting for a change in government policy that would enable standard housing units to be constructed or for the market to offer the maximum profit for some speculative development of offices, shops or luxury flats of no value to people presently without decent homes.

However, a number of our examples show what can be achieved by small groups of people building for themselves in the city: how council tenants in Lewisham built detached houses to their own design; how young, unemployed people in Bristol built small flats for themselves that have gained them a place in the housing market; how co-operative enterprise created a group of low-energy homes in Sheffield.

These people have overcome the tide of dereliction and despair that has resulted from the collapse of traditional industries in the cities and the migration of many people who could afford to move out of the city centres.

Self-build should be encouraged in the cities so that it can form a part of the process of making our towns and cities more humane places to live, places with more greenery and gardens, a more human scale, more diversity and better housing for people with low incomes.

Workspaces

Workspaces are expensive and difficult to come by in some cities. One could take the example of Lightmoor as a model for people building places to live and work in the cities as well as in the countryside. Or small groups of workshops could be purpose built by the people who are going to work in them. Small entrepreneurs will be in a good position to make a go of running a business, having benefited from the confidence-building experience of self-building.

Community buildings

Many communities lack social buildings, meeting places and playgroups. Self-build can offer much higher-quality accommodation than the range of lightweight buildings on the market which are often used for small community buildings at present. The layout and standard of specification can be individually worked out to suit the particular circumstances. Costs can be reduced still further by the users undertaking all or part of the work themselves. Many small voluntary organizations are very short of funds and self-help construction offers them a way of buying the accommodation they need with their limited resources. The one obstacle that such groups have to overcome is that, unlike a self-build housing project, where people's individual commitment is clear and is rewarded with a new house on completion, it is sometimes difficult to get a steady commitment of time from people when the final benefit is less direct. Nevertheless, where there is the will to do it, there is considerable potential for high-quality, affordable buildings, purpose-designed for the needs of particular groups.

Refurbishment

Also relevant is the self-help rehabilitation of existing accommodation. Individuals are doing this all the time of course, but larger-scale projects of this kind are rare in Britain. A block of tenements was gutted and brought back into use as low-cost flats in Stirling, Scotland, and a group is about to start work on a derelict block of pre-war council flats in east London. We have limited our discussion to new construction generally, because working on old buildings is much more difficult and there is always the risk of uncovering expensive and technically complicated problems, whereas with building new you know precisely what is involved right from the outset and the way is clear. In New York, where abandoned tenement buildings are brought back into use on a large scale, one agency involved has concluded that the proper extent of self-help involvement should be limited to gutting out the building at the beginning and carrying out the finishing stages at the end. This allows the professional builders, who are brought in to carry out all the structural and services work, to see at the outset before contracts are let if there are likely to be any problems. This reduces the risk of unforeseen rises in cost. The occupants do that part of the operation that builders find hardest, and that is to finish everything off properly. They can achieve the type and standard of finish that they want.

A new way of thinking

A self-build policy of this kind implies a new way of thinking about how housing works and how buildings are put together. However, there are many vested interests in the housing world that will resist change and there are preconceptions about amateurs in building that have to be overcome.

Many people believe, for instance, that to get high standards of construction and finish, professional builders have to be engaged and skilled tradespeople brought in. To many, the DIY approach means shoddiness. We have found this not to be so. The self-builder is highly motivated and prepared to devote time and patience to achieve results that would be prohibitively expensive to obtain commercially. The key factor is that the self-builder cares about the quality of the work

rather than how quickly or profitably it can be carried out.

It is often held that the building of houses will take longer with people participating fully in the process. We note, however, that the present methods, which leave out consideration of the occupant, are unacceptably long and drawn out. The time council schemes took when we both worked in a London borough architect's department ranged from between four and eight years from scheme inception to handover of the finished buildings. When self-builders became involved, however, they exerted great pressure to speed things up. A sense of personal urgency was introduced that could be applied in no other way.

It may be feared that the self-build process will be difficult to control and the results disorderly. We, on the other hand, welcome a tendency that introduces a sense of vitality into our living environments. In fact, the variety that would emerge from a large output of self-build effort would be much more appropriate to a democratically organized society than the 'architectural consistency' so loved by the planners. It would arise from a proper perception of what is sound and good, not a superficial one.

If it should be thought that self-build is an inefficient way to produce houses compared with the geared-up efforts of the building industry, yes, we admit that in terms of the amount of effort that goes into it, it may be. But the point is that so much of that work comes from a formerly untapped resource, the skill and energy of the general population. And to be obsessed with efficiency would be to leave out of the account a more important factor: that the people who have committed themselves to the effort of housing themselves will have gained so much more. When the community engages in self-build, it builds itself. Not everyone will be able, or will want, to self-build but when those who can and do make a start, society as a whole will benefit. The opportunity to self-build should be a right available to us all.

Afterword

by John Seymour

In an ideal world it would practically never be necessary to build a house. For in an ideal world the population would be stationary (it will *have* to be one day!) and houses will be built to last. There is many a house in Europe that has been continually inhabited for four hundred years—and will probably serve that much longer again provided no fool burns it down. So each generation should inherit its housing from the former generation.

But the world we live in is far from ideal. Houses are not built to last four hundred years—nor, indeed, for a hundred; and it is inconceivable that people will, for much longer, be prepared to live in the kind of ugly, boring, soul-stultifying and grossly inefficient mass housing that has covered the industrial nations of the West like a sort of creeping blight in the last dozen decades.

The age of mass-man will, we hope, one day be over and women and men will consider themselves, and have to be considered, individuals again. And then we will all insist on individual housing—dwelling houses built for us or self-built—built the way we want them and where we want them. We have been so indoctrinated, by school and by mass society, that we cannot believe any more that *we* can do anything—except that one narrow skill which we have selected, or which has been selected for us by society. The joy of this book is that it repudiates this narrow view—we are all of us capable of doing nearly anything that anybody else can do. I learned this very vividly in Africa before the Second World War. The farmer for whom I worked, living right up in the north of that desert country known as South West Africa, wished to build a house for himself. There were no professional builders within at least a thousand miles of him. So not only did he have to lay his own bricks—but he had to *make* his own bricks—and he had to make the charcoal with which he burnt 'em. Cement and lime could not be obtained so he had to quarry limestone and burn it into quicklime and use that for mortar. True, he bought some sawn timber and corrugated iron for the roof and these he had to carry a hundred miles from the rail head in an ox wagon. And *he built a house*. And it is a fine solid house and it will last a thousand years if the termites are not allowed to get at the roof timber and the corrugated iron is replaced every half a century. And everyone else in that country had to do the same, if they wanted a house. There was no other way that they were going to get one.

We often hear that 'an Englishman's home is his castle'. When we consider that the majority of Englishmen (to say nothing of English women) do not even *own* their own homes we realize what a nonsensical statement that is. Surely, if there are such abstractions as the Rights of Man one of them has to be: the right to build one's own home! Even the robin and the wren have that right. Having been born on to this planet surely we must have the *right* to a little piece of it—our fair share of it perhaps? But certainly a piece of it large enough to build a home on. Nobody has a *right* to a house but everybody should have the right to *build* a house.

The authors of this book do not concern themselves with the rights and wrongs of the planning laws of any country, but it might be reasonable to devote a thought or two to the subject here.

Let us consider the case of a farmer who has an acre of ordinary agricultural land in England that he wishes to sell. Without planning permission it might be worth a thousand pounds. If he manages to persuade—or to bribe—some Lord Luck in the person of a *planning officer* to squiggle his name on a piece of paper that thousand-pound field immediately goes up in value *two hundred times*. For if he cuts the acre up into eight plots each plot will be worth £25,000, wherever it is. So the young couple, desperate to build their own home in their own country, have to fork out that amount before they even buy a bag of cement or a concrete block, and they have to pay that money for a piece of land actually worth only £125.

This is not the place for a discussion of the planning laws, nor even of the morality of one man owning more than his fair share of the Earth's surface, but surely it is in order just to consider these figures?

I have seen, indeed slept in, two houses that one of the authors of this book built himself. They are not only both superb houses—but superb *homes*. And, if ever the inhabitants of this planet decide to try true civilization they will have to base it on the civilized *home*, for it cannot be based anywhere else. It certainly won't arise from 'housing units' shoved up by councils, or by speculative builders.

'We have got to have planning laws!'

Maybe, but we don't have to have silly ones.

'We can't all be allowed to build our own houses!'

Why not?

Why not?

Chapter notes

Chapter 1

1-7 Murray Armor, *Building Your Own Home*, Bridport, Devon, Prism Press, 1987. All the extracts in this chapter are from the 9th edition. Murray Armor continually and comprehensively revises, updates and republishs this title. In 1995 the current edition is the 14th, published the the UK by J M Dent and Sons Ltd and the 15th edition is already in preparation.

Chapter 2

1 John McKean, *Learning from Segal: Walter Segal's Life, Work and Influence*, Basel, Birkhäuser, 1989.
2 Walter Segal Self-Build Trust, 57 Charlton Street,London NWI IHU.
3 The Centre for Alternative Technology, Llyngwern Quarry, Machynlleth, Powys SY20 9AZ.
4 Christopher Alexander, *The Timeless Way of Building*, New York, Oxford University Press, 1979.
5 Christopher Alexander *et al.*, *A Pattern Language: Towns, Buildings, Construction*, New York, Oxford University Press, 1977.
6 Alexander, op, cit., p. 7.
7 Ibid., p. 9.
8 Ibid., p. 7.
9 Ibid., p. 14.
10 Ibid., p. 25.
11 Ibid., p. 395.
12 Alexander *et al.*, op. cit.
13 Alexander, op. cit., p. 395.
14 Alexander *et al.*, op. cit.

Chapter 3

1 The Town and Country Planning (General Permitted Development)
Order 1995, London, HMSO, 1995

Chapter 4

1 Ken Kern, *The Owner-Built Home*, New York, Charles Scribner's Sons, 1972.
2 'DIY Plan No. 4 Solar Water Heater', Quarry Publications, the Centre for Alternative Technology, Llyngwern Quarry, Machynlleth, Powys SY20 9AZ.

Chapter 7

1 Nicholas Taylor, *The Village in the City*, London, Maurice Temple Smith Ltd, 1973.
2 Colin Ward, *Anarchy in Action*, London, Freedom Press, 1982.
3 Jonathan Street, *Outlook*, May 1976.
4 Ibid.
5 Model Rules, National Federation of Housing Associations.
6 Councillor Ron Pepper in conversation with Brian Richardson, 1979.

Chapter 8

1 Zenzele Self-Build Association Ltd, 'Bristol Pilot Project—Completion Report', prepared by I. E. Symonds and Partners, Chartered Quantity Surveyors, December 1985.

Chapter 9

1 Tony Gibson, 'Changing the Neighbourhood', Town and Country Planning Association, undated.
2 Tony Gibson, 'Lightmoor isn't the only pebble on the beach BUT it would be a stepping stone', Town and Country Planning Association, undated.
3 Letter written by Margaret Wilkinson to Brian Richardson, 9 December 1988.
4 Tony Gibson, 'Neighbourhood Initiatives Foundation', leaflet, June 1988.
5 Dennis Hardy and Colin Ward, *Arcadia for All: The Legacy of a Makeshift Landscape*, London and New York, Mansell Publishing Ltd, 1984.
6 Colin Ward, *New Town, Home Town, the lessons of experience*, London, Calouste Gulbenkian Foundation, 1993.

Chapter 12

1 VIBA-centrum, Veemarktkade 8, Gebouw B, 's-Hertogenbosch, the Netherlands. Correspondence to Postbus 165, 5201 AD 's-Hertogenbosch.

Chapter 16

1 A magazine specifically for self-builders, *Build It*, features a list of 1000 plots available every month and operates a help-line, 0181 286 3000. Available in bookshops, it is published by Build It Publications Ltd, St James House, St James Road, Surbiton, Surrey, KT6 4BR.

Chapter 18

1 Robin Murrell and Avril Fox, 'Choices', *Environment Now*, Vol. 1, No. 10, 19 November 1988.
2 Nigel Pennick, *Earth Harmony*, London, Century Hutchinson Ltd, 1987, pp. vii-viii.

Chapter 19

1 *Thermal Insulation: Avoiding Risks*, London, HMSO, 1989.
2 *Where the Wind Blows*, *Tapping the Sun* and *Wired up to the Sun*, New Futures series, Centre for Alternative Technology Publications, Machynlleth, Powys, SY20 9AZ.
3 Republished as *Solar Water Heating, a* DIY Guide by CAT Publications as above.
4 John Seymour and Herbert Girardet, *Blueprint for a Green Planet*, London, Dorling Kindersley, 1987.
5 S. R. Curwell and C. G. Marsh, *Hazardous Building Materials: A Guide to the Selection of Alternatives*, London, E. and F. N. Spon Ltd, 1986.
6 Ibid.
7 John Elkington and Julia Hailes, *The Green Consumer Guide*, Victor Gollancz Ltd, 1988.
8 Robin Murrell and Avril Fox, 'Choices', *Environment Now*, Vol. 1, No. 10, 19 November 1988. They have recently expanded these notes into a book *Green Design—A Guide to the Environmental Impact of Building Materials*, London, Architecture Design and Technology Press, 1989. They intend it to fill the gap in information sources which we encountered in looking at this aspect of building and you will find it, if not exhaustive, useful and enlightening.

It is particularly welcome to have the Guide in book form as *Environment Now* has since ceased publication.
9 *Greener Building (Products and Services Directory)* 3rd Edition, Association for Environment-Conscious Building, Windlake House, The Pump Field, Coaley, Gloucestershire, GL11 5DX
10 WOODMARK, The Soil Association, 86 Colston Street, Bristol, BS1 5BB
11 Simon Counsell, *The Good Wood Guide: A Friends of the Earth Handbook*, London, Friends of the Earth, 1990.
12 *Toxic Treatments: Wood Preservative Hazards at Work and in the Home*, London Hazards Centre Trust Ltd, January 1989.
13 Seymour and Girardet, op. cit. pp. 119-121.

14 AURO ORGANIC PAINTS LTD, Unit 1, Goldstones Farm, Ashdon, Essex, CB10 2LZ, and OSTERMANN AND SCHEIWE UK LTD, 26 Swakeleys Drive, Ickenham, Middlesex, UB10 8QD
15 Pat Borer and Cindy Harris, *Out of the Woods, Ecological Designs for Timber-Frame Housing*, The Centre for Alternative Technology and the Walter Segal Self-Build Trust, 1994.
16 AECB op.cit. p.164.

Chapter 20

1 HMSO op.cit. p.37.
2 The Standard Assessment Procedure. The Building Regulations now require that you provide an energy rating for your design, and in an appendix give instructions on calculating it, but the method is not simple. Ask the Building Inspector what your particular proposal entails. If the services of a Building Energy Consultant are necessary the Association for Environment-Conscious Building has an excellent list.

Chapter 23

1 Borer and Harris op.cit. p.171
2 Alexander et al. op.cit. p.25

Chapter 24

1 J. A. Baird and E. C. Ozelton, *Timber Designers' Manual*, second edition, reprinted Oxford, BSP Professional Books, 1989.
2 e.g. Albert Jackson and David Day *The Collins Complete D.I.Y. Manual*, London, Collins, 1986.
3 Albert Jackson and David Day *The Collins D.I.Y. Guide to Plumbing and Central Heating*, London, Collins, 1986. (Extracted from the Complete Manual and available separately.)
4 Albert Jackson and David Day *The Collins D.I.Y. Guide to Wiring and Lighting*, London, Collins, 1986. (Extracted from the Complete Manual and available separately.)
5 e.g. *The Collins D.I.Y. Guide to Plumbing and Central Heating* detailed in note 3.
6 John Brookes *The New Small Garden*, London, Dorling Kindersley, 1989.

Chapter 26

1 Dennis Hardy in a lecture, 'The Persistent Dream', given at the Canon Frome community in 1982.
2 Ibid.
3 Ibid.
4 Dennis Hardy and Colin Ward, *Arcadia for All: The Legacy of a Makeshift Landscape*, London and New York, Mansell Publishing Ltd, 1984, p. vii.
5 Ibid., pp. 291-2.
6 Ibid., pp. 291-2.
7 Ibid., p. 300.
8 Colin Ward in a lecture, 'Planning Reform', at ICA Future Communities Seminar, 9 July 1981.
9 Association of Metropolitan Authorities, submission to Inquiry into British Housing, September 1984, quoted in Colin Ward, *When We Build Again, Let's Have Housing That Works!*, London, Pluto Press Ltd, 1985.
10 *Homes for Today and Tomorrow*, London, HMSO, 1961.
11 John F. C. Turner & Robert Fichter, *Freedom to Build*, (USA & London) New York, Macmillan Co., 1972, London, Collier-Macmillan, 1972.

Chapter 27

1 Colin Ward in a lecture.
2 Richard Norton-Taylor *Whose Land is it Anyway?* Wellingborough, Turnstone Press Ltd, 1982.
3 Ibid.
4 John Seymour, 'Reform for Real', *Resurgence*, issue 131, Nov-Dec 1988.
5 Richard Norton-Taylor, op. cit.
6 Peter Kropotkin, edited by Colin Ward, *Fields, Factories and Workshops Tomorrow*, London, George Allen & Unwin Ltd, 1974.
7 William Morris, *News from Nowhere*, edited by A. L. Morton, London, Lawrence and Wishart, 1973.
8 Ian McHarg, *Design with Nature*, USA, Natural History Publishing Co. 1971.

Build me my home

Further information

There are so many places to go for further information that we can only erect a few signposts. For instance, Borer and Harris's *Out of the Woods* which we discuss in Part Four concludes with a Resource Guide, and this is amplified still further in the CAT publication *Environmental Resource Guide—Details of organizations, consultants, equipment suppliers, courses and publications concerned with minimizing the environmental impact of building.* Both are obtainable from the
Centre for Alternative Technology
Machynlleth, Powys SY20 9AZ

Another main source of environmentally sound building information is the Association for Environment-Conscious Building who publish a quarterly newsletter *Building for a future* and a comprehensive source-list: *Greener Building (products and services directory).* The address is
Windlake House
The Pump Field
Coaley, Gloucestershire GL11 5DX
tel: 01453 890 757

The best route to more information about organizing yourself for self-build is via a charitable trust with which the authors are associated—Jon Broome as initiator and Brian Richardson, a *retired* architect, as a trustee.
The Walter Segal Self Build Trust
57 Chalton Street
London NW1 1HU
tel: 0171 388 9582
fax: 0171 383 3545
The Trust aims to make it a practical proposition for anybody who so chooses to design and build a house for themselves:

- ▷ alone or in company with others
- ▷ self-directed, with such professional support as they deem necessary
- ▷ without the requirement of special craft skills or full-time training
- ▷ whatever their financial status
- ▷ with enjoyment

Its services are available to everyone, especially those in housing need or on low incomes, as individuals or in groups; and to local authorities and housing associations. It provides advice, information, training and support in the following areas:

- ▷ building method
- ▷ land acquisition
- ▷ finance and funding
- ▷ professional services
- ▷ contract management
- ▷ NVQs and training qualifications

The Trust has created a national network and can supply information on completed and current schemes, housing, community and other building types, self-build groups and other self-help groups, experienced architects and contract managers—and other organizations involved in issues of self-build, training, environment and employment.

As well as being co-publishers with CAT of *Out of the Woods* (see above), the Trust sells an information pack *You Build: a guide to building your own home* for people in housing need and those organizations working in the 'social housing' field. *You Build* is also the title of a quarterly bulletin, free to members subscribing to the Trust. It has mobile exhibition material and can sell or hire an historic video made by the Lewisham self-builders in the early eighties with the BBC Open Door team called *The house that Mum and Dad built* In collaboration with CAT it engages in self-build training courses.

Not unnaturally, it tends to emphasize the benefits of timber frame construction as described in Part Four of this book; but as Walter Segal himself was more concerned with realizing the creative potential of self-builders than with any particular building technique, so the Trust helps people to make their own best judgement.

Another organization operating in England and Wales and giving advice to self-builders seeking to benefit from government grant, and whose technical bias is more towards conventional brick and block construction, is the Community Self Build Agency
Unit 26, Finsbury Business Centre
40 Bowling Green Lane
London EC1R ONE
tel: 0171 415 7092
fax: 0171 837 7612

North of the border the Walter Segal Self Build Trust has helped to set up Community Self Build Scotland. The contact address is
Robert Chalmers
CSBS
Bonnington Mill Business Centre
72 Newhaven Road
Edinburgh EH6 5QG
tel: 0131 467 4675
fax: 0131 555 2471

North America
Several schools based in North America offer worthwhile courses and workshops for prospective self-builders. These include:
The Owner Builder Center
1250 Addison Street 209
Berkeley, CA 94702
(510) 841 6827

Real Goods' Institute for Solar Living
555 Leslie Street
Ukiah, CA 95482
tel: (800) 762 7325
fax: (707) 468 9394
email: isl@realgoods.com

Yestermorrow School
RR 1, Box 97-5
Warren, VT 05674
(802) 496 5545

Two key references for finding supplies and information on all aspects of home construction are *The Real Goods Solar Living Sourcebook*, 8th edition (Chelsea Green Publishing, 1994, 670 pages, available in the UK through Green Books) and *The Millennium Whole Earth Catalog* (Harper San Francisco, 1994; 384 pages).

One book of particular interest to self-builders is Richard Manning's *A Good House* (Grove Press, 1993), which takes a personal and environmentally conscious look at the whole construction experience. A more practical guide is John Connell's *Homing Instinct: Using Your Lifestyle to Design and Build Your Home* (Warner Books, 1993; 416 pages). Connell is the founder of the Yestermorrow School (see above), and his book is an excellent resource for anyone contemplating self-building.

Bibliography

The details of some of our favourite books that we think you would enjoy too have been given in the Chapter Notes, notably:

Building Your Own Home Murray Armor
The Timeless Way of Building Christopher Alexander
A Pattern Language Christopher Alexander et al
Learning from Segal John McKean
The Owner-Built Home Ken Kern
Earth Harmony Nigel Pennick
Arcadia for All Dennis Hardy and Colin Ward
Anarchy in Action Colin Ward
When We Build Again Colin Ward
Homes for Today and Tomorrow Colin Ward
Fields, Factories and Workshops Tomorrow Kropotkin, ed. Colin Ward
News from Nowhere William Morris
Blueprint for a Green Planet John Seymour and Herbert Girardet
Design with Nature Ian McHarg
Whose Land is it Anyway? Richard Norton-Taylor
Freedom to Build John Turner
and of course the recently published *Out of the Woods* by Pat Borer and Cindy Harris.

Other recent additions have been *Eco-Renovation* by Edward Harland (Green Books, Devon, 1993) which deals with the sister-subject of self-build: rehabilitation; and *How Buildings Learn: What happens after they're built* by Stewart Brand (Penguin Books, New York and London, 1994) in which he advocates making adaptable buildings that the occupants can continually adjust to meet their changing needs—a particularly relevant notion for self-builders.

Index

Reader response—an invitation

So far, this has been a one-way flow of information and opinion. While you have been reading, you will no doubt have come across things that you disagree with, things that you would like to know more about and things that you do not understand because we have been obscure.

We would be glad to enter into a dialogue. Please let us know

what we have left out of this book that you think should be included;

what factual mistakes you have noticed that should be corrected;

what erroneous or misguided opinions we have expressed;

what topic you would wish us to clarify or explore further.

Please reply to us via our publisher, who has kindly agreed to forward correspondence:

Green Books,
Foxhole, Dartington,
Totnes, Devon TQ9 6EB

If what you have read has stimulated your interest in the Segal approach to building and you want to be kept in touch with developments in this area, please also contact:

The Walter Segal Self Build Trust,
57 Chalton Street,
London NW1 1HU